Origin and Evolution of Biological Energy Conversion

Origin and Evolution of Biological Energy Conversion

Edited by

Herrick Baltscheffsky

Editor
Herrick Baltscheffsky
Department of Biochemistry
Arrhenius Laboratories
Stockholm University
S-106 91 Stockholm
Sweden

This book is printed on acid-free paper. ∞

Library of Congress Cataloging-in-Publication Data

Origin and evolution of biological energy conversion / edited by
  Herrick Baltscheffsky.
      p.   cm.
    Includes bibliographical references and index.
    ISBN 1-56081-614-7 (alk. paper)
    1. Energy metabolism.   2. Evolution.   I. Baltscheffsky, Herrick.
  QP176.075   1996
  574. 19--dc20   96-3922
                                                                CIP

Printed in the United States of America

ISBN 1-56081-614-7 VCH Publishers, Inc.

Printing History:
10   9   8   7   6   5   4   3   2   1

Published jointly by

VCH Publishers, Inc.   VCH Verlagsgesellschaft mbH   VCH Publishers (UK) Ltd.
333 7th Avenue   P.O. Box 10 11 61   8 Wellington Court
New York, New York 10001   69451 Weinheim, Germany   Cambridge CB1 1HZ
                                                              United Kingdom

Contents

3. Iron and Sulfur in the Origin and Evolution of Biological Energy Conversion Systems   43

Richard Cammack

4. The Evolution of Electron Transfer Proteins in Photosynthetic Bacteria and Denitrifying Pseudomonads   71

T. E. Meyer, J. J. Van Beeumen, R. P. Ambler, and M. A. Cusanovich

5. From Dihydrogen to Dioxygen: Evolution of Biological Oxidation–Reduction   109

Robert J. P. Williams

Contributors

R. P. AMBLER. Institute of Molecular and Cell Biology, University of Edinburgh, Edinburgh EH8 9YL, UK

HERRICK BALTSCHEFFSKY. Department of Biochemistry, Arrhenius Laboratories, Stockholm University, S-106 91 Stockholm, Sweden

J. J. VAN BEEUMEN. Laboratory of Microbiology and Cell Biology, State University of Ghent, Ghent, Belgium, and University of Edinburgh, Edinburgh EH8 9YL, UK

RICHARD CAMMACK. Centre for the Study of Metals in Biology and Medicine, King's College, London W8 7AH, UK

JOSE CASTRESANA. European Molecular Biology Laboratory, D-69012 Heidelberg, Germany

W. A. CRAMER. Department of Biological Sciences, Purdue University, West Lafayette, IN 47907-1392, U.S.A.

M. A. CUSANOVICH. Department of Biochemistry, University of Arizona, Tucson, AZ, 85721, U.S.A.

THORSTEN FRIEDRICH. Institut für Biochemie, Heinrich-Heine Universität Düsseldorf, 40225 Düsseldorf, Germany

P. N. FURBACHER. Department of Biological Sciences, Purdue University, West Lafayette, IN 47907–1392, U.S.A.

DESMOND HIGGINS. European Molecular Biology Laboratory, D-69012 Heidelberg, Germany

MATHIAS LÜBBEN. European Molecular Biology Laboratory, D-69012 Heidelberg, Germany

T. MATTIOLI. Département de Biologie Cellulaire et Moléculaire, CEA. de Saclay, 91191 Gif-sur-Yvette, France

T. E. MEYER. Department of Biochemistry, University of Arizona, Tucson, AZ, 85721, U.S.A.

NATHAN NELSON. Department of Biochemistry, Tel Aviv University, Ramat Aviv, 69 978 Tel Aviv, Israel

W. NITSCHKE. BIP/CNRS, 13402 Marseille, France

A. W. RUTHERFORD. SBE, CNRS URA 2096, Département de Biologie Cellulaire et Moléculaire, CEA de Saclay, 91191 Gif-sur-Yvette, France

MATTI SARASTE. European Molecular Biology Laboratory, D-69012 Heidelberg, Germany

VLADIMIR P. SKULACHEV. Department of Bioenergetics, A. N. Belozersky Institute of Physico-Chemical Biology, Moscow State University, Moscow 119899, Russia

G.-S. TAE. Department of Biology, Dankook University, Choongnan, South Korea

LINCOLN TAIZ. Department of Biology, Thiman Laboratories, University of California, Santa Cruz, CA 95064, U.S.A.

HANNS WEISS. Institut für Biochemie, Heinrich-Heine Universität Düsseldorf, 40225 Düsseldorf, Germany

ROBERT J. P. WILLIAMS. Inorganic Chemistry Laboratory, Oxford University, Oxford OX1 3QR, UK

MATTHIAS WILMANNS. European Molecular Biology Laboratory, D-69012 Heidelberg, Germany

Introduction

This volume is a collection of current knowledge about the origin and evolution of biological energy conversion. The eleven chapters cover those major metabolic systems for which significant information about the evolutionary aspects of bioenergetic reactions appears to exist. The first five chapters deal with various basic questions, and examine recent results that have provided new openings. The next six chapters treat the evolution of the most essential parts of photosynthetic and respiratory energy conversion.

Fortunately, because of an accelerating rate of experimental results obtained during the last few years, an overall view of the field is now rapidly developing. Expanding frontiers in chemistry, biology, physics, and geology, such as those found in new contact areas between biochemistry, molecular biology, biophysics, and biogeochemistry, contribute significantly to the emerging picture of how the bioenergetic reactions and systems may have originated and evolved. The reader, from the undergraduate student to the advanced researcher, is invited to share with the authors the present excitement in this broad and fascinating field. A certain amount of overlap can be found between different chapters, some confirmatory and some conflicting, which illustrates the dynamic activity in the growing evolutionary network. The content of differences in points of view also illuminates the considerable amount of "educated guesses", which naturally tends to increase with the distance back in time.

The volume is at the forefront of research and is written by invited specialists. The general backgrounds provided and references given allow an easy access to basic broad and introductory information, for example, in cited books and review articles.

Life as a "flow of energy, matter and information," from the book *The Logic of Life* by François Jacob (translated from *La logique du vivant: une histoire de l'hérédité*), may be regarded as one of the best, and shortest, descriptions of life from a scientific viewpoint. He emphasized the fundamental requirement for energy to drive a multitude of energy-requiring reactions in all living matter, including those required for building its informational macromolecules, DNA and RNA (deoxyribo- and ribonucleic acid).

How did life originate and evolve? Questions about the origin of the universe, sun, earth, life, and man have been asked for thousands of years, both by philosophers such as Aristotle and in the codified texts of the great religions. Early scientific approaches to fundamental problems concerning the origin and evolution of life included the work of, for example, Cuvier, Lamarck, Erasmus Darwin (grandfather of Charles), and Berzelius, but the breakthrough occurred when Charles Darwin in 1859 published *The Origin of Species*. A wealth of experiments and observations supported the incisive conclusions presented in this monumental work.

Since the 1930s, results from investigations on the molecular aspects of biological energy conversion have provided increasingly detailed knowledge about mechanisms involved in the biological transformation, coupling, conversion and conservation of energy. Similarly, the evolutionary picture of these processes started to emerge in the 1960s. Fritz Lipmann, the "father of bioenergetics," played a most significant role in the early breakthroughs. One may describe the present rate of progress as breathtaking. However, it should be emphasized that many large areas still show a very extensive lack of knowledge.

Three examples of discoveries, all from 1995, will illustrate the present pace of development and its possible repercussions. The first concerns an exotic tungsten-containing enzyme, which was purified and crystallized from the hyperthermophilic archaeon (archaebacterium) *Pyrococcus furiosus*, known to grow optimally at 100°C. The three-dimensional structure of this tungstopterin enzyme, aldehyde ferredoxin oxidoreductase, was determined at 2.3 Å resolution and published in March of 1995 [1]. This result with the extremely thermostable protein led to important information both about the binding sites for the tungsten cofactor and a closely situated Fe_4S_4 cluster, and about general aspects of protein thermostability. It is the first hyperthermophilic enzyme to be structurally characterized in three dimensions at atomic resolution. The detailed bioenergetic and evolutionary implications remain a question for the future. In the kingdom Archaea, with its remarkable number of unique redox systems, the evolution of energy conversion is still essentially a *terra incognita*, but the structure determination can well be a most significant step to change this situation. Additional steps toward this end may soon be taken, when current efforts in several laboratories to determine the complete nucleotide sequences of different archaeal genomes become successful.

The second example concerns the two first complete nucleotide sequences of bacterial genomes, from the kingdom Bacteria (Eubacteria). The first, from *Haemophilus Influenzae* Rd, was presented in July and the second, from *Mycoplasma genitalium* in October of 1995 [2, 3]. Determination of these two first complete genome sequences from free-living organisms has extended expectations about deeper knowledge of basic evolutionary patterns, as compared to that obtained from earlier complete sequences from several viral and organellar genomes. Ongoing active genome projects covering various archaeal, bacterial, and eukaryotic species should contribute to the clarification of the major pathways in bioenergetic evolution.

The third example is cytochrome c oxidase. The first high-resolution three-dimensional structures obtained, at 2.8 Å, with this multisubunit protein were published recently. The August 24, 1995, issue of *Nature* contained the structure of the four-subunit enzyme from *Paracoccus denitrificans* [4] and the August 25, 1995, issue of *Science* contained the structure of metal sites of the oxidized 13-subunit enzyme from bovine heart [5]. The new structural data give additional insights into the mechanisms involved in electron transport and proton pumping.

These examples show three areas where new structural data may soon produce a rich harvest of new evolutionary knowledge. The fact that it was two parallel discoveries rather than a single one in two of the three examples, accentuates the rapid pace of progress in areas within or significantly close to the evolution of biological energy conversion. Although we are still only in the beginning of the long road to a clear understanding of its origin and evolution, the recent developments in our field lead to optimism. Even if we realistically keep in mind various problems which may always blur the evolutionary picture, such as, for example, lateral gene transfer, an awesome growth of the relevant sphere of knowledge may be expected, perhaps only five to ten years from now. This would seem to add to the importance of presenting this collection of the current status of the field.

At this very time, the links are strengthening between the three interconnected areas enzyme evolution, enzyme mechanism, and enzyme structure. Evolutionary knowledge increasingly supports knowledge on structure and mechanism, properties which, as may be recalled, arose through evolution in action.

I wish to express may wholehearted thanks to the contributors of the different chapters for their willingness to participate in this endeavor and for the great ability and enthusiasm they have shown, as well as for their patience and support. It is also my pleasure to gratefully acknowledge VCH Publishers for never failing encouragement and patience.

Herrick Baltscheffsky
Stockholm, Sweden

References

1. Chan, M. K., Mukund, S., Kletzin, A., Adams, M. W. W., and Rees, D. C. *Science* 1995; *267*, 1463–1469.
2. R. D. Fleischmann et al. *Science* 1995; *269*, 496–512.
3. C. M. Fraser et al. *Science* 1995; *270*, 397–403.
4. Iwata, S., Ostermeier, C., Ludwig, B., and Michel, H. *Nature* 1995; *376*, 660–669.
5. Tsukihara, T. et al. *Science* 1995; *269*, 1069–1074.

Energy Conversion Leading to the Origin and Early Evolution of Life: Did Inorganic Pyrophosphate Precede Adenosine Triphosphate?

Herrick Baltscheffsky

1.1 Introduction

The further back in time we look, the more dimly we may expect to perceive various details of the molecular events that led to the emergence of life on Earth and to the subsequent evolution of all known living organisms from a common origin. One fruitful approach is to extrapolate backward from what is known about present or fossilized forms of life. Another well-known approach is to try to simulate conditions assumed to have existed on the abiotic, early Earth and to obtain the compounds essential for life. Increased confidence in suggested schemes of very early evolution is gained when different approaches produce results that appear to reinforce each other. Such successful outcomes have given in recent years several plausible chemical pathways of molecular evolution leading from early conditions and events before the appearance of life to its presumed origin about 3.8 Ga (1 gigayear = 1 billion years) ago [1] and to the subsequent biological evolution.

Energy sources for the molecular evolution of those systems, which became the early converters of biologically useful free energy, may have been chemical, photic and/or thermal. These alternatives are briefly discussed before turning to the main topics, which include some suggested alternatives for primordial energy conversion and the question of whether inorganic pyrophosphate (PP_i)

preceded adenosine triphosphate (ATP) in the energy conversion processes leading to the origins and the early evolution of life on Earth. It will be shown that experiment and theory along different lines have provided an increasingly coherent picture of possible early events related to these topics.

1.2 Different Kinds of Energy Sources

1.2.1 Chemical Energy Sources

Chemical energy is generally assumed to have been obtained already in a "prebiotic" world either from energy-rich compounds, such as, for example, inorganic or organic phosphoanhydrides, acyl phosphates, and thioesters [2–7] or from sufficiently exergonic oxidation–reduction reactions [8]. Several kinds of chemical energy conversion reactions may well have occurred on the primitive Earth and participated in the pathways of molecular evolution leading to the origin of life.

1.2.2 Photic Energy Sources

Photons from our own star, the sun, cover an extensive spectrum of wavelengths as light quanta with energies sufficient for driving energy-requiring reactions of biological [9, 10] and potential "prebiological" significance. The very complex photosynthetic systems of living cells, with chlorophyll [9, 10] or carotenoid [11] absorbing the light and transforming its energy to a chemical and/or chemiosmotic form, are assumed to have been preceded by much simpler systems with similar and, at very early stages, also more or less different pigments. In addition, alternative sources of light energy, in the depths of oceans, have recently been suggested [12]. The wavelengths in this case are in the infrared, in regions where certain bacteriochlorophylls absorb, and where, of course, the energy is both photic and thermal!

1.2.3 Thermal Energy Sources

At an earlier stage the thermal connection was essentially the production of energy-rich phosphoanhydrides by heating of orthophosphate, or other low-energy phosphates under anhydrous conditions [13]. Recently a most important extension has been reported, in the continuous formation of inorganic pyrophosphate ($H_4P_2O_7$), as well as tri- and tetrapolyphosphate, from hot volcanic magma, which on cooling gives these energy-rich products in partial hydrolysis of the magmal mother compound P_4O_{10} [14].

1.3 Suggested Energy Conversion Leading to the Origin of Life

1.3.1 Three Chemical Prenucleotide "Worlds"

Before molecular evolution had reached the stage of RNA "world," or more generally, that of mono-, oligo- and polynucleotides, less complex compounds must have participated in evolution. The early flows of energy, matter, and information leading to the origin of life may well have involved simple systems and molecules that are still involved in the metabolism of living cells. Restricting the discussion to the early flow of energy, I shall consider three distinct proposals for early "worlds" of prebiological and prenucleotide energy conversion: the thioester world [7], the iron–sulfur world [15,16], and the inorganic pyrophosphate world (PP$_i$ world) [17].

In contrast to the thioester world, which presupposes a heterotrophic origin of life, the PP$_i$ world may have operated independently of whether the origin of life was autotrophic or heterotrophic (recently an intensely debated subject [18,19]). A chemoautotrophic origin is visualized in the iron-sulfur world [19].

1.3.2 Molecular Links Between Different Proposed "Worlds"

One could discuss at length the question of which of the three proposed "worlds" may have been operating first or have been most central or most probable. Detailed arguments for the plausibility of each have been presented [7, 15, 17, 20–22], as have molecular links between the different suggested worlds. For example, the well-known cellular metabolite acetyl phosphate could have been a prebiological link between a thioester world and a PP$_i$ world [5,7]. In Figure 1.1 the possibility of contacts and interactions between the

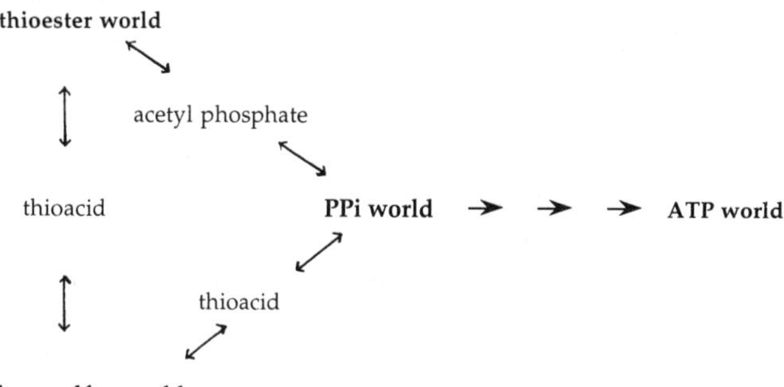

Figure 1.1 Three prenucleotide "worlds" and possible molecular links between them. Suggested evolution from the PP$_i$ world to the ATP world. The background for acetyl phosphate and thioacid as links between "worlds" is given in Refs. [5], [7], and [21].

three suggested prenucleotide worlds is shown, with emphasis on proposed links between them. The particular pathway of molecular evolution, which had been suggested earlier between a PP_i world and an "ATP world" in a subsequently emerging nucleotide world, emphasizing the central significance of this universal cellular carrier of biologically useful chemical energy, is treated below in greater detail. However, the recent [22, 23] and current search for clear molecular support for the hypothesis that PP_i preceded ATP in the energy conversion leading to the origin and early evolution of life may be mentioned at this point.

1.4 PP_i, High-Molecular-Weight Polyphosphate, and ATP

1.4.1 PP_i (Inorganic Pyrophosphate), an Early Energy Carrier?

About three decades ago we discovered, in isolated bacterial chromatophores, the photophosphorylation of inorganic orthophosphate to inorganic pyrophosphate [24, 25] according to:

$$2\,P_i \leftrightarrow PP_i + H_2O$$

PP_i was the first and is still the only alternative to ATP as a primary energy-rich chemical product of photophosphorylation. The reaction, catalyzed by an integrally membrane-bound PP_i synthase, has been considered a possible example of a retained, very early mode of light-induced formation of an energy-rich phosphate compound [26]. The new roles of PP_i in a photosynthetic system, both its formation at the expense of light energy and its function as energy donor [27, 28], corroborated earlier examples from intermediary metabolism indicating that PP_i may have served as a very early energy donor [3, 29].

Comparing the rates of light-induced formation of PP_i and ATP at saturating light intensities, in chromatophores isolated from anaerobically light grown *Rhodospirillum rubrum*, shows 5- to 10-fold higher rates of ATP formation than of PP_i formation. However, at low intensities of light the rate of PP_i formation is about twice as high as that of ATP formation [30], and titration with antimycin, which inhibits electron transport at the level of cytochrome bc_1, at concentrations that almost completely inhibit ATP formation leave the formation of PP_i at a level about as high as 30% of the maximum. This agrees with the thermodynamic difference between PP_i and ATP, the former being only one half to two thirds as energy-rich as the latter, allowing operation at a lower energy level. Independent support for the capability of the PP_i systems to work at lower energy levels than those with ATP came from the recent demonstration that the H^+/PP_i stoichiometry of the *R. rubrum* PPase, nearly 2, is appreciably lower than that of H^+/ATP, which was found to be nearly 3.6 [31]. Is it reasonable to extrapolate backward the expression "PP$_i$ is the poor man's ATP" to "PP$_i$ is early life's ATP"?

What of availability and stability of PP$_i$ on an abiotic Earth? Given the general assumption that its volcanic activity was much higher than at present the continuous production of PP$_i$ from volcanic magma [14] should have secured its availability under extended time periods. The stability question may be answered by referring to the existence of pyrophosphate in mineral form, as canaphite, a calcium sodium pyrophosphate containing four molecules of crystal water [32]. So, although PP$_i$ is thermodynamically hot, it is kinetically cold (even in presence of crystal water), and thus remarkably stable, here as a mineral.

To sum up this part, biological, chemical, and geological properties of PP$_i$ seem to be consistent with the proposal that PP$_i$ may have been active as an early energy carrier.

1.4.2 High Molecular Weight Polyphosphate (Poly-P)

Early work on high molecular weight polyphosphate metabolism [4] led to the view that this linear homopolymer may have been of both prebiological and early biological significance as energy and phosphate donor. A poly-P kinase is known [33, 34] to catalyze the reaction

$$n\text{ATP} \leftrightarrow \text{poly-P}_n + n\text{ADP}$$

where the more favored conversion is from poly-P to ATP. This reaction clearly shows the potential of poly-P to serve as a donor of biologically useful chemical energy.

Investigations and discussion of poly-P and of enzymes catalyzing its reactions [35–37] have gained increased impact recently by the successful cloning of the genes for some of the enzymes involved and the characterization of the proteins polyphosphate kinase [38, 39] and exopolyphosphatase [40, 41].

Although little may be said at present about the stability and availability of linear poly-P on the abiotic or the prebiotic [20] Earth, the relative simplicity of its structure makes it, as a homopolymer, a plausible precursor to proteins, RNA, and DNA [42] and as an inorganic phosphoanhydride type energy-rich compound, to ATP [4].

1.4.3 ATP (Adenosine Triphosphate)

The relative complexity of ATP is a prerequisite for its great versatility in the living cell. In addition to being able to act as an energy and phosphate donor, it can also pyrophosphorylate and adenylate. Little wonder then that PP$_i$, which shares only the two first mentioned capabilities with ATP, occupies a less central position than ATP in current cellular metabolism.

Two well-known distinctions or concepts concerning early evolution may be useful in this context. The first is that the "self-organization" of matter is rather generally assumed to have given an early "prenucleotide world" evolving to a "nucleotide world." This transformation to the nucleotide level of complexity may tentatively be assumed to have coincided with an evolution of ATP metabolism from PP_i metabolism, particularly significant consequences given that ATP at this early stage may already have started to reach its current position as the central molecular energy currency of living cells and that ATP as a donor of the adenylate moiety became a basic prerequisite for various biosynthetic pathways. The second is the sequence from early chemical evolution over progenotes [43] or preprokaryotes [44] to prokaryotes, the earliest known forms of life. A takeover by ATP from PP_i may well have started at the progenote (preprokaryote) level and gone toward its present degree of completion in prokaryotes, soon after the origin of life.

1.5 Evolution from PP_i Metabolism to ATP Metabolism?

1.5.1 PP_i as a Present Metabolic Alternative for ATP

Several reactions in living cells are known to involve PP_i, in many cases as an alternative for ATP [37,45]. Highly conserved elements of active sites have been reported for PP_i- and ATP-dependent 6-P-fructo-1-kinases, suggesting that PP_i, in a PP_i-dependent enzyme from one bacterium and ATP in an ATP-dependent enzyme from another, bind to the "same" site [46]. Most of the cases where PP_i and ATP may substitute for each other concern soluble enzyme reactions, in contrast to the energy conversion reactions involving PP_i and ATP in bacterial chromatophore and in plant vacuolar membranes.

Since 1992 [47] the genes for at least three vacuolar proton pumping PPases have been sequenced. The corresponding amino acid sequences indicate that the proteins have many membrane transversing hydrophobic stretches. No bacterial PP_i synthase gene has yet been fully sequenced, but the protein shows immunological cross-reaction with vacuolar PPase [48]. The indication that the integrally membrane-bound and proton pumping PP_i synthase, similarly to the vacuolar PPases, in its energy conversion capacity is bound to the membrane as a homodimer [49] brings into focus the question of whether there has been a direct evolution from a membrane-bound PPase (PP_i synthase) to a membrane-bound ATPase (ATP synthase). More information on sequence and preferably also three-dimensional structure is required before this question can be evaluated more clearly.

1.5.2 Current Evolutionary Indications

Two findings that support an evolution from enzymes metabolizing energy-rich inorganic phosphoanhydrides to enzymes metabolizing phosphoanhydrides of nucleotides are the recent discoveries that an exopolyphosphate phosphatase belongs to the same superfamily as several ATP metabolizing enzymes [50] and that another exopolyphosphate phosphatase also has guanosine pentaphosphate phosphatase activity [51]. At present the indications for a direct evolutionary connection between the metabolisms of poly-P and ATP are thus stronger than those for PP_i and ATP [22, 23]. However, the rate at which new structural information is now forthcoming also with respect to enzymes involved in PP_i, poly-P, and ATP metabolism, is cause for optimism that, in a not too distant future, some of the many remaining questions in this rapidly growing research area will be answered.

Acknowledgment

The author gratefully acknowledges support from Carl Tryggers Stiftelse för Vetenskaplig Forskning.

References

1. Schidlowski, M. *Nature* 1988; *333*, 313–318.

2. Miller, S. L., Parris, M. *Nature* 1964; *204*, 1248–1250.

3. Lipmann, F. Projecting Backward from the Present Stage of Evolution of Biosynthesis. In *The Origins of Prebiological Systems and of Their Molecular Matrices*, Fox, S. W. (ed.), Academic Press, New York, 1965; pp. 212–226.

4. Kulaev, I. S. *The Biochemistry of Inorganic Polyphosphates*, John Wiley & Sons, New York, 1979; 255 pp.

5. Baltscheffsky, H., Baltscheffsky, M. Molecular Origin and Evolution of Early Biological Energy Conversion. In *Early Life on Earth*, Bengtson, S. (ed.), Columbia University Press, New York, 1994; pp. 81–90.

6. Weber, A. L. *BioSystems* 1982; *15*, 18–189.

7. de Duve, C. *Blueprint for a Cell. The Nature and Origin of Life.* Patterson, New York, 1991; 275 pp.

8. Wächtershäuser, G. *System. Appl. Microbiol.* 1988; *10*, 207–210.

9. Hall, D. O., Rao, K. K. *Photosynthesis,* Cambridge University Press, Cambridge, 1995; 211 pp.

10. Harris, D. A. *Bioenergetics at a Glance.* Blackwell, Oxford, 1995; 116 pp.

11. Váró, G., Lanyi, J. K. *Biochemistry* 1991; *30*, 5016–5022.

12. Nisbet, E. G., Cann, J. R., Van Dover, C. L. *Nature* 1995; *373*, 479–480; *376*, 26–27; Björn, L. O. *Nature* 1995; *376*, 26; Allen, J. F. *Nature* 1995; *376*, 26.

13. Ponnamperuma, C., Chang, S. The Role of Phosphates in Chemical Evolution. In *Molecular Evolution 1. Chemical Evolution and the Origin of Life*, Buvet, R., Ponnamperuma, C. (eds.), North Holland, Amsterdam, 1971; pp. 216–222.

14. Yamagata, Y., Watanabe, H., Saitoh, M., Namba, T. *Nature* 1991; *352*, 516–519.

15. Wächtershäuser, G. *Prog. Biophys. Mol. Biol.* 1992; *58*, 85–201.

16. Kaschke, M., Russell, M. J., Cole, W. J. *Orig. Life* 1994; *24*, 43–56.

17. Baltscheffsky, H. Chemical Origin and Early Evolution of Biological Energy Conversion. In *Chemical Evolution: Origin of Life*, Ponnamperuma, C., Chela-Flores, J. (eds.), A. Deepak, Hampton, 1993; pp. 13–23.

18. de Duve, C., Miller, S. L. *Proc. Natl. Acad. Sci. USA* 1991; *88*, 10014–10017.

19. Wächtershäuser, G. *Proc. Natl. Acad. Sci. USA* 1994; *91*, 4283–4287.

20. de Duve, C. *Vital Dust. Life as a Cosmic Imperative*, Basic Books, New York, 1995; 362 pp.

21. Keller, M., Blöchl, E., Wächtershäuser, G., Stetter, K. O. *Nature* 1994; *368*, 836–838.

22. Baltscheffsky, H., Baltscheffsky, M. Energy, Matter and Self-Organization in the Early Molecular Evolution of Bioenergetic Systems. In *Chemical Evolution; Self-Organization of the Macromolecules of Life*, ChelaFlorers, J., Chadha, M., Negron-Mendoza, A., Oshima, T. (eds.), A. Deepak, Hampton, 1995; pp. 83–89.

23. Baltscheffsky, H., Baltscheffsky, M. Energy-Rich Phosphate Compounds and the Origin of Life. In *Evolutionary Biochemistry and Related Areas of Physicochemical Biology*, Poglazov, B. F., Kurganov, B. I., Kritsky, M. S., Gladilin, K. L. (eds.), Bach Institute and ANKO, Moscow, 1995; pp. 191–199.

24. Baltscheffsky, H., von Stedingk, L.-V., Heldt, H. W., Klingenberg, M. *Science* 1966; *153*, 1120–1122.

25. Baltscheffsky, H., von Stedingk, L.-V. *Biochem. Biophys. Res. Commun.* 1966; *22*, 722–728.

26. Baltscheffsky, H. *Acta. Chem. Scand.* 1967; *21*, 1973–1974.

27. Baltscheffsky, M. *Nature* 1967; *216*, 241–243.

28. Baltscheffsky, M. *Arch. Biochem. Biophys.* 1969; *133*, 46–53.

29. Siu, P. M. L., Wood, H. G. *J. Biol. Chem.* 1962; *237*, 3044–3051.

30. Nyrén, P., Nore, B. F., Baltscheffsky, M. *Biochim. Biophys. Acta* 1986; *851*, 276–282.

31. Sosa, A., Celis, H. *Arch. Biochem. Biophys.* 1995; *316*, 421–427.

32. Rouse, R. C., Peacor, D. R., Freed, R. L. *Am. Mineral.* 1988; *73*, 168–171.

33. Kornberg, A., Kornberg, S. R., Simms, E. S. *Biochim. Biophys. Acta* 1956; *20*, 215–227.

34. Kornberg, S. R. *Biochim. Biophys. Acta* 1957; *26*, 294–300.

35. Harold, F. M. *Bacteriol. Rev.* 1967; *30*, 1229–1242.

36. Kulaev, I. S., Vagabov, V. M., Shabalin, Y. A. New Data on Biosynthesis of Polyphosphates in Yeast. In *Phosphate Metabolism and Cellular Regulation in Microorganisms*, Torriani-Gorini, A, Rothman, F. G., Silver, S., Wright, A., Yagil, E. (eds.), American Society for Microbiology, Washington, D.C., 1987; pp. 233–238.

37. Wood, H. G., Clark, J. E. *Annu. Rev. Biochem.* 1988; *57*, 235–260.

38. Ahn, K., Kornberg, A. *J. Biol. Chem.* 1990; *265*, 11734–11739.

39. Akiyama, M., Crooke, E., Kornberg, A. *J. Biol. Chem.* 1992; *267*, 22556–22561.

40. Akiyama, M., Crooke, E., Kornberg, A. *J. Biol. Chem.* 1993; *268*, 633–639.

41. Andreeva, N. A., Okorokov, L. A. *Yeast* 1993; *9*, 127–139.

42. Kornberg, A. *J. Bacteriol.* 1995; *177*, 491–496.

43. Woese, C. R., Fox, G. E. *Proc. Natl. Acad. Sci. USA* 1977; *74*, 5088–5090.

44. Baltscheffsky, H., Jurka, J. On Protocells, Preprokaryotes, and Early Prokaryotes. In *Molecular Evolution and Protobiology*, Matsuno, K., Dose, K., Harada, K., Rohlfing, D. L. (eds.), Plenum Press, New York, 1984; pp. 207–214.

45. Baltscheffsky, M., Baltscheffsky, H. Inorganic Pyrophosphate and Inorganic Pyrophosphatases. In *Molecular Mechanisms in Bioenergetics*, Ernster, L. (ed.), Elsevier, Amsterdam, 1992; pp. 331–348.

46. Ladror, U. S., Gollapudi, L., Tripathi, R. L., Latshaw, S., Kemp, R. G. *J. Biol. Chem.* 1991; *266*, 16550–16555.

47. Sarafian, V., Kim, Y., Poole, R. J., Rea, P. A. *Proc. Natl. Acad. Sci. USA* 1992; *89*, 1775–1779.

48. Nore, B. F., Sakai-Nore, Y., Maeshima, M., Baltscheffsky, M., Nyrén, P. *Biochem. Biophys. Res. Commun.* 1991; *181*, 962–967.

49. Nyrén, P., Nore, B. F., Strid, Å. *Biochemistry* 1991; *30*, 2883–2887.

50. Reizer, J., Reizer, A., Saier, M. H., Jr., Bork, P., Sander, C. *TIBS* 1993; *18*, 247–248.

51. Keasling, J. D., Bertsch, L., Kornberg, A. *Proc. Natl. Acad. Sci. USA* 1993; *90*, 7029–7033.

Evolution of Convertible Energy Currencies of the Living Cell: From ATP to $\Delta\bar{\mu}_{H^+}$ and $\Delta\bar{\mu}_{Na^+}$

Vladimir P. Skulachev

All known forms of contemporary life require (1) nucleic acids and nucleotides, (2) proteins, and (3) an insulating membrane. In this chapter, I address the question of why only these three components proved to play the key roles in the major processes occurring in living cells. To do this the bioenergetic aspect is considered, an approach that is reasonable if we assume that the energy supply was as critical for the origin of life as it is for its maintenance today.

2.1 Light as the Primary Energy Source. Adenine-Mediated Photosynthesis Supported by Ultraviolet Light

The evolution from a mixture of organic compounds to the first living cell was apparently so lengthy that the energy source that supported this process had to be continuously available during the entire period. The sun seems to be the only candidate for such an energy source. In this respect, the possible role of the middle ultraviolet region of sunlight is especially interesting. Corresponding quanta are of the highest energy charge among those that do not cause immediate and irreversible decomposition of light-absorbing organic molecules.

There are some reasons to assume that ultraviolet (UV) radiation reached the surface of our planet when the primordial living systems were formed, or, alternatively, the Earth was "infected" by living cells originating from another area of the Unisphere, where UV light was available.

2.1.1 Why Adenine Could Be Selected as a UV Pigment

Trying to model the gas phase of the ancient Earth atmosphere Sagan came to the conclusion that there was a "window" in the region 240 to 290 nm transparent to UV owing to the presence of CH_4, NH_3, H_2O, CO, CO_2 (all absorbing shorter than 240 nm), and formaldehyde, which absorbed longer than 290 nm [1]. Purines and pyrimidines have absorption maxima in this "window."

Adenine seems a good candidate to be the UV-absorbing pigment. As Ponnamperuma et al. have shown, UV irradiation of a $10^{-4} M$ solution of hydrocyanic acid results in formation of adenine and guanine [2]. Thermal polymerization of hydrocyanic acid in a solution of aqueous ammonia also produced adenine, but the concentration of hydrocyanic acid had to be as high as $1.5 M$ [3]. Electron irradiation of 5 MeV of methane, ammonia, hydrogen, and water was shown to produce purines and pyrimidines. Under these conditions, the substance synthesized in highest yield was adenine [4]. Besides this observation, Ponnamperuma et al. [2] indicated other features favorable for selection of adenine as the UV chromophore: (1) the highest UV absorption coefficient in comparison to other natural purines and pyrimidines; (2) the highest stability to the decomposing action of UV; and (3) the longest life time of the excited state. All these properties derive from the greatest resonance energy of adenine [2, 5, 6].

It is remarkable that the same energy source, that is, UV light, can, according to data of Ponnamperuma et al. [2, 7] be used for nonenzymatic formation of adenosine from adenine and ribose (the quantum yield for a 1-h irradiation was as high as 1×10^{-5}). Ultraviolet irradiation of an aqueous solution containing adenosine and ethyl metaphosphate produced AMP, ADP, and ATP, in a quantum yield of 1×10^{-4}. A measurable amount of ATP was found when ADP was irradiated by UV light in the presence of ethyl metaphosphate [2, 7]. Synthesis of inorganic pyrophosphate was demonstrated by Barltrop and co-workers when adenine was added to a solution of inorganic phosphate, Mg^{2+}, and HBr_3 [8].

Calculations made some time ago by the Russian chemists Blumenfeld and Temkin [9] attracted our attention to the fact that the dissociation constants for protonated aromatic amine (aniline) and aliphatic methyl amine are 3.5×10^{-10} and 5×10^{-4}, respectively, which corresponds to a free energy difference for protonation as high as $8.4 \, \text{kcal} \times \text{mol}^{-1}$ [10].

Taking these facts into account, I suggested [11] that the mechanism of phosphorylation catalyzed by the adenine-containing compounds under UV irradiation includes addition of a phosphoryl group to the "aliphatic" amino group of the excited adenine. Relaxation of the excited adenine to the ground state is postulated to energize the phosphoamide bond (aliphatic-to-aromatic transition). Phosphorolysis of the phosphoadenine may give inorganic pyrophosphate.

2.1.2 Why ATP Proved to Be the Primary Biological Energy Currency

If ADP, rather than free adenine, is used as the chromophore, ATP might be formed by means of transfer of a phosphoryl group from the adenine head to the pyrophosphate tail of the nucleotide. For such a process, it may be essential that, in the "scorpion" conformation of ADP, the distance between the adenine amino group and the terminal phosphate exactly corresponds to one more (the third) phosphate residue [12].

The phosphoryl group, when transferred from adenine to the phosphate tail of the nucleotide, proved to be stabilized to some degree because more energized aromatic phosphoamide was replaced by less energized phosphoanhydride (Fig. 2.1A).

2.1.3 Why Are Some Coenzymes Nucleotides?

The dinucleotide coenzymes, such as NAD, NADP, and FAD, contain an adenine residue that is coplanar to the coenzyme redox heterocycle, that is, nicotinamide or flavin. Energy transfer between adenine and the redox heterocycle has been shown (for review, see Ref. [11]). This means that UV excitation of the anenine moiety of the dinucleotide coenzyme can change the reactivity of its redox moiety.

In CoA the active SH− group is attached to the rather long flexible "tail," allowing it to interact with the adenine "head."

In thiamine pyrophosphate the amino group of the pyrimidine part of the coenzyme was found to be crucial for the catalysis carried out by the thiazole ring [13]. Again, UV excitation of the pyrimidine ring might affect properties of the active site (thiazole). Perhaps the primary function of the purine- and pyrimidine-containing coenzymes was to catalyze chemical reactions energized by UV irradiation.

Interestingly, an N_6-substituted derivative of adenine was recently shown to be enzymatically formed from NAD^+. This derivative is cyclic adenosine diphosphate ribose (cADPR) [14], in which a high-energy bond exists between the amino group of the adenine residue and one of the ribose residues (Fig. 2.1B). This compound was shown to be involved in the cGMP-mediated Ca^{2+} release from an intracellular Ca^{2+}-storing compartment [15,16]. One may speculate that cADPR is a representative of a family of substances that appeared as a result of the chemical reactions mediated by the UV-excited adenine. According to the above concept, summarized in Section 2.1.1, ribosylation should be facilitated by UV light.

2.1.4 Nucleic Acids

Within the framework of the above logic, one may assume that RNA could be formed from purine and pyrimidine nucleotides at the expense of UV energy.

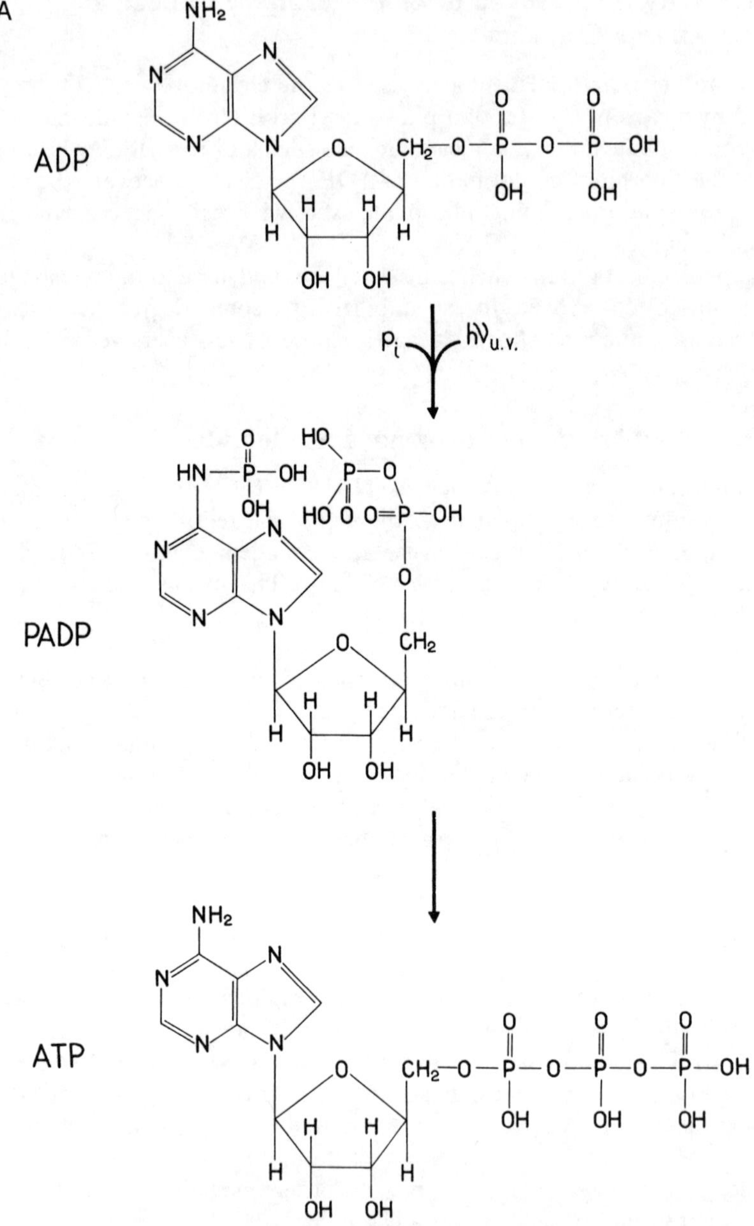

Figure 2.1   Adenine as an energy transducer. (A) Ultraviolet light-supported, adenine-mediated ADP phosphorylation. PADP, "scorpion" conformation of ADP phosphorylated at the amino group. (B) Conversion of NAD^+ to cyclic adenosine diphosphate ribose (cADPR), a high-energy compound with the covalently modified adenine amino group.

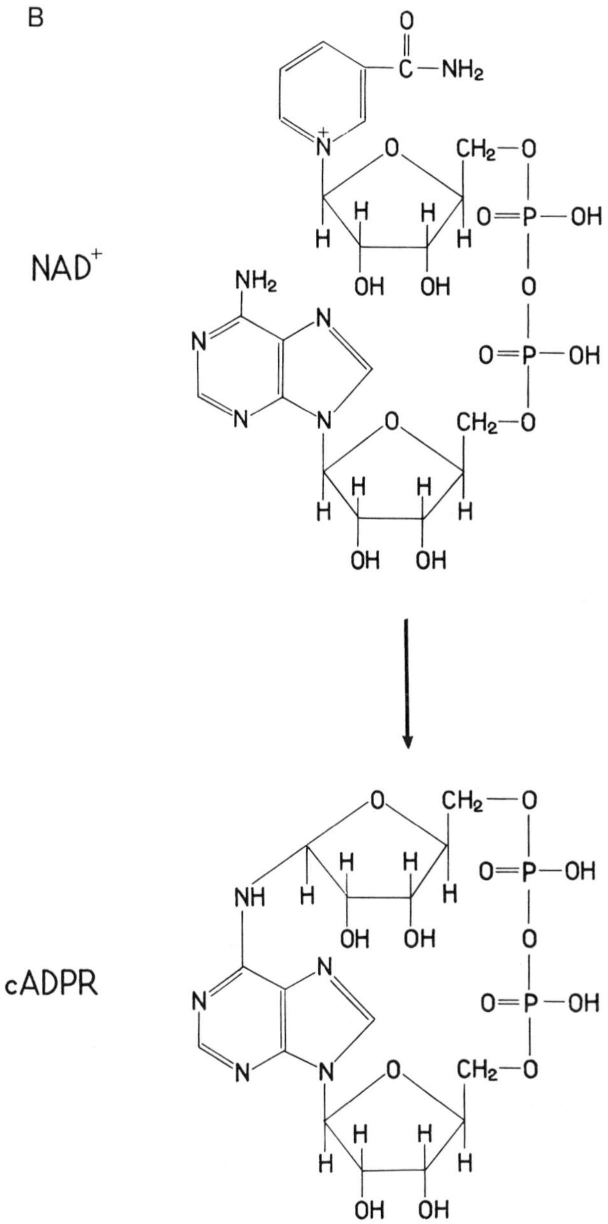

Figure 2.1 (*Continued*).

Apparently RNA was the first biopolymer to perform some catalytic functions. The ribozyme phenomenon might be a reminiscence of such an ancient RNA function. If this were the case, then DNA looks like a later RNA derivative specialized in encoding of RNA and proteins.

An interesting possibility is that oligoribonucleotides in the primordial cells operated as UV — harvesting antenna transferring the excitation energy to ADP, which formed a complex with the terminal nucleotide residue of oligonucleotide. Phosphorylation of the adenine residue in ADP could result in decomposition of the complex. Moreover, oligonucleotides could play the role of catalysts in ADP phosphorylation, assuming that it was a complex composed oligonucleotide, ADP, inorganic phosphate, and Mg^{2+} ions that served as the substrate of the phosphorylation. In this contest, a work by Murphy et al. [16a] may be mentioned. The authors succeeded in direct demonstration of electron transfer over a distance greater than 40 angstroms between metallointercalators tethered to the 5″ termini of a 15-base pair oligonucleotide duplex. Such an effect seems to be consistent with the possibility of antenna function of oligonucleotides.

2.2 Primary Energy Stores at Night

2.2.1 Inorganic Polyphosphates

In modern photosynthetic organisms, photosynthetic products (carbohydrates) are accumulated during the day for utilization at night. These processes are catalyzed by rather long chains of enzymes involved in (1) endergonic synthesis of carbohydrates from CO_2 and H_2O and (2) decomposition of carbohydrates resulting in ATP formation.

One may assume that a simpler process for light energy storage was used at the early stages of evolution. The ATP-driven synthesis of inorganic polyphosphates seems a good candidate for such a role.

Reversal of this process might buffer the ATP level at night (see Ref. [17] for a discussion of ATP-inorganic polyphosphate interconversion in microorganisms).

The ATP-buffering role of inorganic pyrophosphate has been postulated for some contemporary photosynthetic bacteria [18].

2.2.2 Carbohydrates

Accumulation of poly- and pyrophosphates solves only one problem, that is, storage of high-energy phosphate. On the other hand, storage of light energy in the form of carbohydrates also means storage of organic material that can be utilized in the dark not only to produce ATP but also to form pyruvate, acetyl-CoA, and di- and tricarboxylates, that is, key components of intermediary metabolism. It is not surprising, therefore, that in the course of evolution carbohydrates substituted for polyphosphates as the storage material.

At present, two main types of carbohydrate decomposition are known. One of them results in ethanol and CO_2, whereas the other gives lactic acid (or another carboxylic acid) that dissociates to the lactate anion and H^+. The former has an advantage because there is no risk of acidification of the cell interior when fermentation proceeds. On the other hand, a disadvantage consists in that lipid membranes are quite permeable for the fermentation products, ethanol and CO_2. These small neutral molecules immediately escape from the cell when formed. Lactate is a charged compound and, therefore, nonpenetrating. It might accumulate in the cell at night to be used as a substrate for gluconeogenesis during the day. However, for lactate to accumulate, the second glycolytic product, H^+ ion, must equilibrate with the outer medium; otherwise, a severe acidic pH shift occurs in the cytoplasm, resulting in denaturation of cellular proteins, etc. Unfortunately, H^+ is also nonpenetrating for the lipid membranes so that a special mechanism had to develop to facilitate transmembrane flux of H^+.

2.3 H^+-ATPase and Its H^+ Channel as Mechanisms for the Efflux of Glycolysis-Produced H^+

2.3.1 H^+ Channel

A well-known artifact in experiments on natural energy-coupling membranes is that membrane permeability for H^+ strongly increases when the catalytic part of the H^+-ATP synthase, that is, factor F_1, dissociates from its membrane anchor, factor F_0. Since the resulting H^+ conductance is sensitive to F_0 inhibitors, such as dicyclohexylcarbodiimide (DCCD), it was concluded that F_0 operates as an H^+ carrier or channel. It was also found that DCCD specifically modifies one dicarboxylic amino acid residue in the c subunit of factor F_0, an indication that this subunit is involved in H^+ transport (reviewed in Ref. [18]).

An example of a situation in which the c-type subunits function in vivo without other subunits organizing a transmembrane channel is the gap junctions responsible for the cell-to-cell exchange of low molecular weight solutes in animal tissues. It was found that the protein that forms the animal gap junction pore is very homologous in sequence and functionally similar to the DCCD-sensitive subunit of the membrane sector of the yeast vacuolar H^+-ATPase [19]. The cDNA coding for gap junction protein of the arthropod *Nephrops norvegicus* was recently expressed in the yeast *Saccharomyces cerevisiae*, in which the endogenous gene coding for the vacuolar ATPase c subunit has been inactivated. The expression was shown to restore the vacuolar ATPase activity and the cell growth [19].

In *E. coli*, the arsenite-extruding ATPase was shown to be composed of the membrane subunit B and the cytosol-exposed catalytic subunit A [20–23]. On the other hand, in *Staphylococci* the arsenite pump contains B but not A [23–25] and operates, most probably, as a $\Delta\Psi$-dependent, ATP-independent, arsenite-specific pore (or carrier). It was assumed that arsenite anion is extruded via this pore electrophoretically [22, 23].

These observations indicate that in some cases the membrane sectors of ion-transport ATPases can operate as an ion channel independently of the catalytic subunits of the enzyme. In this connection, one may suggest that the F_0 part of the F_0F_1 complex facilitated the H^+ efflux from primitive cells employing lactate-producing glycolysis as the dark mechanism of ATP formation (see Fig. 2.2B2).

A serious limitation of an H^+ channel (or H^+ uniporter) as a mechanism for H^+ unloading of the cell undergoing glycolysis consists in the fact that the H^+ efflux charges the membrane. The formed $\Delta\Psi$(inside negative) must stop large-scale H^+ efflux.

To avoid this difficulty, the membrane must be discharged. This may be done by making the membrane permeable to another cation present in the medium, say K^+ or Na^+, which substitutes for H_{in}^+ when it leaves the cell. To this end, an additional cation channel (carrier) could be used or, alternatively, factor F_0 could be rather nonspecific, transporting not only H^+ but also other small cations. In this context, one may mention that (1) the gap junction is nonspecific for small molecules [18] and (2) F_0 of *Propionigenum modestum* [26] and, apparently, of *E. coli* growing under some specific conditions [27], can transport Na^+.

Thus the chain of events involved in ATP synthesis in the dark could be described by Eqs. (2.1) to (2.3).

$$\text{Carbohydrate} + \text{ADP} + \text{P}_i \rightarrow \text{lactate}^- + \text{H}_{in}^+ + \text{ATP} \tag{2.1}$$

$$\text{H}_{in}^+ \rightarrow \text{H}_{out}^+ \tag{2.2}$$

$$\text{C}_{out}^+ \rightarrow \text{C}_{in}^+ \tag{2.3}$$

where C^+ is for K^+ and Na^+.

2.3.2 H^+-ATPase

For the mechanism shown by Eqs. (2.1) to (2.3), acidification of the medium would be the factor that limits H^+ efflux. To overcome this limitation, an active (i.e., energy-consuming) H^+ pump is required. Since ATP was assumed to be the only convertible energy currency at that time, the pump in question should represent H^+-ATPase composed of the already presnt H^+ channel (F_0) and the newly formed catalytic part (F_1) (see below, Fig. 2.2B3). Thus reaction (2.4) is substituted for reaction (2.2):

$$n\text{H}_{in}^+ + \text{ATP} \rightarrow n\text{H}_{out}^+ + \text{ADP} + \text{P}_i \tag{2.4}$$

Apparently n was higher than 1 to minimize the ATP expenditure for H^+ pumping. Reaction (2.4) describes events taking place in the cells of modern plants when they used glycolysis as a dark mechanism for ATP synthesis. Accumulating H^+ ions are expelled from the cells by the plasma membrane H^+-ATPase, which is strongly activated when the pH_{in} levels drops below 7.0 (for review, see Ref. [18]).

In this context, the H^+-pumping inorganic pyrophosphatase (H^+-PPase) should be mentioned. Such an enzyme, discovered in bacterial chromatophores by Baltscheffsky et al. [28], proved to be composed of two or four identical 56-kDa subunits. Hence, it is much simpler than H^+-ATPase (a minimal version of the bacterial type enzyme consists of 12 subunits of six different kinds [18]). Analysis of sequences revealed a strong similarity between a stretch containing amino acid residues 227 to 245 of a vacuolar H^+-PPase and a transmembrane α-helix of the c-subunit of the F_0 sector of H^+-ATPase. In just this helix a crucial dicarboxylic amino acid residue is localized that is attacked by DCCD [29].

One may speculate that in evolution the c-subunit became not only a component of H^+-ATPase, but also, by means of gene fusion, a part of the H^+-PPase polypeptide. H^+-PPase could well be employed, like H^+-ATPase, as a mechanism to pump glycolytic H^+ from the cell. Inorganic polyphosphate, an energy store in primordial cells, could be the source of PP_i. However, this mechanism most probably proved a dead end of evolution because the majority of contemporary living cells maintain $[PP_i]$ at a very low level which is favorable for numerous biosynthetic reactions supported by hydrolysis of ATP to AMP and PP_i (for details, see Ref. [18]).

2.4 Photosynthesis

2.4.1 Bacteriorhodopsin

The high energy charge of UV quanta had an advantage and simultaneously a disadvantage for primary living systems. On one hand, UV irradiation could activate numerous chemical reactions that otherwise would be highly improbable. On the other hand, it could cause, for the same reason, many unfavorable side processes that were to destabilize and even decompose the primitive cells. In this respect, visible light is much less dangerous, so that substitution of the UV type photosynthesis with the long wavelength type seems to have been a very progressive event of biological evolution. Such an event seems even more important (in fact, inevitable) if we assume that oxygen accumulated in the atmosphere in an abiogenic fashion (e.g., owing to UV-induced water photolysis). The appearance of atmospheric oxygen and then ozone had to result in a decrease in the UV irradiation on the surface of the Earth. To survive, primitive living cells had to invent a photosynthetic mechanism activated by the still available visible light.

The new photosynthetic process, like the old one, had to result in formation of ATP which already occupied the central position on the metabolic map, playing the role of convertible energy currency. However, now the adenine moiety of the nucleotide could not be used as a chromophore, so that new pigments had to be selected that had high absorption coefficients in the visible region of the sun's spectrum.

An attractive possibility is that the new photosynthetic machinery was organized in a way allowing a transmembrane charge separation to occur. If $\Delta\Psi$ produced by this mechanism was of the same direction as that generated by H^+-ATPase, that is, the cell interior negative, the flux of H^+ via H^+-ATPase could be reversed. As a result, H^+-ATPase began operating as an H^+-ATP synthase that catalyzed ATP synthesis coupled to the downhill H^+ influx instead of ATP hydrolysis coupled to the uphill H^+ efflux. To avoid ΔpH formation, that is, acidification of the cytoplasm caused by the H^+ influx, the light-driven charge separation should represent, in fact, H^+ efflux [eqs. (2.5) and (2.6)].

$$n\text{H}^+_{\text{in}} + \text{visible light} \rightarrow n\text{H}^+_{\text{out}} \tag{2.5}$$

$$n\text{H}^+_{\text{out}} + \text{ADP} + \text{P}_{\text{i}} \rightarrow n\text{H}^+_{\text{in}} + \text{ATP} \tag{2.6}$$

The process shown by Eq. (2.5) is now catalyzed by bacteriorhodopsin, a retinal-containing protein found in some archaea (see below, Fig. 2.2C). The efficiency of this mechanism is rather low, that is, one H^+ per photon of about 1.5 eV energy charge. Because the electric potential that can be maintained by a biological membrane does not exceed 0.3 V, this means that the bacterio-rhodopsin photoelectric generator operates with efficiency as low as 20%. It is not surprising, therefore, that bacteriorhodopsin photosynthesis is a very rare case among contemporary organisms. In fact, it is operative in only one group of modern microorganisms, that is, extremely halophilic and thermophilic archaea. On the other hand, bacteriorhodopsin is a rather simple and very stable device for exporting H^+ at the expense of visible light, a fact pointing to its very ancient origin. It is a single 26-kDA polypeptide resistant to heating up to 130°C, and to many oxidants and SH−reagents (it contains no cysteine) as well as to variations in salt concentration from zero up to saturation. It can be deprived of the water-exposed N- and C-terminal sequences and split in the middle of the polypeptide chain without serious impairment of the H^+ pumping.

The simplicity of bacteriorhodopsin resulting in its extreme stability seems to answer the question of why bacteriorhodopsin, rather than more efficient but much more complex chlorophyll-based systems, is used by halobacteria living under extreme conditions (saturated salt solutions, high temperature).

Bacteriorhodopsin is localized in special areas of the halobacterial membrane ("purple sheets") where there are no other proteins. In the sheet it forms trimers, but the monomeric form of bacteriorhodopsin is also quite active.

The chromophore (a retinal residue) is firmly (covalently) bound to the protein via a Schiff base. Light is known to cause all-*trans* → 13-*cis* isomerization of the retinal. This simple process does not include any unstable intermediates of high chemical reactivity.

The key event in H^+ pumping is, most probably, translocation, owing to isomerization, of the protonated Schiff base from the more hydrophobic to the less hydrophobic environment inside the bacteriorhodopsin molecule embedded in the membrane. This should be accompanied by a decrease in the pK value of the Schiff base and, hence, in deprotonation of the Schiff base.

The released H^+ ion moves to the outer medium via a rather hydrophilic region of the bacteriorhodopsin molecule, localized between the outer membrane surface and retinal which is situated midway from one membrane surface to the other. The part of bacteriorhodopsin separating retinal from the cytoplasmic surface of the membrane is very hydrophobic. This prevents H^+ from being transported in the wrong direction, that is, from outside to inside the cell (for reviews, see Refs. [18], [30–32]).

The retinal isomerization causes one more event besides the Schiff base deprotonation, a conformational change of the protein, namely, the appearance of a cleft in the hydrophobic half of the bacteriorhodopsin molecules facing the cytoplasm [33]. This should facilitate the H^+ transfer from the cytoplasm to the deprotonated Schiff base [32, 33].

As a result of reprotonation of the Schiff base by the cytoplasmic H^+, 13-*cis* → all-*trans* isomerization of retinal takes place, the cleft disappears, and bacteriorhodopsin returns to its ground state [18, 30–32].

Interestingly, the light-induced conformational change (most probably, the formation of a cleft on the cytoplasmic surface of the protein) was used in a later step of evolution in proteins operating as very sensitive photoreceptors (photon counters). In halobacteria, this function is performed by sensory rhodopsins I and II, which are involved in the attractant and repellent action of long and short wavelength light. Sensory rhodopsins look like bacteriorhodopsin mutants lacking Asp-96. In bacteriorhodopsin, this dicarboxylic amino acid residue strongly facilitates the H^+ transfer from the end of the cleft to the deprotonated Schiff base. The D96N bacteriorhodopsin mutant has a very slow photocycle because "the cleft intermediate" with the deprotonated Schiff base becomes long lived. One may speculate that just this cleft is recognized by some other protein ("bacteriotransducin") initiating, after complexation with "the cleft intermediate," the remainder of the chain of events resulting in amplification of the light signal and, in the very end, in a change in the sense of rotation of the bacterial flagella.

In some protozoa as well as in animals, retinal-containing proteins are responsible for photoreception. Visual rhodopsins of vertebrates and invertebrates, like sensory rhodopsins of halobacteria, do not contain a group equivalent to AsP-96 and they form long-lived intermediates of changed conformation. Again, an event (cleft formation?) occurs on the cytoplasmic surface of the protein that is essential for the interaction with the transducin [32].

Some receptor proteins using chemical compounds instead of a photon as the signal were shown to be composed, like bacterial and animal rhodopsins, of seven transmembrane α helical columns. Again, the domain exposed to the cytoplasm proved to be of crucial importance. It seems probable that these receptors evolved from rhodopsins in such a way that a signal molecule causes a change in the protein conformation, that is, appearance of "the cleft intermediate" [32, 33].

As to $\Delta\bar{\mu}_{H^+}$ generators, bacteriorhodopsin proved to be a dead end of evolution, apparently owing to low efficiency of the system, where effective absorption of *one* photon results in the transport of *one* H^+ ion.

2.4.2 Chlorophyll-Containing Systems

2.4.2.1 *Noncyclic Electron Flow in Green Sulfur Bacteria*

All known types of chlorophyll-based photosynthesis are more efficient, more complicated, and, hence, more vulnerable than the bacteriorhodopsin H^+ pump.

The increase in efficiency is the result of two facts: (1) The photons used are of longer wavelength and, hence, of lower energy than those absorbed by bacteriorhodopsin. It is noteworthy that long wavelength light can penetrate deeper into ocean water than short wavelength light does. This means that organisms employing chlorophyll-based photosynthesis can occupy more extensive areas compared with those employing bacteriorhodopsin. (2) In chlorophyll-based photosynthesis, a single photon supports the transport of more than one H^+. Moreover the light energy can be utilized not only to generate $\Delta\bar{\mu}_{H^+}$ but also to form "reducing power," that is, to carry out electron transfer from a donor of more positive redox potential to an acceptor of more negative redox potential. Such a function cannot be accomplished by bacteriorhodopsin, which is not involved in redox reactions.

In contemporary green sulfur bacteria, generally considered as a very primitive group [34–36], such as *Chlorobium*, H_2S is oxidized to inorganic sulfur and $2H^+$ on the outer surface of the cytoplasmic membrane. Electrons are transported across the membrane to its inner surface to reduce NAD^+ to NADH and then to a metabolite A equilibrated with the pair $NADH/NAD^+$. As a result, AH_2 is formed (see below, Fig. 2.2D1). Because the redox potential of H_2S/S is more positive than that of $NADH/NAD^+$, such an oxidoreduction is energy-consuming as soon as the stronger reductants, NADH and then AH_2, are formed by means of oxidation of the weaker reductant, H_2S [Eq. (2.7)]:

$$H_2S + A + 2H_{in}^+ + 2 \text{ photons} \rightarrow S + AH_2 + 2H_{out}^+ \tag{2.7}$$

This means that effective absorption of one photon causes production of one H_{out}^+, uptake of one H_{in}^+, rerduction of $1/2A$, and oxidation of $1/2H_2S$.

Combining Eqs. (2.7) and (2.4) (see p. 18) one can obtain simultaneously (1) photophosphorylation and (2) production of reducing power which can be used in reductive biosyntheses.

The process is initiated by photoexcitation of the bacteriochlorophyll a dimer, $(BChl)_2$, which is localized close to the outer membrane surface. The excited dimer is then oxidized by monomeric bacteriochlorophyll c, BChl. The latter transfers one electron to the next acceptor, a FeS cluster. $(BChl)_2$, BChl, and FeS are attached to a specific membrane protein. The whole system is called "the photosynthetic reaction center complex."

The complex reduces cytoplasmic NAD^+ via a flavoprotein localized on the cytoplasmic surface of the membrane.

Reduction of the photooxidized dimer, $(BChl)_2^+$, is carried out by the electron removed from the outer H_2S. This process is mediated by a quinone and cytochromes b and c (for reviews, see Refs. [18] and [37]).

It should be stressed that the primary product of the photoexcitation, that is, excited $(BChl)_2$, is a very reactive compound of negative redox potential. It might be oxidized not only by a bacteriochlorophyll monomer but also by any oxidant that can come in contact with excited $(BChl)_2$. To avoid such side reactions, the excited dimer $\rightarrow$ monomer electron transfer (so-called "primary charge separation") occurs on the picosecond time scale, the fastest redox process of biological origin. Such a high rate becomes possible because the dimer and the monomer are situated very close to each other and oxidoreduction does not require any movement of these two reactants. This allows the short-lived singlet excitation state of the dimer to be used to reduce the monomer.

2.4.2.2 Cyclic Electron Flow in Purple Bacteria

This process can be considered as a variation of the noncyclic process discussed in the preceding section. Let us assume that (1) a redox mediator that functions (in its oxidized form) as the $(BChl)_2$ oxidant is employed, when reduced, as the $(BChl)_2^+$ reductant, and (2) both oxidized and reduced forms of the redox mediator easily traverse the hydrophobic barrier of the membrane. The consequence of such a situation is that a cyclic photoinduced electron transfer substitutes for the noncyclic one.

In contemporary bacteria, the role of the penetrating redox mediator is performed by a quinone derivative, that is, ubi-, mena-, or plastoquinone. *Rhodospirillum rubrum* and *Rhodobacter sphaeroides* employ ubiquinone (CoQ). In these bacteria the $(BChl)_2^+$ reducing mechanism is similar to that of the green sulfur bacteria, that is, electron transfer from $CoQH_2$ to $(BChl)_2^+$ is mediated by a c-type cytochrome (in fact, cytochrome c_2). The $CoQH_2$– cytochrome c_2 oxidoreduction is catalyzed by the low-potential heme of a b-type cytochrome (b_1) and a nonheme iron called FeS_{III} (see Fig. 2.2D2). The $(BChl)_2$ oxidation mechanism, again as in green sulfur bacteria, includes the bacteriochlorophyll monomer. However, the final step differs from those of the green bacteria because $CoQH_2$, rather than NADH, is the final product. After the BChl monomer, bacteriopheophytin and tightly bound ubiquinone (Q_A) are involved. From Q_A an electron is transferred to another CoQ molecule (Q_b).

To be completely reduced, CoQ_B should accept two electrons and two protons. Protons are transferred from the cytosol. As to electrons, one of them comes from CoQ_A whereas the other is furnished by the high-potential cytochrome b heme (b_h). Then b_h is reduced by b_1, which is localized closer to the outer membrane surface than b_h.

The above mechanism allows *two* H^+ ions to be transported across the membrane per photon. Such high efficiency requires b_h and b_1 to be involved in the cyclic electron transfer (so-called Q cycle mechanism). However, at least in a model system (the reaction center proteoliposomes) CoQ was shown to mediate some cyclic electron flow even in the absence of cytochrome b. This process is half as effective as that in the native system, that is, H^+/photon $= 1$.

In this connection, one may assume that in primitive photosynthetic bacteria employing the cyclic electron transport system, just this simple and less effective mechanism was operative. This may explain the arrangement of the reaction center prosthetic groups that form two symmetric transmembrane branches. Each branch is localized below $(BChl)_2$, being composed of BChl, BPheo, and Q. The "right-hand" branch leading to Q_A is active whereas the "left-hand" branch leading to Q_B does not function in contemporary bacteria. Perhaps originally both branches were operative so that Q_A and Q_b were functionally similar, that is, $Q_A H_2$ and $Q_B H_2$ were formed and diffused to the opposite membrane side. In a later step of evolution, the two-heme cytochrome b appeared so that one electron was furnished to Q_B by "the right-hand" branch and another by b_h. As to the "left-hand" branch, it proved to be unnecessary and once was inactivated because of an occasional mutation. Being inactive, the "left-hand" branch was, nevertheless, preserved in the course of further evolution as its removal from the reaction center complex would apparently cause some damage in "ideal" structural organization of the complex which allows it to carry out the transmembrane electron transfer at the highest rate.

Besides this structural role, the "left-hand" BChl is involved in deactivation of the $(BChl)_2$ triplet state. As previously mentioned, it is the singlet excited state of $(BChl)_2$ that is employed by the photosynthetic apparatus to initiate light-induced electron transport. The long-lived triplet state that sometimes spontaneously arises when $(BChl)_2$ is excited should be deactivated as fast as possible to prevent $(BChl)_2^*$ from being attacked by O_2 with subsequent decomposition of this prosthetic group. The triplet $(BChl)_2$, via the "left-hand" BChl, excites carotenoid which is localized very close to this bacteriochlorophyll molecule. Carotenoid, if oxidized by O_2, can be easily replaced by another (native) carotenoid molecule. (For reviews on the reaction center complexes of purple bacteria, see Refs. [18], [38], [39]).

2.4.2.3 *Noncyclic Electron Flow in Cyanobacteria and Chloroplasts*

The appearance of cyanobacteria is generally assumed to be a rather late event in biological evolution, although evidence has been obtained by geologists indicating that cyanobacteria existed as early as 3.2 billion years ago (for discussion, see Refs. [36] and [40]).

The cyanobacterial photoredox chain represents, in fact, a combination of two simpler mechanisms described in the preceding sections, that is, the noncyclic chain of the green bacteria and the cyclic chain of the purple bacteria. The chain in question includes two types of reaction center complexes known as photosystems I and II (see Fig. 2.2D3).

Photosystem I resembles the complex from the green sulfur bacteria. Like this complex, it includes $(Chl)_2$, Chl, and FeS which reduces a nicotinamide nucleotide (in this case, $NADP^+$) via flavin. Photosystem I has been studied in more detail than the green bacterial complex. In particular, it has been shown that after the chlorophyll monomer, vitamin K_1 is operating as the electron carrier reducing the FeS_x cluster. Between FeS_x and flavin three more FeS clusters are functioning, that is, FeS_B, FeS_A, and ferredoxin (Fd).

Photosystem II is almost identical to the reaction center complex of purple bacteria. Specific features differing in these two mechanisms are that (1) plastoquinone (PQ) instead of CoQ or menaquinone is used as the electron acceptor and (2) water, not $CoQH_2$, is the electron donor for $(ChlII)_2{}^+$. The photolysis of water is catalyzed by a complex containing four Mn, a tyrosine residue of a photosystem II protein subunit being the electron carrier between the Mn–protein complex and $(ChlII)_2{}^+$.

PQH_2 produced by photosystem II is oxidized by a Q cycle system similar to that in the purple bacteria. The only principal difference is that electrons come in $(ChlI)_2{}^+$ rather than to $(ChlII)_2{}^+$.

The overall result of operation of the cyanobacterial system consists in the transmembrane translocation of $3\,H^+$, reduction of one $NADP^+$, and oxidation of one H_2O to form $1/2\,O_2$ per two photons (one photon is absorbed by photosystem I and the other by photosystem II).

The obvious advantage of the cyanobacterial system in comparison to that of the green sulfur bacteria is that, instead of H_2S, the most available and low-energy reductant H_2O is used to reduce nicotinamide nucleotide and then to form glucose from CO_2.

Massive production of molecular oxygen was an important consequence of cyanobacterial-type photosynthesis. Initially O_2 was, most probably, a rather dangerous side product that could oxidize reduced intermediates of the photoredox chain. This could result in futile cycles and, even worse, in decomposition of these intermediates as well as the photosynthetic pigments. It is not surprising, therefore, that in one and the same photosynthetic membrane, O_2-consuming enzymes appeared to reduce O_2 back to H_2O.

Later, however, these respiratory enzymes were organized in a H^+-motive fashion. This system became the main light-independent mechanism of energy production (see the next section).

There is no doubt that cyanobacteria or related microorganisms were evolutionary precursors of plant chloroplasts. However, in the plant cell the photosynthetic and respiratory functions were separated in space, the latter being transferred to mitochondria. (For reviews on the noncyclic photoredox chain of cyanobacteria and chloroplasts, see Refs. [18], [41], [42]).

2.5 H^+-Motive Respiration

The H^+-motive respiratory chain of contemporary organisms can include one, two, or three energy coupling sites. Among them, the second is almost identical to the Q cycle of the purple and cyanobacteria or of chloroplasts, whereas the first and third look like new developments in biological evolution.

The first energy coupling site (so-called NADH-CoQ reductase, or complex I) catalyzes transfer of two electrons from NADH to CoQ coupled to the export of $4 H^+$ from the bacterial cell (or mitochondrion) to the outer medium (or cytosol). This is a very complicated multisubunit enzyme containing FMN and at least four FeS clusters. The mechanism of its operation remains obscure.

The formed $CoQH_2$ is oxidized by $CoQH_2$-cytochrome c reductase (also called complex III) containing two-heme cytochrome b, FeS_{III}, and cytochrome c_1 as the electron carriers catalyzing the Q cycle.

Reduced cytochrome c transfers an electron to cytochrome oxidase (complex IV) which passes it to O_2. Cytochrome oxidase contains two hemes (a and a_3) and two coppers (Cu_A and Cu_B). Transfer of each electron from cytochrome c to oxygen by cytochrome oxidase is coupled to translocation of $2 H^+$ through the membrane (see Fig. 2.2D4).

Interestingly, in cyanobacteria one and the same complex III is involved in the photoredox chain and in the respiratory chain. In the light, photosystems II and I serve as PQ reductant and PQ oxidant, respectively. On the other hand, in the dark, NADH-PQ reductase reduces PQ and cytochrome oxidase oxidizes PQH_2.

Paracoccus denitrificans and *Rhodospirillum rubrum* exemplify bacteria possessing the complete respiratory chain which includes all three energy coupling sites. This is also the case in mitochrondria. On the other hand, in some other bacteria shortened versions of the respiratory chain are operative, sometimes because the respiratory chain reductant and oxidant are of such redox potentials that the entire chain cannot be used. For example, bacteria oxidizing Fe^{2+} by O_2 employ only the terminal step of the respiratory chain to form $\Delta\bar{\mu}_{H^+}$, as the redox potential of the Fe^{2+}/Fe^{3+} pair is too positive to reduce the initial and even the middle segments of the chain. On the other hand, anaerobic bacteria reducing fumarate by the NAD-linked substrates use the first energy coupling site as the only one because fumarate accepts electrons from the CoQ level. These shortened redox chains look like secondary simplifications of the complete respiratory chain (complexes $I + III + IV$). The simplifications in question most probably arose in the course of adaptation to specific ecological niches. It is remarkable that the imago of *Ascaris*, living in the intestine under almost anaerobic conditions and employing fumarate as the oxidant, lacks the terminal segment of the respiratory chain, whereas its larva living aerobically contain the complete respiratory chain.

However, some cases of shortened respiratory chains can hardly be explained in this way. For example, *E. coli* has only two energy coupling sites, that is, NADH-CoQ reductase and $CoQH_2$ oxidase, with no complex III and, hence, no Q cycle (see Fig. 2.2D5). Perhaps *E. coli* lost the Q cycle because of mutations. As a result, the shortened respiratory chain of *E. coli* proved to be less effective but, at the same time, less vulnerable than the complete one. It is known that complex III is attacked by some antibiotics, hydrophobic inhibitors, and SH- reagents. Thus, one may suggest that the loss of complex III was of adaptive importance for *E. coli*. (For review on the H^+-motive respiratory chain, see Refs. [18], [43], [44], [46–48]).

2.6　Na^+　Energetics

Substitution of Na^+ for H^+ as a coupling ion may represent one more mechanism of adaptation to unfavorable conditions [45]. Sometimes the H^+ cycle cannot be used owing to (1) high H^+ conductance of the membrane, (2) low $[H^+]$ in the medium, (3) damage in the $\Delta\bar{\mu}_{H^+}$ generator(s), or (4) damage in the $\Delta\bar{\mu}_{H^+}$ consumer(s).

In the available literature, one can find examples when in each of the above cases the H^+ cycle appears to be replaced by the Na^+ cycle. The latter is composed of the $\Delta\bar{\mu}_{Na^+}$ generators (the Na^+-motive respiratory chain enzymes, Na^+-decarboxylases, or Na^+-ATPases) and $\Delta\bar{\mu}_{Na^+}$ consumers (Na^+-driven ATP synthase or reverse electron transfer systems, Na^+, solute-symporters and flagellar Na^+ motors) spanning the Na^+-impermeable membrane.

The Na^+-motive respiratory chain including Na^+–NADH–Q reductase and Na^+–QH_2 oxidase was shown to be induced in *E. coli* and *Bacillus FTU* when these microorganisms were grown with a protonophorous uncoupler or at high pH [49, 50]. In *Bacillus FTU* the same effect could be obtained when the growth medium was supplemented with micromolar cyanide, which specifically inhibits the H^+-motive terminal oxidase rather than the Na^+-motive oxidase [49, 51]. In *E. coli* induction of the Na^+-motive respiratory chain was found to be accompanied by the appearance of Na^+-coupled oxidative phosphorylation and the Na^+-motive ATPase activity [27].

In the anaerobic *Streptococcus faecalis*, the Na^+, K^+-ATPase was induced by uncoupler, high pH, or a mutation in H^+-ATPase [52–55]. In *Str. faecalis* as well as in *E. coli* and *Bacillus FTU*, induction of the Na^+ energetics required rather high concentrations of Na^+ in the growth medium.

An interesting example of the same kind was recently described in the thermophilic anaerobe *Clostridium fervidus*. Here it was found that at high temperature the membrane H^+ conductance was very much higher than the Na^+ conductance. In the *Cl. fervidus* membrane, a Na^+-ATPase was found that hydrolyzes glycolytic ATP and forms $\Delta\bar{\mu}_{Na^+}$. The latter is used by Na^+, amino acid and Na^+, glucose symporters to accumulate these metabolites inside the bacterial cell [56].

The *C. fervidus* Na^+ cycle is constitutive. There is no evidence that it can, under some conditions, be replaced by the H^+ cycle. When such a replacement occurred in other bacteria (the inducible H^+ and Na^+ cycles), the induction can, or cannot, require de novo syntheses of the enzymes involved. In *E. coli* and *Bacillus FTU* the Na^+-motive NADH-Q reductase seems to be similar to that of *Vibrio alginolyticus*, which is very sensitive to heptyl hydroxyquinoline N-oxide and Ag^+, whereas the H^+-motive reductase is not. In *V. alginolyticus* the Na^+-motive reductase contains flavin, like the H^+-motive enzyme [57]. In *E. coli* the role of the Na^+-motive terminal oxidase was shown to be performed by cytochrome *d* and the H^+-motive one by cytochrome *o* [58]. On the other hand, functions of the Na^+-driven ATP-synthase seem to be carried out by one and the same F_0F_1-type enzyme so that the H^+-to-Na^+ switch occurs most probably by means of a posttranslocational modification.

"Protonic" bacteria growing in Na^+-containing media were shown to utilize Na^+ and K^+ gradients as the $\Delta\bar{\mu}_{H^+}$ buffer. They take up K^+ and extrude Na^+ in a $\Delta\bar{\mu}_{H^+}$-dependent fashion when the energy source is in excess. To this end, electrophoretic K^+ influx and Na^+/H^+ antiport are employed. When energy sources are exhausted, the fluxes of K^+ and Na^+ are reversed so that electrogenic K^+ efflux generates $\Delta\Psi$, and influx of Na^+ in exchange for H^+ generates ΔpH [59–62]. Moreover, ΔpNa may be used in the same bacteria to support uphill uptake of some metabolites [18]. It seems probable that utilization of a Na^+ gradient as a $\Delta\bar{\mu}_{H^+}$ buffer and then as a driving force for metabolite accumulation were intermediary steps of evolution on the way from the H^+ cycle to the Na^+ cycle.

An opposite point of view suggests that the Na^+ cycle was the primary one, being inherent in marine bacteria. As to the H^+ cycle, it appeared later when bacteria spread to fresh water in rivers and lakes containing very low Na^+ concentrations [63].

Indeed, employment of the Na^+ cycle by some contemporary marine bacteria can hardly be accounted for by their adaptation to the low $\Delta\bar{\mu}_{H^+}$ conditions. For example, *Propionigenium modestum* uses Na^+-motive methyl-malonyl decarboxylases as the only mechanism producing all the energy required for this bacterium [64]. *P. modestum* lives at neutral pH and normal temperature in medium containing hardly any uncouplers. Thus, there is no reason to suggest that the H^+ cycle cannot, in principle, be used in this bacterium. It employs the Na^+ cycle because Na^+-motive decaryboxylases were developed by living cells whereas H^+-motive decarboxylase were not. The

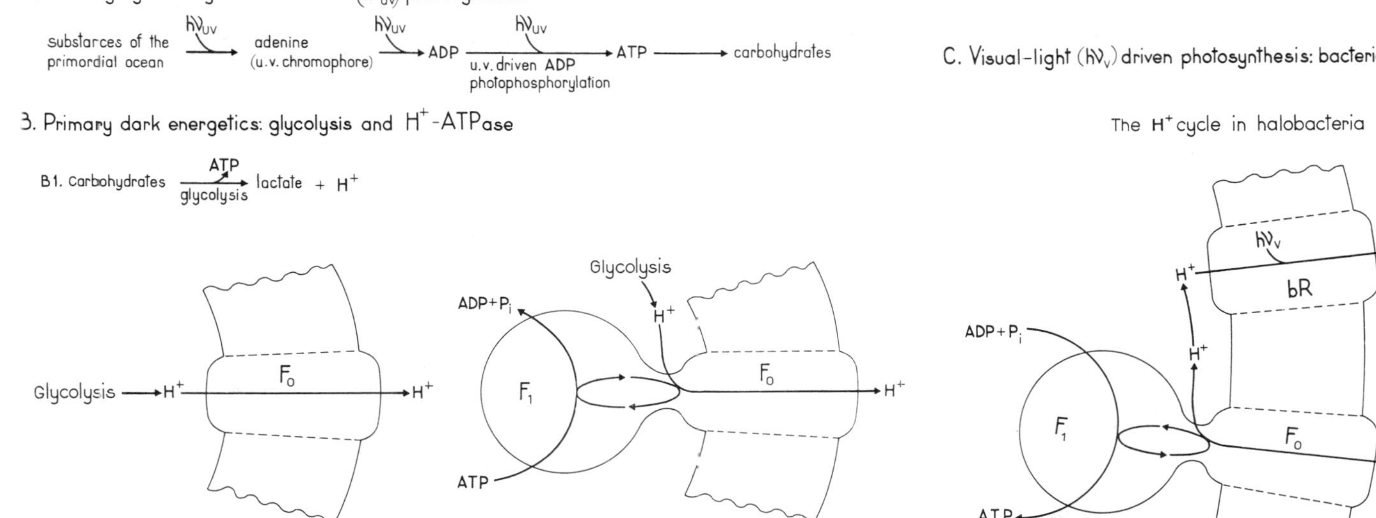

Figure 2.2   Main steps of evolution of the bioenergetic mechanisms. F_0 and F_1, H^+ channel and catalytic part of H^+-ATPase, respectively; bR, bacteriorhodopsin; Q, ubi-, plasto-, or menaquinone; a, a_3, b, c, c_1, d, f, o, o_3, cytochromes or their hemes; $(BChl)_2$, bacteriochlorophyll dimer; BChl, bacteriochlorophyll monomer; FeS (see D1), iron–sulfur cluster(s) in the photosynthetic reaction center complex of the first type (RCI) inherent in the green sulfur bacteria; FeS_{III}, iron–sulfur cluster of the bc complex (complex III); b_h and b_1, high- and low-potential hemes of cytochrome b; BPheo, bacteriopheophytin; PC, plastocyanin; K_1, vitamin K_1; FeS_x, $FeS_{A,B}$, iron–sulfur clusters of the cyanobacterial reaction center complex I; RCII, the reaction center complex of the second type; NADH-Q red., NADH-CoQ reductase; FeS_I 1, 2, 3, 4, the iron–sulfur clusters of the respiratory chain complex I; Cu_A and Cu_B, the cytochrome oxidase coppers. The number of H^+ ions pumped per ATP molecule hydrolyzed or synthesized (B3–D5) as well as numbers of Na^+ ions transferred by the Na^+-linked pumps (D6) are not indicated.

(Continued)

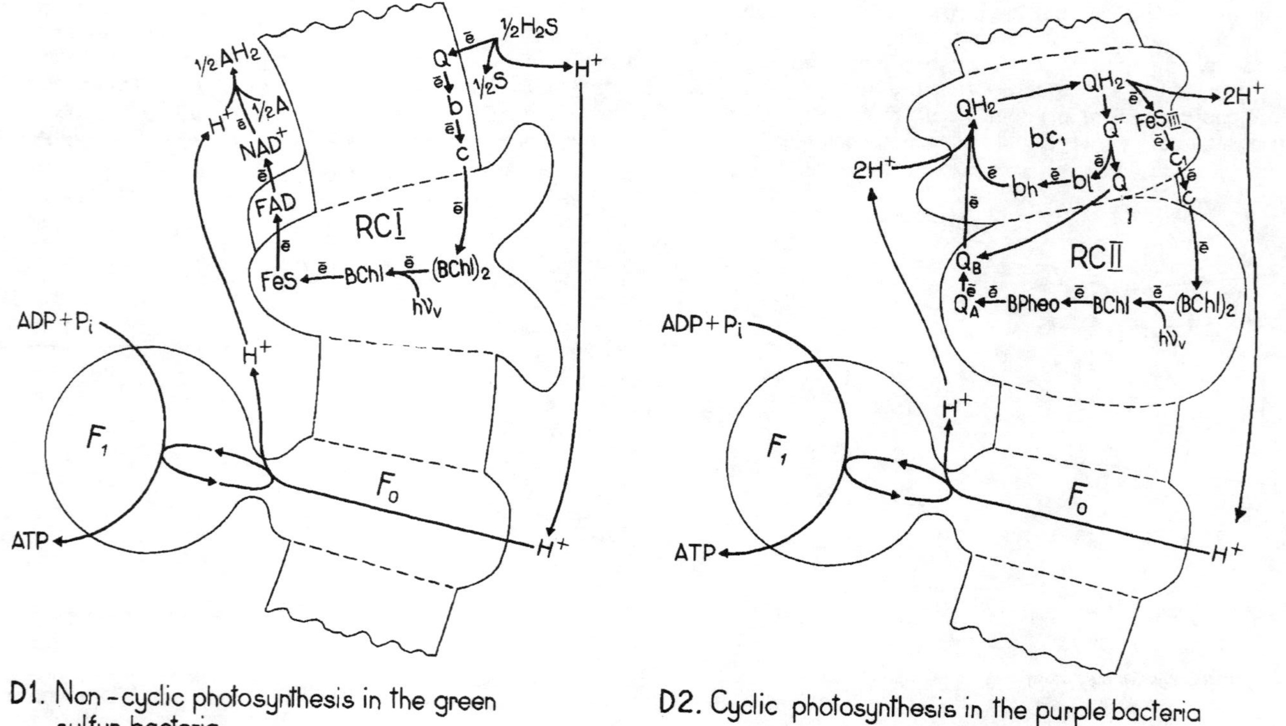

Figure 2.2 (*Continued*).

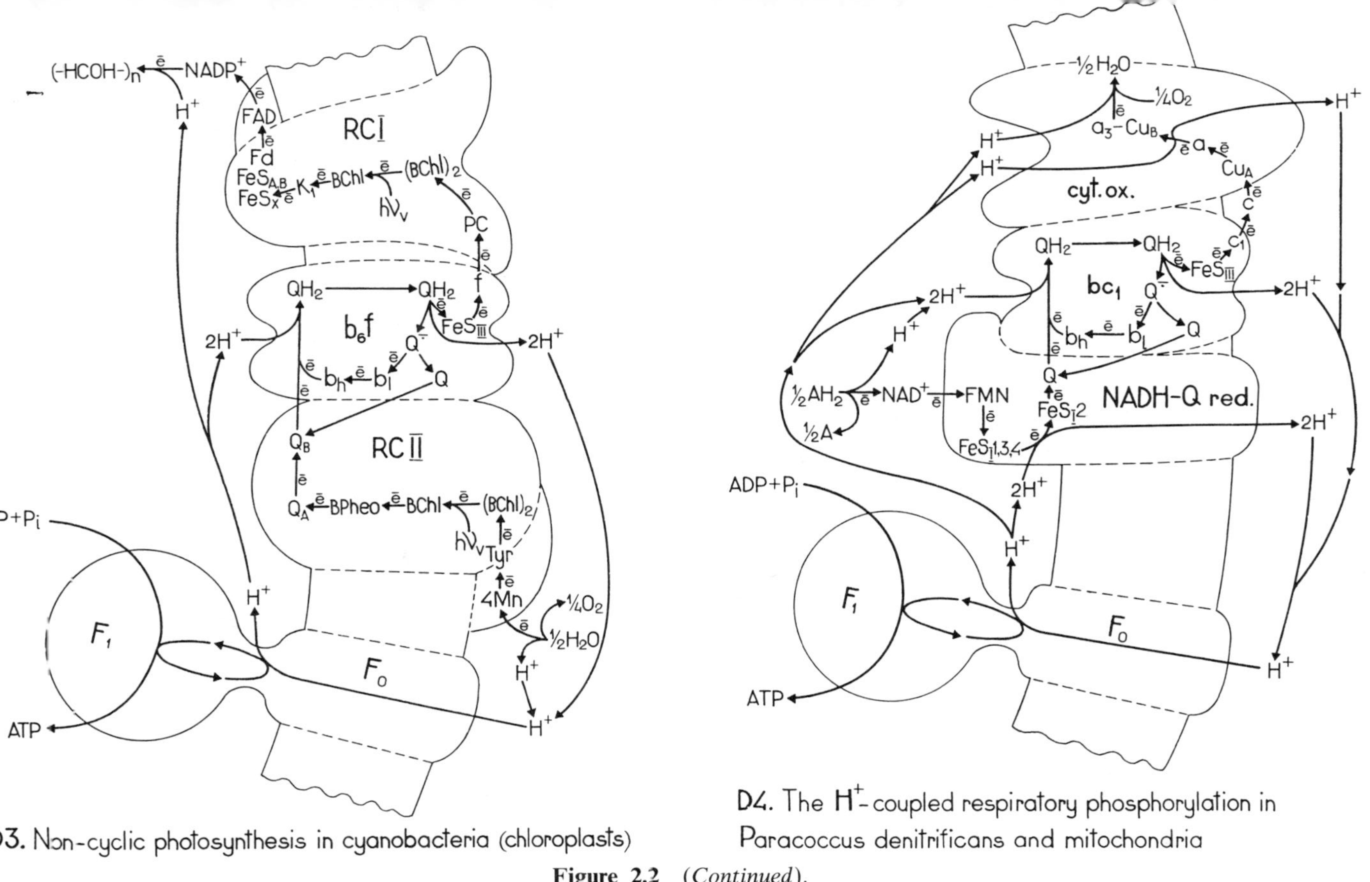

D3. Non-cyclic photosynthesis in cyanobacteria (chloroplasts)

D4. The H$^+$-coupled respiratory phosphorylation in Paracoccus denitrificans and mitochondria

Figure 2.2 (*Continued*).

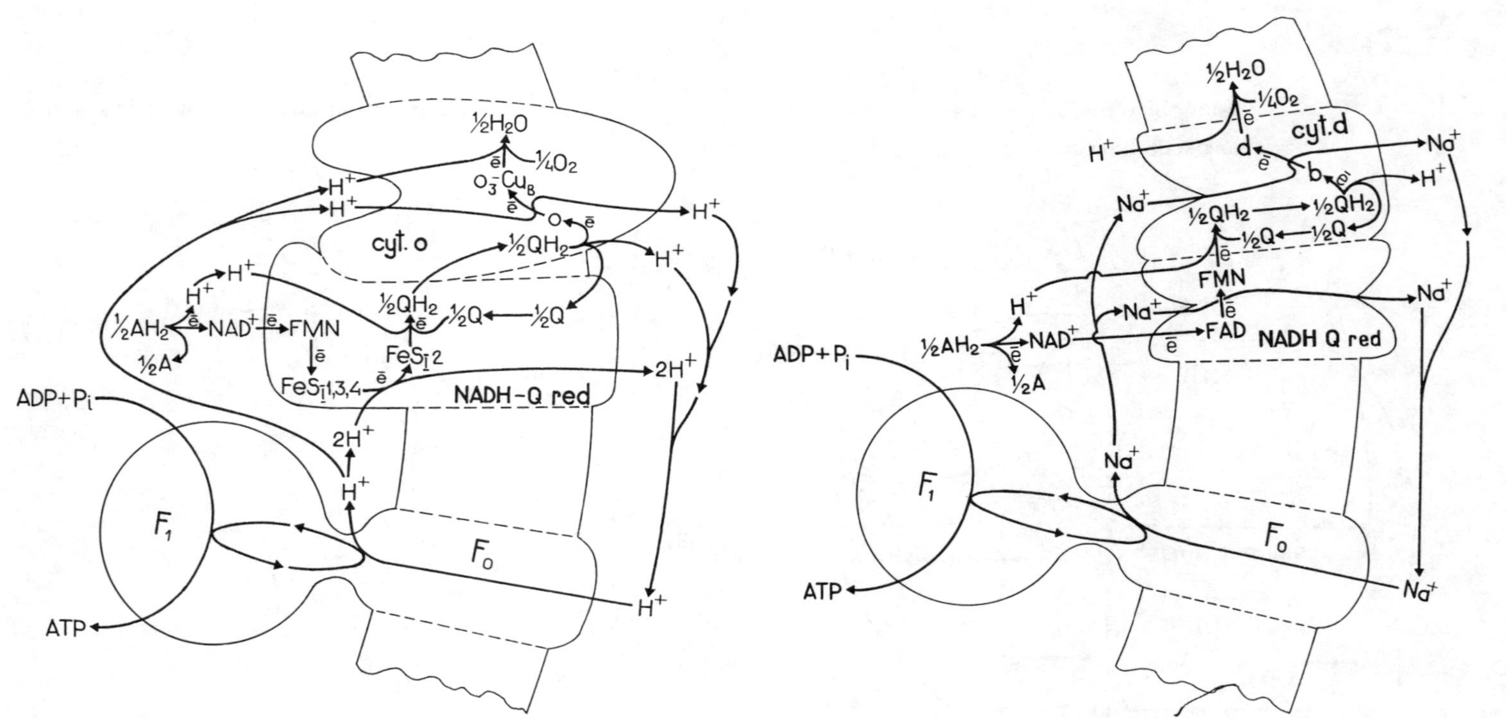

D5. The H⁺-coupled oxidative phosphorylation in E. coli   D6. The Na⁺-coupled oxidative phosphorylation in E. coli

Figure 2.2 (*Continued*).

use of Na^+ as the coupling ion seems reasonable for *P. modestum*, as it is a marine bacterium.

Unfortunately, there is no consensus among geologists concerning the pH and the salt concentration in the ancient ocean. A common view suggests that the ocean was acidic and K^+-rich (see, e.g., Ref. [65]). At the same time, some others assume that it was alkaline and contained a high Na^+ level (hypothesis of the "soda ocean") [66–68]. If the former point of view is right, there was a chance for the H^+ cycle to precede the Na^+ cycle. The idea of the Na^+ cycle as a precursor of the H^+ cycle looks more realistic if the "soda ocean" hypothesis is valid. However, even in this case I prefer to assume that the H^+ cycle appeared first. It is noteworthy in this context that in the animal cell, the "old" membrane of mitochondria employs the H^+ cycle, whereas in the "young" outer cell membrane the Na^+ cycle operates, being composed of the $\Delta\bar{\mu}_{Na^+}$ generator (Na^+, K^+-ATPase) and $\Delta\bar{\mu}_{Na^+}$ consumers (Na^+, metabolite symporters). Moreover, all the logic of the above consideration favors an assumption that the H^+ cycle came first. This logic is illustrated by Figure 2.2.

References

1. Sagan, C. *The Origin of Pre-Biological Systems*, Oparin, A. I. (ed.), Mir, Moscow, 1966; pp. 211–223 (in Russian).

2. Ponnamperuma, C., Sagan, C., Mariner, R. *Nature* 1963; *199*, 222–226.

3. Oro, J., Kimball, A. P. *Arch. Biochem. Biophys.* 1961; *94*, 217–220.

4. Ponnamperuma, C., Lemmon, R. M., Mariner, R., Calvin, M. *Proc. Natl. Acad. Sci. USA* 1963; *49*, 737–740.

5. Pullman, B., Pullman, A. Comparative Effects of Radiation, John Wiley & Sons, New York, 1960; p. 111.

6. Pullman, B., Pullman, A. *Nature* 1962; *196*, 1137–1141.

7. Ponnamperuma, C. *The Origin of Pre-Biological Systems*, Oparin, A. I. (ed.), Mir, Moscow, 1966; pp. 224–238 (in Russian).

8. Barltroop, J. A., Grubb, P. W., Hess, B. *Nature* 1963; *199*, 759–761.

9. Blumenfeld, L. A., Temkin, M. I. *Biophysika* 1962; *7*, 731–734 (in Russian).

10. Pauling, L. *The Nature of the Chemical Bond*, Goskomisdat, Moscow, 1947; p. 206 (in Russian).

11. Skulachev, V. P. *Energy Accumulation Processes in the Cell.* Nauka, Moscow, 1968 (in Russian).

12. Szent-Györgyi, A. *Bioenergetics,* Academic Press, New York, 1957.

13. Golbik, R., Neef, H., Hübner, G., König, S., Seliger, B., Meshalkina, L., Kochetov, G. A., Schellenberger, A. *Bioorg. Chem.* 1991; *19*, 10–17.

14. Lee, H. C., Walseth, T. F., Bratt, G. T., Hayes, R. N., Clapper, D. L. *J. Biol. Chem.* 1989; *264*, 1608–1615.

15. Meszaros, L. G., Bak, J., Chu, A. *Nature* 1993; *364*, 76–79.

16. Galione, A., White, A., Willmott, N., Turner, M., Potter, B. V. L., Watson, S. P. *Nature* 1993; *365*, 456–459.

16a. Murphy, C. J., Arkin, M. R., Jenkins, Y., Ghatlia, N. D., Bossmann, S. H., Turro, N. J., Barton, J. K. *Science* 1993; *262*, 1025–1029.

17. Baltscheffsky, H. Chapter 1, this volume.

18. Skulachev, V. P. *Membrane Bioenergetics*, Springer-Verlag, Berlin, 1988.

19. Holzenburg, A., Jones, P. C., Franklin, T., Pali, T., Heimburg, T., Marsh, D., Findlay, J. B. C., Finbow, M. E. *Eur. J. Biochem.* 1993; *213*, 23–30.

20. Rosen, B. P., Weigel, W., Karkaria, C., Gangola, P. *J. Biol. Chem.* 1988; *263*, 3067–3070.

21. Hsu, C. M., Rosen, B. P. *J. Biol. Chem.* 1989; *264*, 17349–17354.

22. Rosen, B. P., Dey, S., Dou, D., Ji, G., Kaur, P., Ksenzenko, M. Yu, Silver, S., Wu, J. *Ann. NY Acad. Sci.* 1992; *671*, 257–272.

23. Silver, S., Ji, G., Bröer, S., Dey, S., Dou, D., Rosen, B. P. *Microbiol. Rev.* 1993; *8*, 637–642.

24. Ji, G., Silver, S. *J. Bacteriol.* 1992; *174*, 3684–3694.

25. Rosenstein, R., Peschel, P., Wieland, B., Götz. F. *J. Bacteriol.* 1992; *174*, 3676–3683.

26. Laubinger, W., Dimroth, P. *Biochemistry* 1989; *28*, 7194–7198.

27. Avetisyan, A. V., Bogachev, A. V., Murtasina, R. A., Skulachev, V. P. *FEBS Lett.* 1993; *317*, 267–270.

28. Baltscheffsky, H., von Stedingk, L.-V., Heldt, H. W., Klingenberg, M. *Science* 1966; 153, 1120–1124.

29. Nyrén, P., Sakai-Nore, Y., Strid, A. *Plant Cell Physiol.* 1993; *34*, 375–378.

30. Henderson, R., Baldwin, J. M., Ceska, T. A., Zemlin, F., Beckmann, F., Downing, K. H. *J. Mol. Biol.* 1990; *213*, 899–929.

31. Ebrey, T. G. *Thermodynamics of Membrane Receptors and Channels*, Jackson, M. B. (ed.), CRC Press, Boca Raton, FL 1993; pp. 353–387.

32. Skulachev, V. P. *Quart. Rev. Biophys.* 1993; *26*, 177–199.

33. Subramaniam, S., Gerstein, M., Oesterhelt, D., Henderson, R. *EMBO J.* 1993; *12*, 1–8.

34. Gibson, J., Ludwig, W., Stackebrandt, E., Woese, C. R. *Syst. Appl. Microbiol.* 1985; *6*, 152–156.

35. Woese, C. R., Stackebrandt, E., Macke, T. J., Fox, G. E. *Syst. Appl. Microbiol.* 1985; *6*, 143–151.

36. Woese, C. R. *Microbiol. Rev.* 1987; *51*, 221–271.

37. Golbeck, J. H. *Proc. Natl. Acad. Sci. USA* 1993; *90*, 1642–1646.

38. Michel, H., Deisenhofer, J. In *Encyclopedia of Plant Physiology: Photosynthesis III*, Vol. 19, Staehelin, L. A. and Arntzen, C. J. (eds.), Springer-Verlag, Berlin, 1986; pp. 371–381.

39. Deisenhofer, J., Michel, H. *EMBO J.*, 1989; *8*, 2149–2170.

40. Alge, D., Peschek, G. A. *Biochem. Mol. Biol. Int.* 1993; *29*, 511–525.

41. Nitschke, W., Rutherford, A. W. *TIBS* 1991; *16*, 241–245.

42. Mathis, P. *Biochim. Biophys. Acta.* 1990; *1018*, 163–167.

43. Sled, V. D., Friedrich, T., Leif, H., Weiss, H., Meinhardt, S. W., Fukumori, Y., Calhoun, M. W., Gennis, R., Ohnishi, T. *J. Bioenerg. Biomembr.* 1993; *25*, 347–356.

44. Yagi, T. *Biochim. Biophys. Acta* 1993; *1141*, 1–17.

45. Skulachev, V. P. *TIBS* 1984; *9*, 483–485.

46. Trumpower, B. L. *Microbiol. Rev.* 1990; *54*, 101–129.

47. Poole, R. K. *Bacterial Energy Transduction*, Anthony, C. (ed.), Academic Press, London, 1988; pp. 231–291.

48. Saraste, M., Holm, L., Lemieux, L., Lübben, M., van der Oost, J. *Biochem. Soc. Trans.* 1991; *19*, 608–612.

49. Skulachev, V. P. *Molecular Mechanisms in Bioenergetics*, Ernster, L. (ed.), Elsevier, Amsterdam, 1992; pp. 37–73.

50. Avetisyan, A. V., Dibrov, P. A., Semeykina, A. L., Skulachev, V. P., Sokolov, M. V. *Biochim. Biophys. Acta* 1991; *1098*, 95–104.

51. Semeykina, A. L., Skulachev, V. P. *FEBS Lett.* 1990; *269*, 69–72.

52. Kinoshita, N., Unemoto, T., Kobayashi, H. *J. Bacteriol.* 1984; *158*, 844–848.

53. Kakinuma, Y., Harold, F. M. *J. Biol. Chem.* 1985; *260*, 2086–2091.

54. Kakinuma, Y., Igarashi, K. *FEBS Lett.* 1990; *261*, 135–138.

55. Kakinuma, Y., Igarashi, K. *J. Bacteriol.* 1990; *172*, 1732–1735.

56. Speelmans, G., Poolman, B., Abee, T., Konings, W. N. *Proc. Natl. Acad. Sci. USA* 1993; *90*, 7975–7979.

57. Unemoto, T., Hayashi, M. *J. Bioenerg. Biomembr.* 1993; *25*, 385–391.

58. Avetisyan, A. V., Bogachev, A. V., Murtasina, R. A., Skulachev, V. P. *FEBS Lett.* 1992; *306*, 199–202.

59. Skulachev, V. P. *FEBS Lett.* 1978; *87*, 171–179.

60. Wagner, G., Hartmann, R., Oesterhelt, D. *Eur. J. Biochem.* 1978; *89*, 169–179.

61. Brown, I. I., Galperin, M. Yu., Glagolev, A. N., Skulachev, V. P. *Eur. J. Biochem.* 1983; *143*, 345–349.

62. Drachev, A. L., Markin, V. S., Skulachev, V. P. *Biochim. Biophys. Acta* 1985; *811*, 197–215.

63. Rosen, B. P. *Annu. Rev. Microbiol.* 1986; *40*, 263–286.

64. Dimroth, P. *Microbiol. Rev.* 1987; *51*, 320–340.

65. Maisonneuve, J. *Sediment. Geol.* 1982; *31*, 1–11.

66. Kempe, S., Degens, E. T. *Chem. Geol.* 1985; *53*, 95–108.

67. Kazmierczak, J., Degens, E. T. *Mitt. Geol. Paläontol. Inst. Universität Hamburg* 1986; *61*, 1–20.

68. Kempe, S., Kazmierczak, J., Degens, E. T. *Origin, Evolution, and Modern Aspects of Biomineralization in Plants and Animals*, Crick, R. E. (ed.), Plenum Press, New York, 1990; pp. 29–43.

Appendix:
Carnitine: An Evolutionary Aspect

Some structural features of contemporary biochemical compounds cannot be accounted for if we ignore their evolution. This is true not only for biopolymers but also for certain low molecular weight substances of biological origin. For instance, we need to take into account something like UV photosynthesis in primordial living systems (see Section 2.1) to explain why ATP, a convertible biological energy currency, contains such a UV chromophore as adenine.

In this section an attempt is made to infer the structure of another low molecular weight compound in light of the evolution of the cell energetics within the framework of the above concept. Let me start with the history of the carnitine studies, which seems instructive for bioenergeticists.

Carnitine (Fig. 2.A1) was discovered by Russian biochemists Gulewitsch and Krimberg in 1905 [1]. At that time many biochemical compounds were described, and so the discovery of yet another substance of biological origin was not a sensation. In the 1930s, when particularly impressive progress was made in the study of the functions of simple biochemical substances, the similarity between carnitine and choline inspired a number of studies which, however, failed to identify the role of carnitine in the living cell.

In the 1940s, the possible function of small molecules as vitamins was extensively investigated and the first indication that carnitine is really a substance of biological importance was obtained. In 1952 Carter et al. [2] established that it is a growth factor for the meal worm *Tenebrio molitor* and carnitine was called vitamin B_T (T for *Tenebrio*). Later it was found that

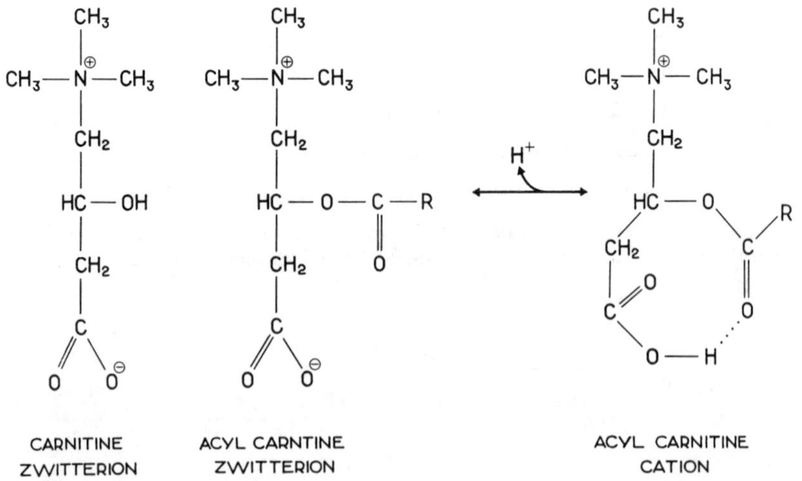

Figure 2.A1   Carnitine and acyl carnitines in zwitterionic (deprotonated) and cationic (protonated) forms. R, hydrocarbon residue of a fatty acid.

carnitine-deficient larvae died "fat" when they were starved, being unable to utilize fat stores to survive [3].

Unfortunately for the studies of carnitine, attempts to demonstrate the vitamin function of carnitine in organisms more important for our life than the meal worm were without success. This fact, certainly, did not stimulate any investigations in carnitine because mankind is not interested in the well-being of worms. Thus, well into the middle of the century, when the majority of low molecular weight substances of the cell had already found their place on the metabolic map, carnitine was still unplaced.

In 1955, 50 years after the discovery of carnitine, Friedman and Fraenkel found that carnitine is reversibly acetylated by acetyl-CoA [4]. Acetyl-CoA holds a central position in metabolism, and one was to expect that acetyl carnitine is an intermediate of acetyl residue transfer to some acetyl acceptor. However, this is not the case. The only reaction of acetyl carnitine utilization proved to be a transfer of acetyl back to CoA. The same was found to be true for fatty acyl carnitine, the second and the last representative of the carnitine derivative family. Thus the impression is that carnitine acylations are a dead end of metabolism.

Also in 1955, Fritz pointed to the stimulating effect of carnitine on the fatty acid oxidation by liver homogenate [5]. Such an influence of carnitine was rather surprising for a compound that is absent from metabolic highways. Therefore that observation awaited an explanation for some period of time. In the 1960s, when membranology studies were launched, carnitine was also investigated from this aspect. It was found that the above-mentioned stimulating effect of carnitine is localized at the mitochondrial level, and Fritz suggested that carnitine is somehow involved in the transport of fatty acids across the membrane of mitochondria [6]. This idea was supported by the finding that (1) fatty acyl carnitine transferase, catalyzing acyl transfer between carnitine and CoA, is localized on both sides of the hydrophobic barrier of the inner mitochondrial membrane, and (2) this membrane is permeable for fatty acyl carnitine, but not for fatty acyl CoA (for review, see Ref. [7]). Thus, the function of carnitine as the fatty acyl carrier was summarized in the following chain of events:

$$acyl - CoA_{out} + Cn_{out} \rightarrow acyl - Cn_{out} + CoA_{out} \qquad (2.A1)$$

$$acyl - Cn_{out} \rightarrow acyl - Cn_{in} \qquad (2.A2)$$

$$acyl - Cn_{in} + CoA_{in} \rightarrow Cn_{in} + acyl - CoA_{in} \qquad (2.A3)$$

where Cn is carnitine.

Two questions remained obscure in the framework of this concept.

First, what happens to Cn_{in}? It was found that carnitine per se is nonpenetrating for micochondria [7, 8]. This means that Cn_{in} cannot escape the matrix, simply moving down ΔpCn.

Second, it is not clear why such a hydrophilic molecule as carnitine was chosen in evolution as a fatty acyl carrier.

Both these problems were solved in the 1970s. For bioenergeticists, this period was highlighted by the struggle over the chemiosmotic hypothesis of Peter Mitchell [9], which culminated in his triumph. In particular, it was found that any uphill transports of solutes into mitochondria are $\Delta\bar{\mu}_{\mathrm{H}^+}$-driven. This is why in 1970 we suggested that the role of carnitine might be connected in some way with $\Delta\bar{\mu}_{\mathrm{H}^+}$ utilization [10]. The fact is that free fatty acids are hydrophobic anions and, hence, should be extruded from mitochondria in a $\Delta\Psi$-dependent fashion. Their protonated forms might return to mitochondria moving down Δp (fatty acid) and $\Delta \mathrm{pH}$. Such a circulation must discharge $\Delta\bar{\mu}_{\mathrm{H}^+}$ (uncoupling). On the other hand, neither carnitine nor acyl carnitine can be an anion. In the deprotonated form, they are zwitterions (the net charge is zero); in the protonated form, cations, with the zwitterions being more hydrophilic (Fig. 2A1). This warrants a supposition that, combining with carnitine, fatty acid may lose its uncoupling activity and acquire the capability to accumulate inside mitochondria where fatty acid β-oxidation is located. Such an accumulation, if it does occur, would mean a symport of fatty acyl and H^+ down total $\Delta\bar{\mu}_{\mathrm{H}^+}$.

$$\mathrm{acyl} - \mathrm{Cn}_{\mathrm{out}} + \mathrm{H}_{\mathrm{out}}^+ \rightarrow \mathrm{acyl} - \mathrm{Cn} \cdot \mathrm{H}_{\mathrm{out}}^+ \tag{2.A4}$$

$$\mathrm{acyl} - \mathrm{Cn} \cdot \mathrm{H}_{\mathrm{out}}^+ \rightarrow \mathrm{acyl} - \mathrm{Cn} \cdot \mathrm{H}_{\mathrm{in}}^+ \tag{2.A5}$$

$$\mathrm{acyl} - \mathrm{Cn} \cdot \mathrm{H}_{\mathrm{in}}^+ \rightarrow \mathrm{acyl} - \mathrm{Cn}_{\mathrm{in}} + \mathrm{H}_{\mathrm{in}}^+ \tag{2.A6}$$

In agreement with this scheme, Levitsky in our group [8, 11] found that the planar phospholipid membrane is permeable for the palmitoyl carnitine cation. As the following experiments showed, a discharge of the mitochondrial membrane potential by uncouplers strongly inhibited the oxidation of palmitoyl carnitine added after the uncoupler. On the other hand, the uncoupler stimulated oxidation of palmitoyl carnitine when the latter was added 3 min before the uncoupler. Apparently palmitoyl carnitine accumulated inside energized mitochondria during these 3 min in an amount sufficient to support active respiration for the duration of the polarographic experiment.

Thus, we could explain why carnitine plays the role of a fatty acyl carrier. All three functional groups of carnitine proved to be necessary for the function of fatty acyl, H^+-symporter, namely, hydroxyl to combine with the fatty acyl residue, carboxyl to reversibly bind the H^+ ion, and quaternary nitrogen to charge protonated acyl carnitine positively.

The above scheme failed to answer but one, the final question: What is the fate of the intramitochondrial carnitine released in reaction (2.A3)? This problem was solved in 1976 by Pande and Parvin [12, 13] and by Ramsay and Tubbs [14], who disclosed stoichiometric carnitine/fatty acyl carnitine antiport across the inner mitochondrial membrane. Later a protein catalyzing the process was isolated and reconstituted into proteoliposomes [17–22].

A general scheme of events resulting in the transport of extramitochondrial fatty acids to the matrix is given in Figure 2.A2. The process starts in the outer mitochondrial membrane where long-chain fatty acyl-CoA synthase is local-

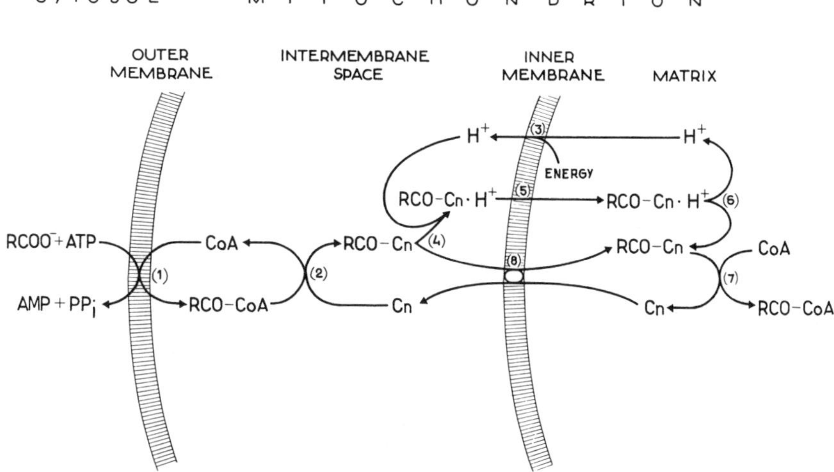

Figure 2.A2 Carnitine-mediated transport of fatty acyls into the mitochondrial matrix. R, hydrocarbon residue of fatty acid; Cn, carnitine. (1) acyl-CoA synthase; (2) outer carnitine acyltransferase; (3) $\Delta\bar{\mu}_{H^+}$ generators of the respiratory chain; (4) protonation of acyl carnitine carboxylic group; (5) $\Delta\bar{\mu}_{H^+}$-driven influx of protonated acyl carnitine; (6) deprotonation of acyl carnitine carboxylic group; (7) inner carnitine acyltransferase; (8) carnitine/acyl carnitine antiporter.

ized. This enzyme esterifies CoA by fatty acids at the expense of the ATP energy (step 1). In the intermembrane space or on the outer surface of the inner membrane, acyl-CoA is attacked by outer fatty acyl carnitine transferase so that acyl is transferred from CoA to carnitine (step 2). The resultant acyl carnitine is protonated by H_{out}^+. The pool of H_{out}^+ is regenerated by respiratory $\Delta\bar{\mu}_{H^+}$ generators that pump protons from the matrix to the intermembrane space (steps 3 and 4). Protonated acyl carnitine cation moves across the inner membrane down $\Delta\Psi$ and ΔpH (step 5) and is deproned in the matrix (step 6). Acyl carnitine transfers its acyl group to intramitochondrial CoA, the process being catalyzed by inner fatty acyl carnitine transferase (step 7). The formed inner carnitine is exchanged for outer acyl carnitine via the carnitine/ acyl carnitine antiporter (reaction 8).

An interesting feature of this system is that the level of the matrix carnitine can regulate distribution of the acyl carnitine flow between the $\Delta\bar{\mu}_{H^+}$-driven (i.e., energy-expending) pathway (5) and the energy-independent antiporter-mediated pathway (8). Apparently, as the cell begins to oxidize fatty acids, the $\Delta\bar{\mu}_{H^+}$-driven pathway is predominant, as intramitochondrial carnitine is below K_m of the antiporter for carnitine, which is rather high, that is, 8.7 mM (for outer carnitine, 0.45 mM) [22]. At this stage, $\Delta\bar{\mu}_{H^+}$ energy is spent to accelerate the fatty acid oxidation process. Large-scale uphill import of acyl-carnitine, followed by utilization of fatty acyls, leads to the accumulation of carnitine in the matrix to a level sufficient for antiporter saturation. Now, $\Delta\bar{\mu}_{H^+}$ consumption at the stage of transport of fatty acyl residue may be decreased, provided antiport (8) is faster than uniport (5). It should be stressed, however, that under energized conditions, the uniport can hardly be completely excluded, for this is a physical property of fatty acyl carnitine cation to cross phospholipid membranes. Such a process must take place in any systems where at least some of the membrane areas are at the phospholipid bilayer.

It should be noted that all the above considerations cannot be applied to acetyl carnitine, which is too hydrophilic for a penetrating cation. As postulated by our group, the main function of acetyl carnitine is that of a buffer of high-energy acetyls just as creatine phosphate buffers high-energy phosphoryls [15, 16].

Such a buffer function, besides the transport function, may also be inherent in fatty acyl carnitine.

An attractive assumption is that both the transport and buffer functions of the carnitine derivatives stem from some other, initial role performed by these compounds in the primordial cell.

Both fatty and acetic acid can easily penetrate through the phospholipid membrane [23, 24], that is, can escape the cell to be diluted by the outer medium. Combining with a hydrophilic carnitine residue, which is always a zwitterion or cation, the acetyl residue becomes nonpenetrating, and, hence, can be retained by the cell. Fatty acyl carnitine is penetrating in the cationic form which cannot leave the cell because the potential difference (the cell interior is negative) is unfavorable for a cation efflux. Further, if fatty acyl carnitine were to escape the cell as a zwitterion, it would be returned to the cytoplasm electrophoretically in its cationic form.

One may suggest that the preservation and buffer functions of fatty acyl carnitine were supplemented with its transport function when mitochondria appeared. Electrophoresis of the protonated fatty acyl carnitine cation to the mitochondrial matrix proved, most probably, the evolutionary primary mechanism of realization of the transport function. This did not require development of any special protein carrier because of the ability of the protonated fatty acyl carnitine to cross phospholipid bilayer.

Later this simple mechanism was replaced (partially, or maybe even completely) by the electroneutral fatty acyl carnitine/carnitine antiport, the process being catalyzed by special transport protein. In such a replacement, a general tendency of biological evolution was revealed that the cell always prefers to deal with a protein-catalyzed (rather than with a spontaneous) process to be able to control this process.

It seems obscure at what stage of evolution we are at present. Perhaps in contemporary systems the role of the spontaneous electrophoretic transport of fatty acyl carnitine, under physiological conditions, is quantitatively negligible. This would be the case if the level of free carnitine in mitochondria were always higher than K_m for the antiporter or if the rate of the carrier-mediated process were much higher than that of electrophoretic transport even when we are below K_m.

References (Appendix)

1. Gulewitsch, W., Krimberg, R. *Hoppe Seylers Z. Physiol. Chem.* 1905; *45*, 326–330.

2. Carter, H. E., Bhattacharyya, P. K., Weidman, K. R., Fraenkel, G. *Arch. Biochem. Biophys.* 1952; *38*, 405–416.

3. Fraenkel, G., Friedman, S. *Vitamins and Hormones*, Vol. 15, Harris, R. S. (ed), Academic Press, New York, 1957; pp. 73–118.

4. Friedman, S., Fraenkel, G. *Arch. Biochem. Biophys.* 1955; *59*, 491–501.

5. Fritz, I. B. *Acta Physiol. Scand.* 1955; *34*, 367–385.

6. Fritz, I. B. *Adv. Lipid Res.* 1963; *1*, 285–334.

7. Bremer, J. *Physiol. Rev.* 1983; *63*, 1420–1480.

8. Levitsky, D. O., Skulachev, V. P. *Biochim. Biophys. Acta* 1972; *275*, 33–50.

9. Mitchell, P. *Nature* 1961; *191*, 144–148.

10. Severin, S. E., Skulachev, V. P., Yagujinsky, L. S. *Biokhimiya* 1970; *35*, 1250–1257 (in Russian).

11. Levitsky, D. O., Skulachev, V. P. *Molek. Biologia* 1972; *6*, 338–345 (in Russian).

12. Pande, S. V., Parvin, R. *J. Biol. Chem.* 1976; *251*, 6683–6691.

13. Pande, S. V., Parvin, R. *J. Biol. Chem.* 1980; *255*, 2994–3001.

14. Ramsey, R. R., Tubbs, P. K. *Eur. J. Biochem.* 1976; *69*, 299–303.

15. Skulachev, V. P. *Energy Transduction in Biomembranes*, Nauka, Moscow, 1972 (in Russian).

16. Skulachev, V. P. *Membrane Bioenergetics*, Springer-Verlag, Berlin, 1988.

17. Indiverti, C., Tonazzi, A., Palmieri, F. *Biochim. Biophys. Acta* 1990; *1020*, 81–86.

18. Palmieri, F., Bisaccia, F., Capobianco, L., Iacobazzi, V., Indiveri, C., Zara, V. *Biochim. Biophys. Acta* 1990; *1018*, 147–150.

19. Indiverti, C., Tonazzi, A., Prezioso, G., Palmieri, F. *Biochim. Biophys. Acta* 1991; *1065*, 231–238.

20. Indiveri, C., Tonazzi, A., Palmieri, F. *Biochim. Biophys. Acta* 1991; *1069*, 110–116.

21. Kaminska, J., Nalecz, K. A., Azzi, A., Nalecz, M. J. *Biochem. Mol. Biol. Int.* 1993; *29*, 999–1007.

22. Indiveri, C., Tonazzi, A., Palmieri, F. *Biochim. Biophys. Acta* 1994; *1189*, 65–73.

23. Wilschut, J., Scholma, J., Eastman, S. J., Hope, M. J., Cullis, P. R. *Biochemistry* 1992; *31*, 2629–2636.

24. Kamp, F., Hamilton, J. A. *Biochemistry 1993; 32*, 11074–11086.

Iron and Sulfur in the Origin and Evolution of Biological Energy Conversion Systems

Richard Cammack

3.1 Early Iron–Sulfur Metabolism

Iron–sulfur proteins are proteins that contain clusters of iron and (with the exception of rubredoxins) sulfide, bound to cysteine thiolates and other ligands from the protein. The simplest iron–sulfur proteins are the ferredoxins and rubredoxins, which are small proteins whose only function is electron transfer. Other iron–sulfur proteins are more complex, have enzymatic activities, and often contain other metal ions or organic prosthetic groups. They are widely distributed in the electron-transfer pathways of bioenergetics, including mitochondrial and bacterial respiratory chains, photosynthesis, methanogenesis, and sulfate reduction. The different types of iron–sulfur proteins that have been identified, numbering over 100, have been reviewed [1].

3.1.1 Ferredoxins as Primitive Proteins

The possible evolution of ferredoxins has been discussed in previous reviews [2–6]. Soon after they were discovered, it was suggested that the ferredoxins were likely candidates for the types of proteins that occurred on the primitive Earth [7,8], for a number of reasons:

1. Iron and sulfide are believed to have been abundant on the early Earth. Iron forms a major part of the Earth's core, while sulfur continues to emerge from volcanic vents.

2. Artificial iron–sulfur clusters are readily synthesized in a mixture of thiols, iron salts, and sulfide [9]. Under anaerobic conditions, the [4Fe–4S] clusters are the most stable, although other structures such as [2Fe–2S] clusters are formed with suitable ligands.

3. The same mixtures of thiols, iron salts, and sulfide may be used to create artificial iron–sulfur proteins by using simple polypeptides containing cysteine [10]. In present-day cells such high concentrations, of free iron particularly, are not found, and it is likely that more elaborate mechanisms for cluster insertion are used [11, 12].

4. They have short polypeptide sequences, containing a high proportion of the more thermodynamically stable amino acids. The exact composition of the amino acids may not be critical, and, indeed, it has been shown that amino acids other than cysteine are capable of binding the clusters (see later).

5. The ferredoxins are stable proteins for their size, as shown by the observation that they occur in extreme thermophilic archaea such as *Pyrococcus furiosus* and *Sulfolobus acidocaldarius*. In these organisms, iron–sulfur proteins and ferredoxins may take the place of conventional dehydrogenases and pyridine nucleotides [13, 14]. The strong covalent bonding of the clusters provides rigidity for the structure of a protein, analogous to the disulfide bridges of extracellular proteins of present-day aerobic organisms. This would be useful when protein replication was not so accurate.

When the sequences of the 2[4Fe–4S] ferredoxins were determined, they were proposed as an early example of gene duplication, consisting of two homologous sequences of about 26 amino acids. However, when the three-dimensional structure was determined it revealed that the duplicated sections of sequence are intertwined around both [4Fe–4S] clusters (Fig. 3.1). Half of the molecule would be incapable of binding a single cluster in the same way. This point is emphasized by the observation that, when a ferredoxin was cleaved proteolytically in the middle, reconstitution with iron and sulfide resulted in the reassembly of two polypeptides to form a 2[4Fe–4S] protein [15]. The structure of the ferredoxins suggests that far from being primitive, they represent a relatively sophisticated design. The ferredoxins with a single [4Fe–4S] cluster are envisaged as evolving from the structure by the *loss* of one of the two clusters [16]. All traces of the ancestral precursor of the ferredoxins have probably been lost from this structure. An alternative approach to find the primitive proto-ferredoxin is the construction of simple, stable [4Fe–4S] proteins by peptide synthesis [17–19].

3.1.2 Iron and Sulfur in Prebiotic Evolution

Wächtershäuser and colleagues [20–21a] have proposed that iron sulfide played a central role in prebiotic chemistry. They point out that the reaction converting iron sulfide to the stable crystalline form, pyrite (FeS_2), is highly

(a) 2[4Fe-4S]

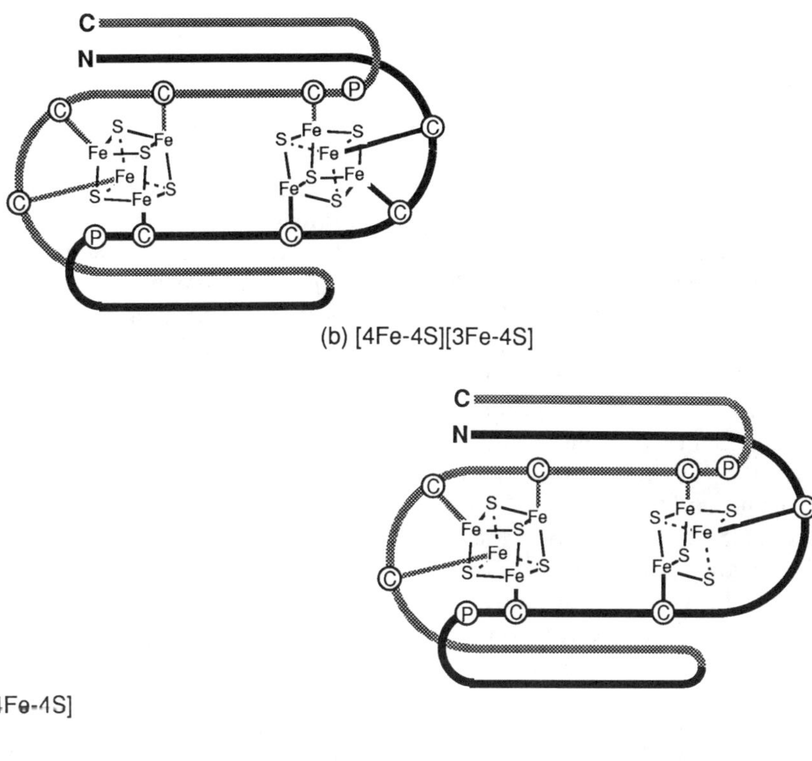

(b) [4Fe-4S][3Fe-4S]

(c) [4Fe-4S]

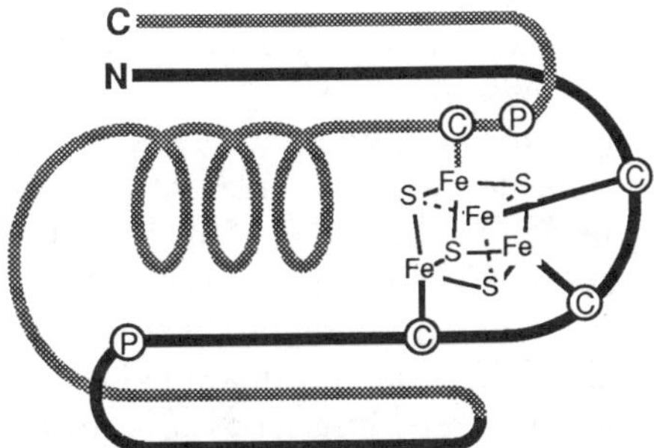

Figure 3.1 Protein folding in the 2[4Fe–4S] ferredoxins. (Reproduced with permission from Ref. [1].)

exergonic:

$$FeS + H_2S \text{ (aq)} = FeS_2 + H_2 \quad \Delta G^{\circ\prime} = -41.9 \, kJ \cdot mol^{-1} \tag{3.1}$$

The feasibility of this reaction has been demonstrated, under anaerobic conditions at pH 7 and 100°C [22]. There is evidence for the existence of pyrite, probably of biogenic origin [23], in palaeozoic sediments [24]. The possibility of forming an enclosing membrane for an early cell from iron sulfides is discussed by Russell et al. [24a].

The pyrite-forming reaction was proposed as a source of strong reductant to "pull" the biosynthetic reactions, so that CO_2 could be used as carbon source. As such it is an alternative to photochemical reactions, or the degradation of preexisting carbon compounds postulated in previous models of evolution. Blöckl et al. [25] have proposed a cycle of reactions in which the reactants are negatively charged carboxylic acids, adsorbed onto the positive charged pyrite surface. The reactions are analogous to the reversed tricarboxylic acid (TCA) cycle used by present-day green anaerobic photosynthetic bacteria, the Chlorobiaceae [26], and, in part, by the methanogens [27]. Reaction (3.1) shows that FeS in the presence of H_2S is a much stronger reductant than NADPH or hydrogen gas, which might obviate the thermodynamic requirement for ATP-driven reactions in the cycle.

The next stage in this proposed line of evolution would be the sequestration of iron–sulfur crystallites into iron–sulfur proteins [28]. It should be noted that the formation of pyrite would compete with the formation of such protein-bound clusters for iron and sulfur. Pyrite itself is quite stable, being insoluble even in hydrochloric acid, so that other forms of iron and sulfide presumably would act as the source.

The TCA cycle, or primitive variants of it, have been a focus of interest in models for prebiotic reaction pathways (e.g., [29]). It is noteworthy that in anaerobic bacteria, several of the enzymes of the TCA cycle (pyruvate: ferredoxin oxidoreductase, 2-oxoglutarate synthase, fumarate reductase, class I fumarase, and aconitase) are iron–sulfur proteins. Ferredoxin, another iron–sulfur protein, serves as reductant. By contrast, in aerobic organisms many of these enzymes either do not exist, or are substituted by enzymes, such as class II fumarases, that do not contain iron–sulfur clusters.

Hydrogen gas is central to the energy metabolism of many microorganisms. It is produced as an end-product by some fermentative organisms, and used as a reductant for various types of respiration in others [30, 31]. Hydrogen production and consumption are catalyzed by hydrogenases, which are iron–sulfur proteins. Most types of iron–sulfur cluster do not produce hydrogen gas, even at extremely low redox potentials, which is an advantage in low-potential energy metabolism such as photosynthesis. In hydrogenases, a particular type of catalytic site is required to react with hydrogen. This site is either the "H-cluster" (an iron–sulfur cluster, structure unknown), or an NiFe center [31a], depending on the type of hydrogenase. Hydrogenases of the hyperthermophiles *Thermococcus* stetteri [31b] and *Pyrococcus furiosus* [31c] are able to reduce elemental sulfur directly, which might have been the principal activity of an ancestral hydrogenase [31c]. Nitrogenases are also capable of producing hydrogen: this appears to be a function of the iron and sulfur rather than molybdenum, since the alternative nitrogenase, which contains neither molybdenum nor vanadium, also catalyzes the reaction [32].

3.2 Iron–Sulfur Clusters and Proteins: The Range of Possibilities

If we consider that iron–sulfur proteins, or similar compounds, could have been an evolutionary development from these autocatalytic substrate cycles, we should examine the chemistry that is possible for such structures. Figure 3.2 shows a number of different arrangements of iron and sulfur that have been found in proteins. Other structures that are almost certainly different, but have not been resolved, are the catalytic H-clusters of Fe-hydrogenases [33] and the

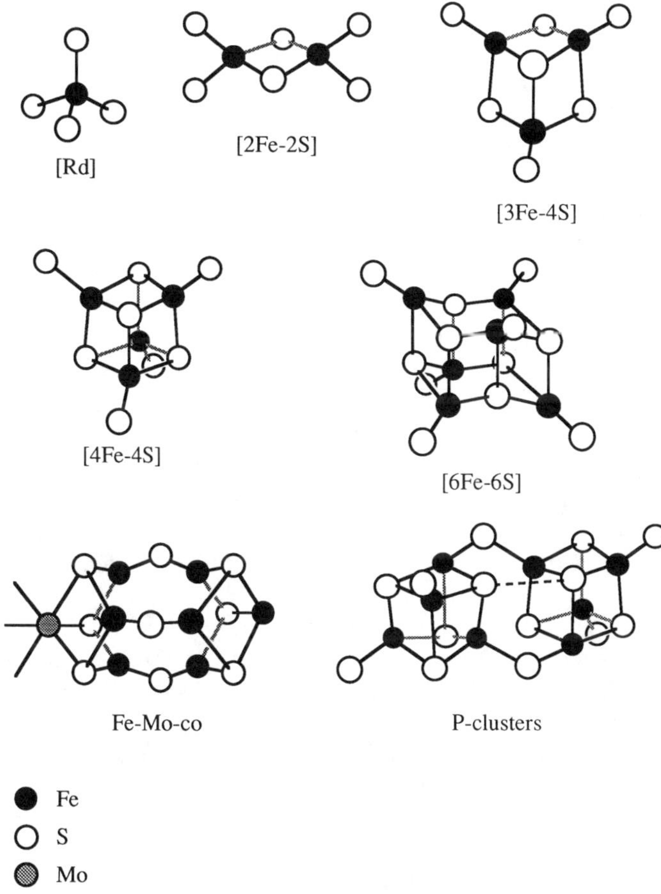

Figure 3.2 Structures of iron–sulfur clusters found in proteins. All have been determined crystallographically [1], except the [6Fe–6S] cluster, which has been observed in synthetic compounds, and has been indicated by spectroscopic results [136].

iron-only nitrogenase [34]. The iron–molybdenum center in nitrogenase contains a molybdenum atom. Other mixed-metal clusters appear to be involved in the nickel-containing carbon monoxide dehydrogenases and the vanadium nitrogenase. These structures require different binding motifs, and generally are found only in large, complex proteins. It seems likely that these evolved later than the ferredoxins.

3.2.1 Redox Potentials

Many iron–sulfur clusters are simple one-electron carriers. As such, their functionality is defined by their midpoint reduction potentials [35]. The simple [4Fe–4S] clusters in ferredoxins have midpoint potentials near the hydrogen potential at pH 7, -420 mV. However, within proteins, their potentials vary from -705 mV for cluster X in photosystem I [36] to $+80$ mV in *Escherichia coli* respiratory nitrate reductase [37]. Similar manipulation of the redox potentials by the protein environment is seen in all electron transfer proteins, including cytochromes, flavoproteins, and molybdenum proteins. Various mechanisms have been proposed for this [38], and they appear to involve relatively minor changes in the protein structure. Such modifications could have evolved in a gradual manner.

Another relatively minor modification of the [4Fe–4S] ferredoxins, which brings about a modification of cluster structure, is the loss of a cysteine which converts them to the [3Fe–4S] type (Fig. 3.1). The modification involves an overall oxidation of the cluster, and leads to a more positive redox potential. Such a modification is seen in one of the four clusters in the respiratory-chain nitrate reductase of *Escherichia coli* [39], where a Trp is found in place of an expected Cys, producing a [3Fe–4S] cluster. This is to be compared with DMSO reductase of the same organism, which contains four [4Fe–4S] clusters with a more negative average midpoint potential [40]. The potential is presumably tuned to that of the substrate.

Other modifications are observed in iron–sulfur proteins with radically different protein sequences and conformation. These types of protein appear to represent different lines of descent. In the high-potential iron–sulfur proteins (HiPIPs), an alternative redox chemistry of the [4Fe–4S] clusters is selected. The higher $[4Fe–4S]^{3+} \leftrightarrow [4Fe–4S]^{2+}$ transition occurs, instead of the usual $[4Fe–4S]^{2+} \leftrightarrow [4Fe–4S]^{1+}$ transition, as found in all other proteins [41]. The resulting midpoint is much higher, between $+100$ and 350 mV [41a]. The way in which this is achieved is dependent on the protein structure, since both transitions are equally feasible in chemical model compounds. The [4Fe–4S] cluster in HiPIP is deeply buried in the protein, and this may prevent the charge compensation that would be needed to stabilize the $[4Fe–4S]^{1+}$ state. In ferredoxins, attempts to oxidize the [4Fe–4S] clusters to the $[4Fe–4S]^{3+}$ state by chemical means usually result in loss of a sulfur ligand and conversion to a [3Fe–4S] cluster [42], although a possible $[4Fe–4S]^{3+}$ intermediate has been detected by NMR [43].

The use of noncysteine ligands is another way to alter the total cluster charge and thus its midpoint potential. This is exploited in the Rieske-type [2Fe–2S] clusters, where the ligands to one iron atom are histidine nitrogens instead of cysteines. This leads to considerably more positive redox potentials [44]. Recently, a histidine nitrogen ligand has been found in a [4Fe–4S] cluster of *Desulfovibrio gigas* hydrogenase [31a].

3.2.2 Catalytic Activity

Modification of the cluster protein structure may lead to clusters with catalytic activities other than electron transfer. As previously mentioned, the reaction with hydrogen gas requires a special type of cluster structure, and presumably a protein structure, which allows the transfer of H^+ ions to the site. The catalytic cofactors of nitrogenases are clearly very specialized for their functions. In addition to these redox-active clusters, some clusters can act as a nucleophiles in nonredox reactions. These involve the exposure of an iron atom so that it can undergo coordination of substrates, as in aconitase. Model iron–sulfur cubane clusters have been synthesized that emulate the binding of oxygen ligands to one iron atom [45].

3.3 Evolution of Complex Electron-Transfer Proteins

3.3.1 Insertion of Iron–Sulfur Clusters into Preexisting Structures

Some iron–sulfur proteins, which show no similarities in sequence or structure to the ferredoxins, could have arisen by insertion of clusters into preexisting proteins. It is envisaged that in the earliest cells, iron–sulfur proteins could have arisen by nonenzymic incorporation of iron and sulfide into the nascent polypeptides. In the cells that evolved later, iron was sequestered to prevent unwanted reactions such as oxygen radical production. Sulfide was likewise not freely available. Mechanisms, which are not well understood, have evolved for insertion of iron–sulfur clusters into proteins. These appear to be catalyzed by specific proteins [11, 45a], which recognize certain features of the apo-iron–sulfur proteins. Recent gene-transfer experiments have shown that when such apo-proteins are expressed in *E. coli*, the correct iron–sulfur clusters are often inserted by the bacterium. For example, expression of the Fe-hydrogenase of *Desulfovibrio vulgaris* in *E. coli* led to the insertion of the [4Fe–4S] clusters, but not the catalytic H-clusters, for which a specific gene product is presumably required [46]. We have found that the proteins of *Pseudomonas putida* benzene dioxygenase can be expressed to extremely high levels in *E. coli*, with their complement of Rieske-type [2Fe–2S] clusters, even though such clusters have not been reported to exist normally in *E. coli* [47].

However, there are cases where the cluster-insertion machinery can make mistakes, and insert iron–sulfur clusters into foreign proteins, where they are not intended. When the zinc-binding mammalian LIM proteins, which contain a pattern of cysteines similar to these iron–sulfur proteins [48], were expressed in *E. coli*, the bacterium erroneously inserted iron–sulfur clusters [49]. This presents a possible mechanism whereby iron–sulfur proteins could have arisen. Iron–sulfur clusters might be inserted into proteins totally unrelated to the ancestral ferredoxins. This might explain the existence of some proteins that do not bear any strong resemblance to other iron–sulfur proteins, such as endonuclease III of *E. coli* [50].

3.3.2 Formation of Chimeric Proteins

The discovery of numerous ferredoxin-like sequences in more complex electron-transfer proteins indicates that the electron-transfer domains of these proteins evolved by fusion of the genes for simple electron-transfer proteins. An example is provided by the enzyme phthalate dioxygenase reductase [51]. The structure, which has been determined crystallographically, contains a domain strikingly similar to plant-type ferredoxins, and another, containing flavin, similar to ferredoxin: NADP reductase. The evolutionary relationships between the electron-transfer pathways of bacterial dioxygenase systems have been discussed by Harayama et al [52].

More elaborate chimeric proteins can be envisaged, in which different types of ferredoxin sequence were combined with those of other redox enzymes. Complexes I and II of the respiratory chain, which are discussed below, can serve as examples.

3.3.3 Stability to Oxygen

Oxygen, produced by photosynthesis, would have been poisonous to the early bacteria. The photosynthetic ancestors of the cyanobacteria must presumably have been the first to develop mechanisms to protect iron–sulfur clusters from oxygen. It is now the case that iron–sulfur proteins from aerobic organisms are fairly stable, whereas their anaerobic counterparts tend to be unstable to oxygen. Various strategies have evolved:

1. Substitution of [4Fe–4S] clusters by [2Fe–2S], which appear to be more stable to oxygen. A possible reason for this is that oxidation of reduced [4Fe–4S] ferredoxins by oxygen leads to a greater production of superoxide than the oxidation of [2Fe–2S] ferredoxins [53, 54]. The [2Fe–2S] clusters are more common in aerobic or aerotolerant organisms. For example, photosynthesis in anaerobic green photosynthetic bacteria involves [4Fe–4S] ferredoxins, whereas [2Fe–2S] ferredoxins are found in cyanobacteria and higher plants [55]. The ferredoxins that interact with pyruvate:ferredoxin reductase in anaerobic bacteria are [4Fe–4S] types, whereas those in the aerobic halobacteria and algae are [2Fe–2S] types (Table 3.1). The

Table 3.1   Pyruvate:Ferredoxin Reductases and Pyruvate Synthases

Source	Function	Electron Carrier	Refs.
Methanogens	Pyruvate synthesis	2[4Fe–4S] ferredoxin	[27]
Halobacteria	Fermentation of peptides	[2Fe–2S] ferredoxin	[127]
Thermophilic archaea	Glucose fermentation	[3Fe–4S][4Fe–4S] ferredoxin	[128]
Thermophilic eubacteria	Glucose fermentation	[4Fe–4S] ferredoxin	[129]
Sulfate-reducing archaea	Electron donor for sulfate reduction	?[4Fe–4S] ferredoxin	[129a]
Green photosynthetic bacteria	Pyruvate synthesis	2[4Fe–4S] ferredoxin	[130]
Sulfate-reducing bacteria	Electron donor for sulfate reduction	[4Fe–4S] ferredoxin	[131, 131a]
Anaerobic fermentative bacteria	Phosphoroclastic ATP synthesis	2[4Fe–4S] ferredoxin	[132]
Anaerobic bacteria	Electron donor for nitrogen fixation	2[4Fe–4S] ferredoxin	[133]
Facultative aerobic bacteria	Electron donor for nitrogen fixation	Flavodoxin	[133]
Cyanobacteria	Electron donor for nitrogen fixation	Plant-type [2Fe–2S] ferredoxin	[134]
Trichomonads	Phosphoroclastic ATP synthesis	Hydroxylase-type [2Fe–2S] ferredoxin	[135]

hydroxyacid dehydratase of anaerobic *E. coli* contains a [4Fe–4S] cluster [56], whereas the counterpart in plants contains a [2Fe–2S] cluster [57].

2. Protective proteins. Nitrogenase from the aerobe *Azotobacter vinelandii* is protected during growth by removal of oxygen by high-affinity oxidases [58]. In the resting state a specific [2Fe–2S] protein binds to the enzyme and appears to protect it from oxygen damage [59].

3. Mechanisms have evolved to reinstall the iron–sulfur clusters if they are lost (as appears to be the case for aconitase [60]). It is possible that stress proteins produced in response to denaturing conditions include enzymes for iron–sulfur cluster insertion.

3.4 Families of Iron–Sulfur Proteins

Computer databases of protein sequences, derived from protein and gene sequencing, are a valuable resource for investigation of the evolution of iron–sulfur proteins. Many amino acid sequences of iron–sulfur proteins have been determined, either by protein or DNA sequencing. For ferredoxins alone, over 100 sequences have been determined [6]. These sequences can be studied in various ways.

1. The cluster-binding sites in iron–sulfur proteins may be identified by searching for characteristic arrangements (or motifs) of cysteines in the sequences. As will be seen, some of these motifs show homologies to the ferredoxins, so it is likely that they evolved from the simpler proteins. Because the structures of a representative range of the ferredoxins have been determined, this gives a good indication of the structure of the cluster-binding sites in the other proteins.
2. Where the sequences are not homologous to proteins of known structure, the iron–sulfur-binding motifs may be determined by searching for invariant residues in alignments of homologous proteins. Other "nonessential" cysteines may be eliminated in this way. Examination of sequences of iron–sulfur proteins of known structure, for example, ferredoxins, allows us to draw some tentative general rules. For example, two adjacent cysteines have not been observed to bind to a cluster.
3. From a survey of the sequences it is possible to identify a number of different families, containing different iron–sulfur cluster-binding motifs.
4. The motifs may be used to search the gene banks for previously unidentified iron–sulfur proteins. A number of open reading frames coding for iron–sulfur proteins have been found in this way.
5. Where a number of sequences of the same protein from different species have been determined, the sequences may be used to derive evolutionary relationships between those species. This presupposes that the proteins have the same functions in different species, which is often not the case for ferredoxins of bacteria. Moreover, an organism will often contain several ferredoxins, some similar, some different [6, 61].

3.4.1 General Iron–Sulfur Motifs

We will consider the iron–sulfur cluster-binding motifs, and their use to detect iron–sulfur proteins and their evolution. Just as there is no method yet devised to predict the folding of a protein from its structure, there is no procedure as yet for identifying all iron–sulfur proteins and rejecting all others. A reliable criterion for retrieving many iron–sulfur protein sequences from a database is to search for cysteines with particular spacings. The amino acids in between are much more variable. This approach will not find all sequences, owing to the insertion of loops or gaps in some of the proteins. A less discriminating search pattern will retrieve more iron–sulfur proteins, but also more sequences of other proteins.

Another caveat is that the insertion of iron–sulfur clusters is, to some extent, a function of the cell in which the protein is expressed. As previously mentioned, when the zinc-binding LIM proteins were expressed in *E. coli*, they were converted into iron–sulfur proteins [49].

Motif 1: –CXXC–

This is found in most iron–sulfur proteins. It therefore seems not unlikely that a primitive ferredoxin could have been constructed from duplication of a polypeptide containing just two cysteines.

The disadvantage of motif 1 as a means of following the ancestry of iron-sulfur proteins is that it is not discriminating. When applied to the database, it also retrieves many other proteins, including cytochromes *c* (haem binding); metallothioneins (Cd/Zn binding); keratins and other extracellular proteins containing structural disulfides; proteins with redox-active disulfides, such as thioredoxins; and binding sites for other metals, such as zinc.

3.4.2 [4Fe–4S] Clusters

Motif 2: –CXXCXXCXXXCP–

This is listed in the "Prosite" list of sequence motifs, and will find most of the ferredoxins with two [4Fe–4S] clusters.

However, some [4Fe–4S] proteins, which are clearly related, do not conform to motif 1. In the photosynthetic bacterial ferredoxins from *Chromatium* and *Chlorobium*, gaps are inserted into the sequence between the second and third C, causing a loop on the surface of the protein. The less restricted sequence

Motif 3: –CXXC(2–8)C(10–50)CP–

retrieves most of these 4Fe ferredoxins, and many enzymes (Table 3.2) (for references, see Ref. [1]). Here the numbers in parentheses indicate the number of amino acids spacing between the cysteines. All of these are proteins containing [4Fe–4S] clusters. It seems likely that the [4Fe–4S] domains of these proteins are descended from the bacterial ferredoxins, although the possibility of convergent evolution cannot be ruled out.

It can be seen from Fig. 3.1 that in the protein structure, the cysteine of the Cys–Pro from one group binds to the cluster that is bound by another group of three cysteines. Therefore for 50% of the [4Fe–4S] clusters, the CP precedes the other three cysteines. The pattern for these is

Motif 4: –CP(10-40)CXXC(2–8)C–

This motif finds most of the [4Fe–4S] proteins retrieved by motif 3, but in addition a few proteins with a single [4Fe–4S] cluster, where a unique CP precedes the other three cysteines, such as the *Rhizobium trifolium fixX*-encoded protein [62].

Another class of proteins that does not conform to motifs 2 and 3 comprises those where the proline of the CP group may be substituted by another amino

Table 3.2   Proteins with [4Fe–4S] Motifs

Protein	Example of Source
Carbon monoxide dehydrogenase	*Methanothrix soehngenii*
Complex I (NADH:ubiquinone reductase)	*E. coli*
Corrinoid iron–sulfur protein	*Rhodospirillum rubrum*
DMSO reductase subunit B	*E. coli*
F420-reducing Ni-hydrogenase	*Methanococcus voltae*
Fe-hydrogenase	*Desulfovibrio vulgaris*
Formate dehydrogenase	*E. coli*
Fumarate reductase	*E. coli*
Glutamate synthase	*Azospirillum brasilense*
Glycerol-3-phosphate dehydrogenase	*E. coli*
Nitrate reductase	*E. coli*
Photosystem I 9-kDa protein	Cyanobacteria and plants
Polyferredoxin	*Methanobacterium thermoautrophicum*
Polysulfide reductase	*Wolinella succinogenes*
Pyruvate:flavodoxin oxidoreductase	*Klebsiella pneumoniae*
Succinate dehydrogenase	*Bacillus subtilis*
Sulfite reductase	*Archaeoglobus fulgidus*
Sulfite reduction protein	*Salmonella typhimurium*
Trimethylamine dehydrogenase	Bacterium W_3A_1

acid. In the structure, the proline stabilizes a turn in the polypeptide. Sequence-directed mutagenesis investigations have shown that replacement of this Pro by other amino acids does not prevent the insertion of [4Fe–4S] clusters [63]. In the more complex proteins, such as *E. coli* nitrate reductase [39], other amino acids, such as glutamate or glycine, substitute for this proline.

Iron–sulfur clusters are often found in cysteine-rich regions, containing more cysteines than are needed to bind the cluster. In the case of *A. vinelandii* ferredoxin I, one of these cysteines (C24) was able to bind to the cluster when one of the cluster-binding cysteines (C20) was deleted by site-directed mutagenesis [64]. The function of these other cysteines might be to act as polarizable groups in the vicinity of the cluster, which could facilitate electron transfer.

Chemical synthesis studies have shown that the [4Fe–4S] and [2Fe–2S] clusters can be maintained in stable form with other terminal ligands, including hydroxyl and carboxylate groups [9]. Synthetic [4Fe–4S] clusters can be synthesized with amino acids of these types as ligands [65]. It was therefore not unexpected to find that in the sequences of some of the iron–sulfur proteins, one of the expected cysteines is missing. This includes proteins such as *Desulfovibrio africanus* ferredoxin III [66], where an aspartate is found at position 14, instead of the cysteine which is expected to bind to the [4Fe–4S] cluster. This cluster undergoes facile conversion to a [3Fe–4S] cluster [67]. In *E. coli* succinate dehydrogenase, one of the cysteines that are expected to bind a [2Fe–2S] cluster is missing and its place is taken by an aspartate [68].

Site-directed mutagenesis studies on a number of iron–sulfur proteins have indicated that, in some cases, substitution of a cluster-binding Cys by Ser still allows the retention of [2Fe–2S] and [4Fe–4S] clusters. Examples of both are seen in fumarate reductase [69].

There are two types of protein known where a [4Fe–4S] cluster is bound to cysteines from different protein subunits. In the Fe-protein of nitrogenase, the cluster is bound between two identical subunits. In photosystem I of plants and cyanobacteria, the low-potential iron–sulfur cluster X is bound between two different polypeptides, A1 and A2, each containing two conserved cysteines that are adjacent to a sequence resembling a "leucine zipper" motif [70]. This latter takes the form:

Motif 5: –LXXXXXXLXXXXXXLXXXXXXL–

Leucine zippers are more familiar in DNA-binding proteins, where they form two-stranded helices between two different polypeptides, allowing many combinations of heteroduplexes from a library of monomer polypeptides. In photosystem I, it is possible that the sequence may be a coincidence in such a hydrophobic protein, but if it does indeed form a helix, a reasonable function to stabilize the cluster structure can be recognized. Another iron–sulfur protein that contains a potential leucine zipper is NADH:ubiquinone reductase, where it is found in subunits 2 and 3.

The proteins retrieved by motifs 2 to 4 (Table 3.2) appear to be the largest family of iron–sulfur proteins, descending from the 2[4Fe–4S] ferredoxins, and having a similar protein structure (Fig. 3.1). Studies of protein structures and sequences indicate that other families exist, which have characteristic sequence motifs.

3.4.3 High-Potential Iron–Sulfur Proteins

These proteins were among the first to be isolated, from anaerobic photosynthetic bacteria and other autotrophic bacteria.

Their sequences do not resemble those of other proteins, but fit the pattern:

Motif 6: –CXXC(9–16)C(14–16)C–

They represent a method to achieve a relatively high redox potential, by exploiting the higher oxidation state of the [4Fe–4S] cluster. The function of the HiPIPs is uncertain, although a capacity for photosynthetic electron transfer has been demonstrated [70a]. However, their high potential fits them for a role similar to that of cytochromes and copper proteins in aerobic bacteria. It may be noted that, in an anacrobic, sulfide-rich environment, copper would be unavailable because of the insoluble nature of copper sulfide.

3.4.4 Other [4Fe–4S] Clusters

A number of enzymes contain [4Fe–4S] clusters that do not show close homologies to other iron–sulfur proteins, and may represent categories of their own. These include glutamine:phosphoribosylpyrophosphate amidotransferase [71] and endonuclease III [50].

3.4.5 [2Fe–2S] Clusters

The plant ferredoxins were the first [2Fe–2S] proteins to be isolated. Similar ferredoxins exist in cyanobacteria and photosynthetic bacteria [72]. Their sequences are characterized by three cysteines in a group, then a distant cysteine:

> Motif 7: –CXXXXCXXC...C–

The structure of these proteins was found to be a stable β-barrel [73]. It appears that, to synthesize stable [2Fe–2S] clusters, a more rigid structure is required than for [4Fe–4S] clusters, and so these could represent a later development than the [4Fe–4S] types. [2Fe–2S] ferredoxins have been found as electron donors to P-450-type monooxygenases of bacteria and mammals. Although differing in sequence and catalytic properties, the proteins have a similar arrangement of cysteines and are likely to be related to the plant-type ferredoxins. Halobacteria, which are aerophilic archaea, also contain a ferredoxin that resembles the cyanobacterial ferredoxin, with extensions at the N and C termini [74]. These proteins have various functions; all three types of [2Fe–2S] ferredoxins are used as electron acceptors for pyruvate:ferredoxin reductase (Table 3.1).

A distinct type of [2Fe–2S] protein has been found in anaerobic bacteria such as *Clostridium pasteurianum* [75] and the photosynthetic prokaryote *Chlorobium* [76]. It appears to be associated with nitrogen fixation. The sequence of the *C. pasteurianum* protein shows a totally different arrangement of cysteines [75], indicating a different structural type.

Yet another, distinct class of [2Fe–2S] proteins comprises the "Rieske" iron–sulfur proteins. These are characterized by a quite different sequence motif [77, 78]:

> Motif 8: –CXH(15–17)CXXH–

in which the two ligands to one iron atom are the δ-nitrogens of histidines [79]. These iron–sulfur proteins have more positive redox potentials, on average, than the conventional ferredoxin-type [2Fe–2S] clusters. The simplest proteins of this type are the ferredoxins associated with class IIB dioxygenase [80]. The more familiar Rieske proteins are those associated with the quinol-oxidizing cytochrome b/c complexes (complex III) of photosynthetic and respiratory bacteria [44, 81].

3.4.6 Rubredoxins

The rubredoxins represent another enigmatic group of nonheme iron proteins, with intermediate redox potentials ($-50\,$mV). The rubredoxins contain no labile sulfide, the single iron site being coordinated by four cysteine sulfurs. One such protein has been shown to be involved in electron transport to a monooxygenase in *Pseudomonas oleovorans* [82]. This protein is unusual in being "double-headed," with two similar iron-binding sites. In anaerobic bacteria the function of the rubredoxins is not clear, although some have been linked to nitrogen fixation [83]. The desulforedoxins of sulfate-reducing bacteria are of exceptionally low molecular mass and have a dimeric structure [84,85].

3.4.7 Sulfite and Nitrite Reductases

Sulfite and nitrite reductases are proteins containing siroheme, and iron and sulfur [86]. Sequence alignments indicated two groups of conserved cysteines that are now known to bind the cluster [87]:

Motif 9: –CXXXXXC...CXXXC–

Sulfite reductase is an enzyme in the bioenergetic pathway of sulfate-reducing bacteria. The structure of the active site comprises a [4Fe–4S] cluster coupled to the iron of siroheme by a bridging cysteine sulfur [87,88]. This structure may be designed to facilitate the six-electron reduction of the substrates, sulfite to sulfide, and nitrite to ammonia.

3.4.8 Mixed-Metal Clusters: Nitrogenases

The structure of the iron–sulfur clusters is further elaborated in those proteins with mixed-metal clusters, containing either molybdenum or vanadium, in the nitrogenases, or nickel, in the carbon monoxide dehydrogenases. The latter are important in the production or consumption of carbon monoxide by methanogens and acetogens.

Nitrogenases contain two proteins, the iron-protein (protein 1, dinitrogenase reductase), with an iron–sulfur cluster, and the nitrogenase itself (protein 2). The recently determined structure of molybdenum nitrogenase shows that it contains two unique types of iron–sulfur cluster [89]. The iron–molybdenum cluster, Fe–Mo-co, is the most elaborate structure so far determined (Fig. 3.2). The P-cluster resembles two [4Fe–4S] clusters and is presumably able to transfer two electrons simultaneously to Fe–Mo-co. This may be a mechanism to overcome the limitation of most iron–sulfur clusters, which transfer only one electron. Each of these iron–sulfur cluster structures requires its own specific binding motif [90]. The ligands to the P-clusters are six cysteine sulfurs and possibly a serine. Fe–Mo-co is a more self-contained complex and is bound by one cysteine; a histidine and homocitrate form ligands to the molybdenum.

In addition to this type I nitrogenase, it has been found that nitrogen-fixing bacteria are able to express two other nitrogenases, one (type II) containing vanadium instead of molybdenum, the other (type III) containing iron as the only metal. The three different types of nitrogenase show strong conservation of sequence, although they are different gene products [90]. Types I and III are found in methanogens, and probably represent an ancestral form. Type II nitrogenases appear to be a more recent development in the proteobacteria [91].

3.4.9 Catalytic Iron–Sulfur Clusters

Aconitase, an essential enzyme of the TCA cycle, is an iron–sulfur protein [92]. The catalytic site, of which the crystal structure has been determined, is a [4Fe–4S] cluster with only three cysteine ligands, leaving the fourth iron atom available to coordinate the substrate [93]. A number of other enzymes containing iron–sulfur clusters, with functions other than electron transfer, have been isolated, mostly from anaerobic bacteria [94]. Some of these proteins, which have homologous aminoacid sequences to aconitase, catalyze reactions that may be classified broadly as hydration–dehydration reactions. For example, the TCA cycle enzyme fumarase is an iron–sulfur protein, containing a [4Fe–4S] cluster [95, 96] in anaerobic *E. coli* (fumarases A and B). A different protein without iron is expressed in *E. coli* (fumarase C), where it is regulated by oxygen concentration [97], and in mitochondria. This suggests that the iron–sulfur protein has advantages as a catalyst under anaerobic conditions, but its oxygen sensitivity is a drawback under aerobic conditions. The dihydroxyacid dehydratase of anaerobic *E. coli* is a [4Fe–4S] protein [95], whereas the equivalent enzyme in plants is a [2Fe–2S] protein, and is less oxygen sensitive [57].

3.5 Iron–Sulfur Proteins of Bioenergetic Pathways

Iron–sulfur proteins are widespread in the electron-transfer pathways of plant-type photosynthesis and mitochondrial-type respiration, and other forms of dissimilatory metabolism found in bacteria, including methanogenesis, and nitrate and sulfate reduction. In the low-potential, anaerobic dissimilatory pathways, iron–sulfur proteins have particularly prominent roles as electron carriers.

3.5.1 Iron–Sulfur Proteins in the Metabolism of the Archaea

The methanogens are strictly anaerobic archaea [98]. In these organisms, CO_2 fixation involves a corrinoid iron–sulfur protein and the iron–sulfur–nickel enzyme, acetyl-Co A synthase ("carbon monoxide dehydrogenase"). Reductant is supplied by nickel-containing hydrogenases. The energetics of methane formation are complex, involving transmembrane potentials of protons and sodium ions [99, 100]. Iron–sulfur proteins are also involved in transmembrane electron transfer. The methanogens also contain complex, soluble iron–sulfur proteins, including the polyferredoxins, in which the tendency for multiplication of [4Fe–4S] clusters finds its most extreme form. These contain up to eight repeated ferredoxin-like sequences, which would bind 16 [4Fe–4S] clusters. This protein appears to be involved in electron transfer from hydrogen in the methanogenesis pathway.

The oxidation and reduction of sulfur compounds provide energy for growth of many lithoautotrophic and phototrophic bacteria, including some branches of the Archaea. These include sulfate-reducing bacteria such as *Archaeoglobus fulgidus* [86] and sulfur oxidizers such as *Sulfolobus acidocaldarius* [101]. Many of the proteins extracted from these organisms, and believed to be involved in energy metabolism, are iron–sulfur proteins.

3.5.2 Pyruvate:Ferredoxin Oxidoreductases

Pyruvate and related 2-oxoacids are strong reducing agents when the reaction is coupled to decarboxylation. In many anaerobic chemolithotrophs and photoautotrophs, the oxoacid:ferredoxin oxidoreductases play a central role in bioenergetic metabolism and carbon fixation. The enzymes differ from the more familiar pyruvate and oxoglutarate dehydrogenases of aerobic metabolism, where the reaction involves lipoate and a flavoprotein, and the electron acceptor is NAD. The latter enzymes probably represent a different line of descent. The ony similarity is the involvement of thiamin diphosphate.

The phosphoroclastic pathway for substrate-level ATP synthesis, involving release of hydrogen by the enzyme hydrogenase, exists in many organisms, including the purely fermentative Clostridia, the cyanobacteria, and some eukaryotic algae, fungi, and protozoa, in which evolutionary relationships can be traced [101a]. As can be seen from Table 3.1, all the different types of low-potential ferredoxin are used as electron carriers in different species. In the sulfate-reducing bacteria, pyruvate, derived from lactate, serves as electron donor for sulfate reduction. Pyruvate also provides the low-potential electron donor to nitrogenase in methanogens, cyanobacteria, and other eubacteria.

In Wächtershäuser's model [25], the formation of carbon–carbon bonds in the autocatalytic cycles involves reactions analogous to the oxoacid synthases of the reductive TCA cycle. The equivalent enzymes in photosynthetic bacteria and methanogens, pyruvate synthase and oxoglutarate synthase, use the strong reducing power of ferredoxin, and are involved in various pathways for CO_2 fixation. These enzymes are used for the reverse TCA cycle in photosynthetic bacteria, where the electron donor is a low-potential 2[4Fe–4S] ferredoxin. In methanogens, the CO_2 fixation pathways use incomplete versions of the TCA cycle [27]. Methanogens such as *Methanobacterium thermoautotrophicum*, which form methane from hydrogen and CO_2, use the malate–succinate branch of the reversed TCA cycle to produce glutamate. This pathway has the effect of fixing further CO_2 in the oxoglutarate synthase reaction. Methanogens such as *Methanosarcina barkeri*, which produce methane from energetically less favorable substrates such as acetate, use the citrate–isocitrate arm of the TCA cycle (Fig. 3.3).

3.5.3 Complex I

The NADH:quinone reductases such as mitochondrial complex I are the most elaborate of the iron–sulfur proteins of bioenergetics. With up to 40 polypeptide subunits, the mitochondrial enzyme rivals the ribosome in size and complexity [102, 103]. The bacterial counterparts, from *E. coli*, *Paracoccus denitrificans*, or *Rhodobacter sphaeroides*, are simpler, but have a similar structure [104]. They all consist of two parts: a hydrophilic, iron–sulfur flavoprotein that contains the NAD-binding site, and a hydrophobic, membrane-bound component, which contains iron–sulfur clusters and is involved in reduction of quinone. The latter component is also believed to contain the proton-pumping mechanism of the enzyme [105]. Unexpected homologies have been observed between the NAD-binding FMN–iron–sulfur domain of complex I and the NAD-reducing hydrogenase of *Alcaligenes eutrophus* [106, 107], between the membrane-bound iron–sulfur protein domain and formate:hydrogen lyase [108] and nickel hydrogenases [108a], and between glucose dehydrogenase and the quinone-binding domain [109, 109a]. Consideration of these sequences suggests that the enzyme evolved by assembly of existing multisubunit proteins [110]. Traces of homology have been detected with the nickel–iron–sulfur hydrogenases [110a]. Gene sequence comparisons indicate that proteins with similar structure to complex I exist in cyanobacteria [111] and the chloroplasts of higher plants [112]. These proteins are presumably involved in respiratory electron transfer in the dark.

It is possible that complex I arose in evolution from proteins that were involved in reduction of NAD for biosynthesis. In the hydrogen-oxidizing bacteria, the nickel- and iron–sulfur-containing hydrogenase, which shows homologies to part of complex I [113], is used for the reduction of NAD for the carbon assimilation pathway. Proteins similar to complex I are found in chemolithotrophic and phototrophic bacteria [114], where they catalyze the reduction of NAD by proton-driven reversed electron transfer.

3.5.4 Succinate Dehydrogenase and Fumarate Reductase

These are similar enzymes of the TCA cycle (Fig. 3.3), which catalyze the same reaction in opposite directions:

$$\text{Succinate} + \text{quinone} \rightleftharpoons \text{fumarate} + \text{quinol}$$

In *E. coli*, the synthesis of these two proteins is controlled by the *fnr* gene, the product of which is an iron-sulfur protein [115]. Succinate dehydrogenase,

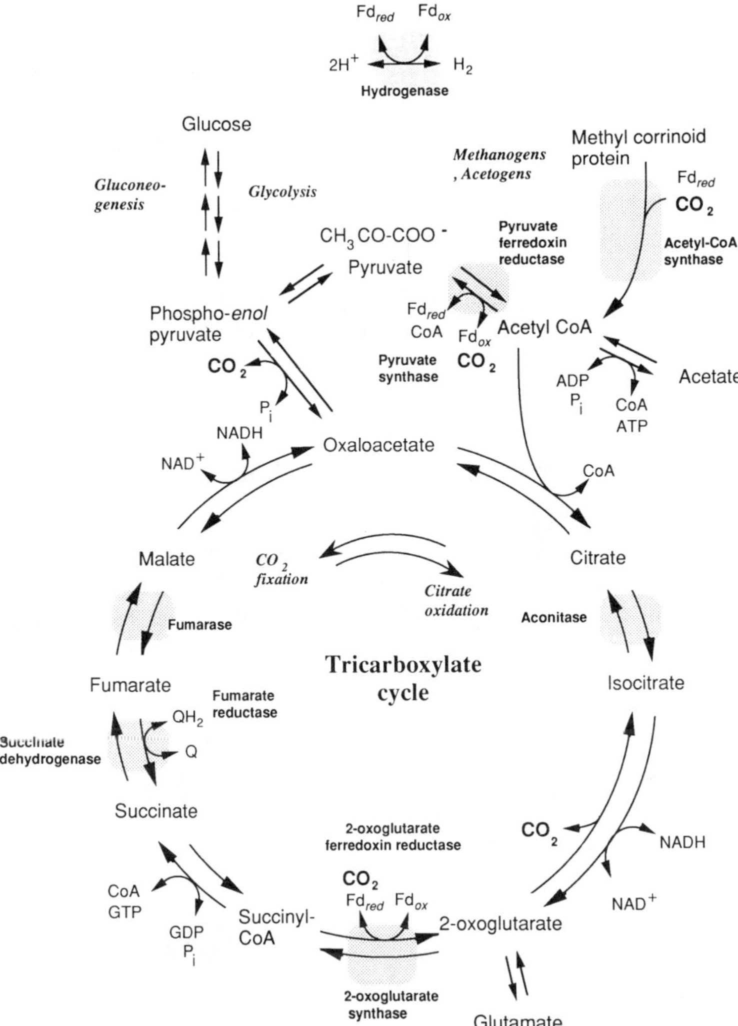

Figure 3.3 Reactions of the reductive and oxidative TCA cycles, showing reactions involved in the fixation of CO_2 by photosynthetic bacteria, the oxidation of carbohydrates by aerobes, and carbon dioxide fixation pathways in the methanogens. Not all of these reactions occur in the same organism. Reactions involving iron–sulfur proteins are shaded.

which oxidizes succinate to fumarate (using ubiquinone), is expressed under aerobic conditions. Fumarate reductase catalyzes the reduction of fumarate to succinate (using menaquinone), and is expressed under anaerobic conditions. Studies with mutants have shown that one protein can substitute, with lower efficiency, for the latter.

Succinate dehydrogenase and fumarate reductase comprise a flavoprotein subunit that reacts with succinate and fumarate, an iron–sulfur subunit, and two membrane anchor subunits [116]. The presence of [2Fe–2S], [3Fe–4S], and [4Fe–4S] clusters was predicted from the sequence [117–119]. The main differences between the two enzymes appear to be in the relative midpoint potentials of the clusters, which are optimized for electron transfer in one direction or the other. Electrochemical studies indicate that succinate dehydrogenase has a "ratchet" effect, which inhibits the reduction of fumarate by the enzyme under strongly reducing conditions [120]. This behavior has been attributed to the redox properties of the flavin. This interesting refinement is an example of the way in which the direction of electron transfer can be controlled by the protein, in addition to the tendencies imposed by redox potentials.

3.5.5 Membrane-Bound Complexes of Anaerobic Bacteria

Bacteria contain iron–sulfur proteins that transfer electrons between soluble substrates and a membrane-bound quinone. Many of them follow the same general structure as succinate dehydrogenase: a subunit containing a catalytic group such as flavin, a molybdopterin or nickel center, an iron–sulfur-containing subunit, and one or more membrane anchor subunits with hydrophobic transmembrane α-helices. In many cases, heme (cytochrome b) is also involved. The soluble substrate-binding subunits protrude substantially out of the membrane. In anaerobic *E. coli*, enzymes of this type include glycerol phosphate dehydrogenase, nitrate reductase, DMSO/TMAO reductase, and

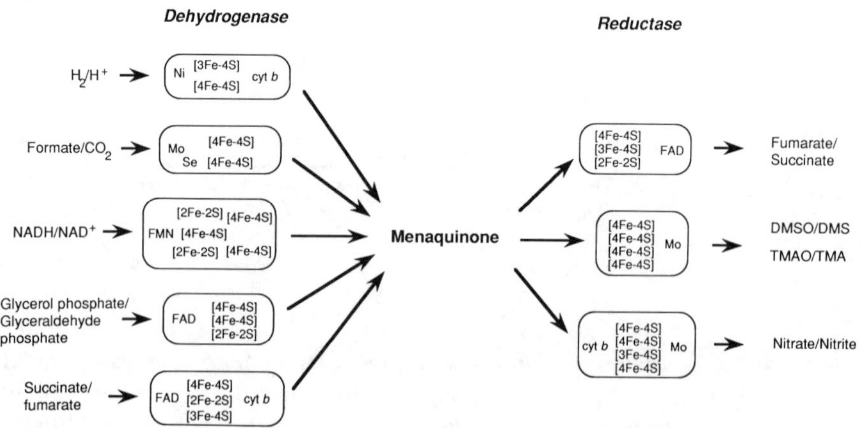

Figure 3.4   Membrane-bound electron-transfer enzymes in anaerobic bacteria.

hydrogenase. They communicate by transferring electrons through mena-quinone (Fig. 3.4).

Other redox proteins, operating at more negative redox potentials than quinone, transfer electrons through specific iron–sulfur clusters. In *E. coli*, a specific hydrogenase and formate dehydrogenase, together with an iron–sulfur subunit, comprise the formate:hydrogen lyase [121]. Formate oxidation may also be coupled through iron–sulfur protein to nitrite reduction [122]. In anaerobic photosynthetic bacteria such as *Rhodospirillum rubrum*, the nickel–iron–sulfur enzyme that catalyzes the oxidation of CO is probably coupled, through another iron–sulfur protein subunit, to the production of H_2 [123]. Such iron–sulfur protein complexes are also involved in the energy-yielding reactions of methanogenesis. The reaction that releases methane from methyl-coenzyme M leaves a mixed disulfide between coenzyme M and HS-HTP (*N*-7-mercaptoheptanoyl-*O*-phosphothreonine). The re-reduction of this com-pound by a membrane-bound iron–sulfur protein [123a], coupled to a nickel-hydrogenase, leads to the formation of a transmembrane proton gradient [124].

3.5.6 ATP-Dependent Electron Transfer Proteins

The nitrogenase iron-protein, or dinitrogenase reductase, catalyzes the hy-drolysis of ATP, a process that is coupled to electron transfer to the iron–molybdenum protein. The structure of the protein reveals that the [4Fe–4S] cluster is situated at one end of the molecule, between the two identical subunits, each of which donates two cysteine ligands. ATP is bound at a site between the two subunits at the other end [125]. Spectroscopic studies indicate a change in the conformation of the cluster when the protein binds ATP or ADP [89]. This type of structure might be considered as an ATP-driven electron "pump."

Genes encoding homologous proteins to the nitrogenase iron-protein have been found in cyanobacteria, purple photosynthetic bacteria, and in the chloroplasts of bryophytes [126]. These genes have been found to encode protochlorophyllide and chlorin reductases, which lead to the formation of chlorophyll and protochlorophyll. Here again, the function of the iron–sulfur protein seems to be to promote a kinetically difficult electron transfer presum-ably accompanied by the hydrolysis of ATP.

Since both type I and type III nitrogenases are found in methanogens, it appears that these proteins originated in the Archaea [91]. At an earlier stage of evolution, it is possible that such proteins could have been more widespread and had other functions. Furthermore, in the presence of a strongly reducing electron donor such as the pyrite precipitation or a photochemical reaction, the reverse of this process could form the basis of a mechanism for redox-driven ATP synthesis, catalyzed by soluble enzymes.

3.6 Concluding Remarks

The value of these observations and conjectures is that they lead to hypotheses that can be tested experimentally. The interconversion of [4Fe–4S] and [3Fe–4S] clusters is already being investigated by sequence-directed mutagenesis. The evolution of complex electron-transfer proteins by combination of the genes for catalytic centers may be investigated by genetic engineering techniques. The substrate cycles, taking place on a crystallizing pyrite surface, proposed by Wächtershäuser, may be simulated chemically, although detection of the products as a surface layer is a challenge to analytical chemistry.

Acknowledgments

I thank G. S. Athwal, M. Seta, and Dr. P. Cunningham for assistance with the database searches.

References

1. Cammack, R. *Adv. Inorg. Chem.* 1992; *38*, 281–322.

2. Cammack, R., Rao, K. K., Hall, D. O. *Physiol. Veg.* 1985; *23*, 649–658.

3. George, D. G., Hunt, L. T., Yeh, L. S., Barker, W. C. *J. Mol. Evol.* 1985; *22*, 20–31.

4. Bruschi, M., Guerlesquin, F. *FEMS Microbiol. Rev.* 1988; *54*, 155–176.

5. Meyer, J. *Trends Evol. Ecol.* 1988; *3*, 222–226.

6. Matsubara, H., Saeki, K. *Adv. Inorg. Chem.* 1992; *38*, 223–280.

7. Eck, R. V., Dayhoff, M. O. *Science* 1966; *152*, 363–366.

8. Hall, D. O., Cammack, R., Rao, K. K. *Nature* 1971; *233*, 136–138.

9. Holm, R. H. *Accts. Chem. Res.* 1977; *10*, 427.

10. Que, L., Anglin, J. R., Robrick, M. A., Davison, A., Holm, R. H. *J. Am. Chem. Soc.* 1974; *96*, 6042.

11. Takahashi, Y., Mitsui, A., Matsubara, H. *Pl. Physiol.* 1991; *95*, 97–103.

12. Allen, R. M., Homer, M. J., Chatterjee, R., Ludden, P. W., Roberts, G. P., Shah, V. K. *J. Biol. Chem.* 1993; *268*, 23670–23674.

13. Conover, R. C., Kowal, A. T., Fu, W., Park, J. B., Aono, S., Adams, M. W. W., Johnson, M. K. *J. Biol. Chem.* 1990; *265*, 8533–8541.

14. Minami, Y., Wakabayashi, S., Wada, K., Matsubara, H., Kerscher, L., Oesterhelt, D. *J. Biochem. (Tokyo)* 1985; *97*, 745–946.

15. Orme-Johnson, W. H. *Biochem. Soc. Trans.* 1972; *1*, 30.

16. Fukuyama, K., Nakahara, Y., Tsukihara, T., Katsube, Y., Hase, T., Matsubara, H. *J. Mol. Biol.* 1988; *199*, 183–194.

17. Smith, E. T., Feinberg, B. A., Richards, J. H., Tomich, J. M. *J. Am. Chem. Soc.* 1991; *113*, 688–689.

18. Davasse, V., Moulis, J. M. *Biochem. Biophys. Res. Commun.* 1992; *185*, 341–349.

19. Ueyama, N., Ueno, S., Nakamura, A., Wada, K., Matsubara, H., Kumagai, S. I., Sakakibara, S., Tsukihara, T. *Biopolymers* 1992; *32*, 1535–1544.

20. Wächtershäuser, G. *Prog. Biophys. Mol. Biol.* 1992; *58*, 85–201.

21. Wächtershäuser, G. *Syst. Appl. Microbiol.* 1990; *10*, 000–000.

21. a. Wachtershauser, G. *Proc. Natl. Acad. Sci. USA* 1994; *91*, 4283–4287.

22. Drobner, E., Huber, H., Wächtershäuser, G., Rose, D., Stetter, K. O. *Nature* 1990; *346*, 742–744.

23. Mann, S., Sparks, N. H. C., Frankel, R. B., Bazylinski, D. A., Jannasch, H. W. *Nature* 1990; *343*, 258–261.

24. Suk, D., Peacor, D. R., Vandervoo, R. *Nature* 1990; *345*, 611–613.

24. a. Russell, M. J., Daniel, R. M., Hall, A. J., Sherringham, J. A. J. *Mol. Evol.* 1994; *39*, 231–243.

25. Blöckl, E., Keller, M., Wächtershäuser, G., Stetter, K. O. *Proc. Natl. Acad. Sci. USA* 1992; *89*, 8117–8120.

26. Buchanan, B. B., Arnon, D. I. *Photosynth. Res.* 1990; *24*, 47–53.

27. Hemming, A., Blotevogel, K. H. *Trends Biochem. Sci.* 1985; *10*, 198–200.

28. Wächtershäuser, G. *Microbiol. Rev.* 1988; *52*, 452–484.

29. Gest, H. *FEMS Microbiol. Lett.* 1981; *12*, 209–215.

30. Wu, L. F., Feng, D. F., Mandrand-Berthelot, M. A. *FEMS Microbiol. Rev.* 1993; *104*, 243–270.

31. Adams, M. W. W. *FEMS Microbiol. Rev.* 1990; *75*, 219–237.

31. a. Volbeda, A., Charon, M. H., Piras, C., Hatchikian, E. C., Frey, M., Fontecilla-Camps, J. C. *Nature* 1995; *373*, 580–587.

31. b. Pusheva, M. A., Slobodkin, A. J., Bonch-Osmobovskaya, E. A. *Mikrobiologiya (Russian)* 1991; *60*, 5–11.

31. c. Ma, K., Schicho, R. N., Kelly, R. M., Adams, M. W. W. *Proc. Natl. Acad. Sci. USA* 1993; *90*, 5341–5344.

32. Eady, R. R. *Adv. Inorg. Chem.* 1991; *36*, 77–102.

33. Adams, M. W. W. *Biochim. Biophys. Acta* 1990; *1020*, 115–145.

34. Joerger, R. D., Bishop, P. E. *CRC Crit. Rev. Microbiol.* 1988; *16*, 1–14.

35. Cammack, R. In *Charge and Field Effects in Biosystems*, Allen, M. J., Usherwood, P. N. R. (eds.), Abacus Press, Tunbridge Wells, 1984; pp. 41–51.

36. Chamorovsky, S. K., Cammack, R. *Photobiochem. Photobiophys.* 1982; *4*, 195–200.

37. Augier, V., Asso, M., Guigliarelli, B., More, C., Bertrand, P., Santini, C. L., Blasco, F., Chippaus, M., Giordano, G. *Biochemistry* 1993; *32*, 5099–5108.

38. Moore, G. D., Pettigrew, G. W., Rogers, N. K. *Proc. Natl. Acad. Sci. USA* 1986; *83*, 4998–5000.

39. Blasco, F., Iobbi, C., Giordano, G., Chippaux, M., Bonnefoy, V. *Mol. Gen. Genet.* 1989; *218*, 249–256.

40. Cammack, R., Weiner, J. H. *Biochemistry* 1990; *29*, 8410–8416.

41. Carter, C. W., Jr. *J. Biol. Chem.* 1977; *252*, 7802–7811.

41. a. Luchinat, C., Capozzi, F., Borsari, M., Battistuzzi, G., Sola, M. *Biochem. Biophys. Res. Commun.* 1994; *203*, 436–442.

42. Thomson, A. J., Robinson, A. E., Johnson, M. K., Cammack, R., Rao, K. K., Hall, D. O. *Biochim. Biophys. Acta* 1981; *637*, 423–432.

43. Nagayama, K., Ohmori, D. *FEBS Lett.* 1984; *173*, 15–18.

44. Trumpower, B. L. *Microbiol. Rev.* 1990; *54*, 101–129.

45. Ciurli, S., Carrie, M., Weigel, J. A., Carney, M. J., Stack, T. D. P., Papaefthymiou, G. C., Holm, R. H. *J. Am. Chem. Soc.* 1990; *112*, 2654–2664.

45. a. Zheng, L. M., Dean, D. R. *J. Biol. Chem.* 1994; *269*, 18723–18726.

46. Voordouw, G., Hagen, W. R., Krüse-Wolters, K. M., Van Berkel-Arts, A., Veeger, C. *Eur. J. Biochem.* 1987; *162*, 31–36.

47. Shergill, J. K., Butler, C. S., White, A. C., Cammack, R., and Mason, J. R. *Biochem. Soc. Trans.* 1994; *22*, 288S.

48. Michelsen, J. W., Schmeichel, K. L., Beckerle, M. C., Winge, D. R. *Proc. Natl. Acad. Sci. USA* 1993; *90*, 4404–4408.

49. Archer, V. E. V., Breton, J., Sanchez-Garcia, I., Osada, H., Forster, A., Thomson, A. J., Rabbits, T. H. *Proc. Natl. Acad. Sci. USA* 1994; *91*, 316–320.

50. Kuo, C. F., Mcree, D. E., Fisher, C. L., O'Handley, S. F., Cunningham, R. P., Tainer, J. A. *Science* 1992; *258*, 434–440.

51. Correll, C. C., Ludwig, M. L., Bruns, C. M., Karplus, P. A. *Protein Science* 1993; *2*, 2112–2133.

52. Harayama, S., Polissi, A., Rekik, M. *FEBS Lett.* 1991; *285*, 85–88.

53. Orme-Johnson, W. H., Beinert, H. *Biochem. Biophys. Res. Commun.* 1979; *36*, 905.

54. Allen, J. F. *Biochem. Biophys. Res. Commun.* 1975; *66*, 36–43.

55. Meyer, T. E., Cusanovich, M. A. *Biochem. Biophys. Acta* 1989; *975*, 1–28.

56. Flint, D. H., Smyk-Randall, E., Tuminello, J. F., Draczynskalusiak, B., Brown, O. R. *J. Biol. Chem.* 1993; *268*, 25547–25552.

57. Flint, D. H., Emptage, M. H. *J. Biol. Chem.* 1988; *263*, 3558–3591.

58. Hill, S., Viollet, S., Smith, A. T., Anthony, C. *J. Bacteriol.* 1990; *172*, 2071–2078.

59. Scherings, G., Haaker, H., Veeger, C. *Eur. J. Biochem.* 1977; *77*, 621–630.

60. Haile, D. J., Rouault, T. A., Harford, J. B., Kennedy, M. C., Blondin, G. A., Beinert, H., Klausner, R. D. *Proc. Natl. Acad. Sci. USA* 1992; *89*, 11735–11739.

61. Fitch, W. M., Bruschi, M. *Mol. Biol. Evol.* 1987; *7*, 381–394.

62. Iisma, S. E., Watson, J. M. *Nucleic Acids Res.* 1987; *15*, 3180–3181.

63. Gaillard, J., Quinkal, I., Moulis, J. M. *Biochemistry* 1993; *32*, 9881–9887.

64. Martin, A. E., Burgess, B. K., Stout, C. D., Cash, V. L., Dean, D. R., Jensen, G. M., Stephens, P. J. *Proc. Natl. Acad. Sci. USA* 1990; *87*, 598–602.

65. Evans, D. J., Leigh, G. J. *J. Inorg. Biochem.* 1991; *42*, 25–35.

66. Bovier-Lapierre, G., Bruschi, M., Bonicel, J., Hatchikian, E. C. *Biochim. Biophys. Acta* 1987; *913*, 20–26.

67. Armstrong, F. A., George, S. J., Cammack, R., Hatchikian, E. C., Thomson, A. J. *Biochem. J.* 1989; *264*, 265–273.

68. Yao, Y., Wakabayashi, S., Matsuda, S., Matsubara, H., Yu, L., Yu, C.-A. In *Iron-Sulfur Protein Research*, Matsubara, H. (ed.), Japan Scientific Societies Press/Springer-Verlag, Tokyo/Berlin, 1986; pp. 240–244.

69. Werth, M. T., Cecchini, G., Manodori, A., Ackrell, B. A. C., Schroder, I., Gunsalus, R. P., Johnson, M. K. *Proc. Natl. Acad. Sci. USA* 1990; *87*, 8965–8969.

70. Webber, A. N., Malkin, R. *FEBS Lett.* 1990; *264*, 1–4.

70. a. Schoepp, B., Parot, P., Menin, L., Gaillard, J., Richaud, P., Vermeglio, A. *Biochemistry*; 1995; *34*, 11736–11742.

71. Onate, Y. A., Vollmer, S. J., Switzer, R. L., Johnson, M. K. *J. Biol. Chem.* 1989; *264*, 18386–18391.

72. Grabau, C., Schatt, E., Jouanneau, Y., Vignais, P. M. *J. Biol. Chem.* 1991; *266*, 3294–3299.

73. Fukuyama, K., Hase, T., Matsumoto, S., Tsukihara, T., Katsube, Y., Tanaka, N., Kakudo, M., Wada, K., Matsubara, H. *Nature* 1980; *286*, 522–524.

74. Hase, T., Wakabayashi, H., Matsubara, H., Kersher, D., Oesterhelt, D., Rao, K. K., Hall, D. O. *J. Biochem.* 1978; *83*, 1659.

75. Meyer, J. *Biochim. Biophys. Acta* 1993; *1174*, 108–110.

76. Evans, M. C. W., Smith, R. V., Telfer, A., Cammack, R. In *Proceedings of the First European Biophysics Congress,* Baden, Australia, Broda, E., Locker, A., Springer-Lederer, H. (eds.), Vienna Medical Academy, Vienna, 1971: pp. 115–119.

77. Mason, J. R., Cammack, R. *Annu. Rev. Microbiol.* 1992; *46*, 277–305.

78. Neidle, E. L., Hartnett, C., Ornston, L. N., Bairoch, A., Rekik, M., Harayama, S. *J. Bacteriol.* 1991; *173*, 5385–5395.

79. Gurbiel, R. J., Batie, C. J., Sivaraja, M., True, A. E., Fee, J. A., Hoffman, B. M., Ballou, D. P. *Biochemistry.* 1989; *28*, 4861–4871.

80. Mason, J. R., Cammack, R. *Annu. Rev. Microbiol.* 1992; *46*, 277–305.

81. Prince, R. C. *Biochim. Biophys. Acta* 1983; *723*, 133–138.

81. Lode, E. T., Coon, M. J. *J. Biol. Chem.* 1971; *246*, 791–802.

83. Rey, L., Hidalgo, E., Palacios, J., Ruiz-Argueso, T. *J. Mol. Biol.* 1992; *228*, 998–1002.

84. Bruschi, M., Moura, I., Le Gall, J., Xavier, A. V., Sieker, L. C. *Biochem. Biophys. Res. Commun.* 1979; *90*, 596–605.

85. Archer, M., Huber, R., Tavares, P., Moura, I., Moura, J. J. G., Carrondo, M. A., Sieker, L. C., Legall, J., Romao, M. J. *J. Mol. Biol.* 1995; *251*, 690–702.

86. Dahl, C., Kredich, N. M., Deutzmann, R., Truper, H. G. *J. Gen. Microbiol.* 1993; *139*, 1817–1828.

87. Crane, B. R., Siegel, L. M., Getzoff, E. D. *Science* 1995; *270*, 59–67.

88. McRee, D. E., Richardson, D. C., Richardson, J. S., Siegel, L. M. *J. Biol. Chem.* 1986; *261*, 10277–10281.

89. Rees, D. C., Chan, M. K., Kim, J. *Adv. Inorg. Chem.* 1993, *40*, 89–119.

90. Pau, R. N. *Trends Biochem. Sci.* 1989; *14*, 183–186.

91. Normand, P., Bousquet, J. *J. Mol. Evol.* 1989; *29*, 436–447.

92. Beinert, H., Kennedy, M. C. *FASEB J.* 1993; *7*, 1442–1449.

93. Lauble, H., Kennedy, M. C., Beinert, H., Stout, C. D. *Biochemistry* 1992; *31*, 2735–2748.

94. Emptage, M. H. In *Metal Clusters in Proteins*, Que, L. (ed.), American Chemical Society, Washington, D. C., 1988; pp. 343–371.

95. Flint, D. H., Emptage, M. H., Guest, J. R. *Biochemistry* 1992; *31*, 10331–10337.

96. Ueda, Y., Yumoto, N., Tokushige, M., Fukui, K., Ohyanishiguchi, H. *J. Biochem.* 1991; *109*, 728–733.

97. Liochev, S. I., Fridovich, I. *Proc. Natl. Acad. Sci. USA* 1992; *89*, 5892–5896.

98. Woese, C. R., Wolfe, R. S. *Nature* 1986; *19*, 273–274.

99. Thauer, R. K. *Biochim. Biophys. Acta* 1990; *1018*, 256–259.

100. Becher, B., Müller, V., Gottschalk, G. *FEMS Microbiol. Lett.* 1992; *91*, 239–244.

101. Oshima, T., Ohba, M., Wakagi, T. *Origins Life* 1984; *14*, 665–680.

101. a. Hardy, I., Müller, M. *J. Mol. Evol.* 1995; *41*, 388–396.

102. Finel, M. *J. Bioenerg. Biomembr.* 1993; *25*, 357–366.

103. Walker, J. E. *Q. Rev. Biophys.* 1992; *25*, 253–324.

104. Sled, V. D., Friedrich, T., Leif, H., Weiss, H., Meinhardt, S. W., Fukumori, Y., Calhoun, M. W., Gennis, R. B., Ohnishi, T. *J. Bioenerget. Biomembr.* 1993; *25*, 347–356.

105. Weiss, H., Friedrich, T. *J. Bioenerget. Biomembr.* 1991; *23*, 743–754.

106. Tran-Betcke, A., Warnecke, U., Bocker, C., Zaborosch, C., Friedrich, B. *J. Bacteriol.* 1990; *172*, 2920–2929.

107. Pilkington, S. J., Skehel, J. M., Gennis, R. B., Walker, J. E., *Biochemistry* 1991; *30*, 2166–2175.

108. Sauter, M., Böhm, R., Böck, A. *Mol. Microbiol.* 1992; *6*, 1523–1532.

108. a. Albracht, S. P. *J. Biochim. Biophys. Acta* 1993; *1144*, 221–224.

109. Friedrich, T., Strohdeicher, M., Hofhaus, G., Preis, D., Sahm, H, Weiss, H. *FEBS Lett.* 1990; *265*, 37–40.

109. a. Friedrich, T., Vanheek, P., Leif, H., Ohnishi, T., Forche, E., Kunze, B., Jansen, R., Trowitzsch-Kienast, W., Hofle, G., Reichenbach, H., Weiss, H. *Eur. J. Biochem.* 1994; *219*, 691–698.

110. Friedrich, T., Weidner, U., Nehls, U., Fecke, W., Schneider, R., Weiss, H. *J. Bioenerg. Biomembr.* 1993; *25*, 331–337.

110. a. Albracht, S. P. *J. Biochim. Biophys. Acta* 1993; *1144*, 221–224.

111. Berger, S., Ellersiek, U., Steinmuller, K. *FEBS Lett.* 1991; *286*, 129–132.

112. Arizmendi, J. M., Runswick, M. J., Skehel, J. M., Walker, J. E. *FEBS Lett.* 1992; *301*, 237–242.

113. Friedrich, B., Schwartz, E. *Annu. Rev. Microbiol.* 1993; *47*, 351–383.

114. Dupuis, A. *FEBS Lett.* 1992; *301*, 215–218.

115. Green, J., Guest, J. R. *FEBS Lett.* 1993; *329*, 55 58.

115. a. Khoroshilova, N., Beinert, H., Kiley, P. J. *Proc. Natl. Acad. Sci. USA* 1995; *92*, 2499:–2503.

116. Ackrell, B. A. C., Johnson, M. K., Gunsalus, R. P., Cecchini, G. In *Chemistry and Biochemistry of Flavoenzymes*, Müller, F. (ed.), CRC Press, Boca Raton, FL, 1992; pp. 229–297.

117. Cammack, R. *Chem. Script.* 1983; *21*, 87–95.

118. Darlison, M. G., Guest, J. R. *Biochem. J.* 1984; *223*, 507–517.

119. Wood, D., Darlison, M. G., Wilde, R. J., Guest, J. R. *Biochem. J.* 1984; *222*, 519–534.

120. Sucheta, A., Ackrell, B. A. C., Cochran, B., Armstrong, F. A. *Nature* 1992; *356*, 361–362.

121. Bohm, R., Sauter, M., Bock, A. *Mol. Microbiol.* 1990; *4*, 231–243.

122. Darwin, A., Tormay, P., Page, L., Griffiths, L., Cole, J. *J. Gen. Microbiol.* 1993; *139*, 1829–1840.

123. Ensign, S. A., Ludden, P. W. *J. Biol. Chem.* 1991; *266*, 18395–18403.

123. a. Setzke, E., Hedderich, R., Heiden, S., Thauer, R. K. *Eur. J. Biochem.* 1994; *220*, 139–148.

124. Peinemann, S., Hedderich, R., Blaut, M., Thauer, R. K., Gottschalk, G. *FEBS Lett.* 1990; *263*, 57–60.

125. Dean, D. R., Bolin, J. T., Zheng, L. M. *J. Bacteriol.* 1993; *175*, 6737–6744.

126. Burke, D. H., Hearst, J. E., Sidow, A. *Proc. Natl. Acad. Sci. USA* 1993; *90*, 7134–7138.

127. Kerscher, L., Oesterhelt, D. *Eur. J. Biochem.* 1981; *116*, 595–600.

128. Kerscher, L., Nowitski, S., Oesterhelt, D. *Eur. J. Biochem.* 1982; *128*, 223–230.

129. Blamey, J. M., Adams, M. W. W. *Biochemistry* 1994; *33*, 1000–1007.

129. a. Kunow, J., Linder, D., Thauer, R. K. *Arch. Microbiol.* 1995; *163*, 21–28.

130. Cammack, R., Rao, K. K., Hall, D. O. *BioSystems* 1981; *14*, 57–80.

131. Bruschi, M., Hatchikian, E. C., LeGall, J., Moura, J. J. G., Xavier, A. V. *Biochim. Biophys. Acta* 1976; *449*, 275–284.

131. a. Pieulle, L., Guigliarelli, B., Asso, M., Dole, F., Bernadac, A., Hatchikian, E. C. *Biochim. Biophys. Acta* 1995; *1250*, 49–59.

132. Uyeda, K., Rabinowitz, J. C. *J. Biol. Chem.* 1971; *246*, 3111–3119.

133. Wahl, R. C., Orme-Johnson, W. H. *J. Biol. Chem.* 1987; *262*, 10489–10497.

134. Schmitz, O., Kentemich, T., Zimmer, W., Hundeshagen, B., Bothe, H., *Arch. Microbiol.* 1993; *160*, 62–67.

135. Williams, K. P., Leadlay, P. F., Lowe, P. N. *Biochem. J.* 1990; *268*, 69–75.

136. Pierik, A. J., Wolbert, R. B. G., Mutsaers, P. H. A., Hagen, W. R., Veeger, C. *Eur. J. Biochem.* 1992; *206*, 697–704.

The Evolution of Electron Transfer Proteins in Photosynthetic Bacteria and Denitrifying Pseudomonads

T. E. Meyer, J. J. Van Beeumen, R. P. Ambler, and M. A. Cusanovich

4.1 Background

The evolution of bacteria is a nebulous concept for a variety of reasons. Although bacteria are presumably older than any other life form, they have a very simple anatomy, there is virtually no fossil record, they have very rapid generation times, they are promiscuous, and the numbers and types of habitats they occupy have changed since they first evolved. The principal differences among bacteria are metabolic and all such pathways have had to adapt to the changing environment through either simple mutagenesis or gene transfer. The likelihood that any truly primitive characteristics, much less primitive genomes, survived to the present or could be recognized is vanishingly small. Nevertheless, simple hierarchical evolutionary schemes based on 16S rRNA sequences have become fashionable [1–5]. It has been argued that the ribosome is ubiquitous, large, and essential. Thus, it should be representative of the genome as a whole and unlikely to be transferred among species. Proponents have even gone so far as to conclude that 16S rRNA sequences reveal the "true" phylogeny of bacteria [5]. However, we believe that bacterial evolution is far more complicated than such optimistic reports would have us believe and that a more cautious approach to evolutionary studies is warranted.

Studies of 16S rRNA sequences suggest that the photosynthetic bacteria do not represent a valid taxon, but are intermixed with various nonphotosynthetic bacteria such as the coliforms and many other metabolically diverse forms [1]. Furthermore, purple bacteria, cyanobacteria, green sulfur bacteria, *Heliobacterium*, and *Chloroflexus* are widely separated by 16S rRNA analyses that give no insight on how they might be specifically related to one another if at all [1]. Analyses of 16S rRNA suggest two unlikely alternatives, that photosynthesis independently arose several times or was present in the ancestor of the majority of bacteria and was subsequently lost in the nonphotosynthetic families. The purple bacteria alone are divided by 16S rRNA sequences into three major nontraditional groups known as α, β, and γ [1]. These three plus a δ group encompass the gram-negative bacteria [1]. In fact, many authors now incorrectly refer to purple bacteria as a synonym for gram-negative bacteria. Respiration is said to have originated with photosynthetic bacteria [5] although there is no factual basis for such an assertion, which would require that oxygen utilization independently arose in such bacteria as *Halobacterium*. Thus, there are indications that bacteria may not have evolved in a simple divergent manner as previously proposed [1].

We take issue with most, if not all, of the 16S rRNA-based conclusions. First, there is no proof that any particular evolutionary scheme is correct. Furthermore, the "universal phylogenetic tree of life" is not representative of bacterial genomes as can be seen in the mixtures of metabolisms (the principal attribute of bacteria) on various branches of the 16S rRNA tree. One of the most bizarre aspects of the 16S rRNA tree is the mixture of the aerobic *Halobacterium* with the strictly anaerobic methanogens. Because gene transfer has been shown to be involved in bacterial evolution [6, 7], no one gene can be said to be representative of the genome. Therefore, if *Halobacterium* has some attributes of the methanogens on the one hand or of aerobic bacteria on the other, then it is logical to assume that they may have been acquired through gene transfer. However, the list of typical bacterial characteristics of *Halobacterium* is growing. The [2Fe–2S] ferredoxin gene [8] was supposedly transferred from cyanobacteria to *Halobacterium* [1] but what of the gas vesicle proteins [9], DNA photolyase [10], superoxide dismutase [10], denitrification [11], and aerobic respiratory pathways [11]? Does the presence of these characteristics mean that *Halobacterium* is not an "archaeon" or that the supposed distinctions between archaebacteria and other bacteria are no longer as clear as they previously seemed? Were virtually all major metabolic capabilities really present in the common ancestor of archaebacteria and typical bacteria? This is the rationalization favored by the proponents of 16S rRNA based hierarchies. Successful gene transfer does seem to be rare, but given approximately 4 billion years and sufficient selection pressure, there has been ample time for all genomes to become thoroughly mixed. In fact, the bacterial kingdom (in the broadest sense) probably represents one large gene pool [12], access to which is limited in part by habitat, codon usage, and restriction enzymes. Even if we assume for purposes of argument that the ribosome represents a nontransferable core and that most other genes are mobile, then what is the use of an rRNA-based classification of bacteria if it is not representative of the whole genome?

The basic issue is: What are bacteria metabolically capable of doing and how did they acquire that ability? This requires studies of overall metabolism and detailed comparison of the genes that are involved rather than focussing on any single gene. The "winds of evolutionary change" mentioned in regard to the impact of rRNA on microbiology [4] will not come from rRNA as such, but from a variety of properties well characterized at the molecular level including as one part the protein biosynthetic apparatus. The shortcomings of traditional microbiology stem not from a lack of appreciation of 16S rRNA, but from the reliance on vague tests such as gram stain amongst others. These may be overcome, for example, in the case of the gram stain, by a detailed molecular analysis of the structure of the cell wall including the presence or absence of an outer membrane and associated periplasmic space. The gram-negative bacteria were also criticized as an invalid division from the standpoint of 16S rRNA sequences [1]. However, the evolution of the periplasmic space has had tremendous impact on the subsequent development of energy metabolism including soluble cytochromes, photosynthesis, denitrification, and aerobic respiration which are the subjects of the present chapter.

We are cautiously optimistic that continued analyses of sequences and three-dimensional structures for a number of genes will yield useful evolutionary information, but we are convinced that there is no simple solution to bacterial evolution. Based on our own work with electron transfer proteins [13–17], we believe that the photosynthetic bacteria are an important and valid division and that there is strong evidence for a single evolutionary origin of photosynthetic pathways. Photosynthesis has been around for a long time, but may not be as ancient as has been assumed [1]. That is, the majority of bacteria probably did not have photosynthetic ancestors. There is evidence for loss of photosynthesis in some lineages, but not in the majority. Although it is possible that aerobic respiration may have originated with photosynthetic bacteria, it is as likely to have evolved in any species already capable of anaerobic respiration and the requisite genes subsequently may have been transferred to other species. There is a continuum of change from purple sulfur to nonsulfur purple bacteria with little or no support for either the traditional bipartite separation into sulfur and nonsulfur groups or the tripartite division into α, β, and γ based on 16S rRNA. Furthermore, there is a good case for a relationship between electron transfer proteins of phototrophs and pseudomonads, but virtually no support for a close relationship to coliforms as previously claimed based on 16S rRNA [1].

The various kinds of cytochromes c that provide a basis for the above assertions were extensively reviewed by Meyer and Kamen [13], Pettigrew and Moore [18], and Moore and Pettigrew [19] and their distribution in photosynthetic bacteria reported by Bartsch [14, 15]. Ambler [16] and Van Beeumen [17] described the sequences of new types of cytochromes found in these bacteria since the earlier reviews. The following is an extension of the sequence comparisons of electron transfer proteins, which strengthens the relationships among the families of photosynthetic bacteria and with the pseudomonads and presents a new type of evolutionary tree based on well-characterized insertions and deletions in amino acid sequences.

4.2 Types of Soluble Electron Transfer Proteins in Bacteria

Throughout this chapter, purple photosynthetic bacteria are restricted to those that contain bacteriochlorophyll-based photosynthetic pathways. "Purple bacteria" recently have been redefined by 16S rRNA analyses to embrace most of the gram-negative bacteria [1]. We adhere to the stricter definition of photosynthetic purple bacteria but recognize that there are related bacteria that contain many of the same proteins as purple bacteria, hence may have a common genetic origin, but that are unable to grow photosynthetically such as *Paracoccus denitrificans* and *Roseobacter denitrificans*. There are five distinct kinds of soluble electron transfer proteins commonly occurring in purple phototrophic bacteria. Bacterial ferredoxin and cytochrome c' are found in nearly all purple bacteria. Cytochrome c_2 occurs in the nonsulfur purple bacteria, whereas high-potential iron–sulfur protein (HiPIP) and flavocytochrome c are more characteristic of the purple sulfur bacteria [14, 15]. Prototypic species at opposite ends of the purple bacterial spectrum are *Chromatium vinosum* and *Rhodobacter sphaeroides*. Green bacteria such as *Chlorobium* contain bacterial ferredoxin, rubredoxin, cytochrome c-555, and flavocytochrome c as the dominant soluble proteins [14,15]. Cyanobacteria such as *Anabaena* produce cytochrome c-553 (now called cytochrome c_6), a 2-Fe-S ferredoxin, and the copper protein plastocyanin [13].

The pseudomonads are a group of related aerobic and facultatively anaerobic bacteria that include the genera *Pseudomonas*, *Alcaligenes*, *Achromobacter*, and *Azotobacter*. With some exceptions, the denitrifying species appear to have a greater variety and abundance of soluble electron transfer proteins than their strictly aerobic counterparts. The following discussion is intended to include the pseudomonads in general. *Pseudomonas stutzeri* and *Pseudomonas aeruginosa* are prototypic species in that a greater number of electron transfer proteins have been characterized in these than in any other. The pseudomonads commonly contain a variety of electron transfer proteins including: cytochrome c-551 (now called cytochrome c_8), cytochrome c_4, cytochrome c_5, cytochrome cd, bacterial cytochrome c peroxidase (BCCP), bacterial ferredoxin, and the copper protein azurin [13, 16, 17]. Less well known *Pseudomonas* proteins are a diheme cytochrome c-552, and a tetraheme cytochrome c. The above *Pseudomonas* proteins have all been found in one or more species of purple bacteria. Some of these may represent isolated instances of recent gene transfer, but it is unlikely that all would result from this process. If recent gene transfer were a valid explanation for the general occurrence of pseudomonad proteins in phototrophs, then purple bacterial proteins should be equally common in pseudomonads. However, cytochromes c_2 and HiPIP have not yet been documented in pseudomonads. There is one proven occurrence of cytochrome c' in an *Alcaligenes* sp. [20]. This suggests a line of descent from *Pseudomonas* to purple bacteria that is somewhat blurred owing to occasional gene transfers and losses. Moreover, the nearest homologs of *Pseudomonas* cytochrome c_5 in photosynthetic bacteria are *Ectothiorhodospira* cytochrome c-551, *Chlorobium* cytochrome c-555, and cyanobacterial cytochrome c_6. The relationships among these four proteins, which for any one of them is stronger than to any other cytochrome, suggest the order in which the photosynthetic bacterial families may have evolved. What we know of the evolution of each of the photosynthetic bacterial and *Pseudomonas* electron transfer proteins is outlined in the following section and a hypothetical pathway for the evolution of photosynthesis presented.

4.2.1 Cytochrome c_2

Cytochrome c_2 is the nearest bacterial homolog of mitochondrial cytochrome c, the definition of which is based on sequence and structure, not on function, as cytochrome c_2 has several functional roles in the various species in which it is found [13]. It is one of the most thoroughly characterized of the bacterial cytochromes with many sequences [24] and several three-dimensional structures known [21]. Cytochrome c_2 is most commonly found in nonsulfur purple bacteria (family Rhodospirillaceae) with the exception of *Rhodocyclus* species [15] and is present in a number of nonphototrophs, some of which may be close relatives of purple bacteria and some of which may have acquired it through gene transfer. In this regard, there is no proof that cytochrome c_2 first evolved in purple bacteria and that they did not originally obtain it through gene transfer. That is, there is no small cytochrome in purple bacteria that is closer to cytochrome c_2 than any other and that could have been the direct precursor. A sequence alignment of representative species of cytochrome c_2 based on comparison of three-dimensional structures [21] is shown in Figure 4.1. Several important residues are boxed (the heme binding site C16, C19, and H20; the sixth heme ligand M109; aromatics in the N- and C-terminal helices, 12 and 126; and residues that H-bond the heme or its ligands 38, 47, 58, 74, and 82).

The cytochromes c_2 have diverged to such an extent that back and parallel (in sum, convergent) mutations balance divergent mutations; thus matrices of sequence identities have limited utility in quantifying relationships [22]. The approximately $40 \pm 15\%$ average identity among the cytochromes c_2 apparently represents the percentage of residues necessary to maintain both structure and function [22]. The Gaussian distribution of most data within the matrix of sequence identities defines the extent of convergence. Only identities greater than approximately 55% (which is outside the normal distribution) may be used to construct valid evolutionary trees for this family of proteins. There are relatively few data that meet this criterion; thus matrix trees are practically uninformative for the cytochromes c_2.

Insertions and deletions are less subject to convergent mutations than are amino acid substitutions (although not free of them); thus they can be used to quantify relationships for highly divergent sequences provided that they occur in the proteins in question and are precisely located. Dickerson [23] was the first to construct a cytochrome tree using a limited number of insertions and

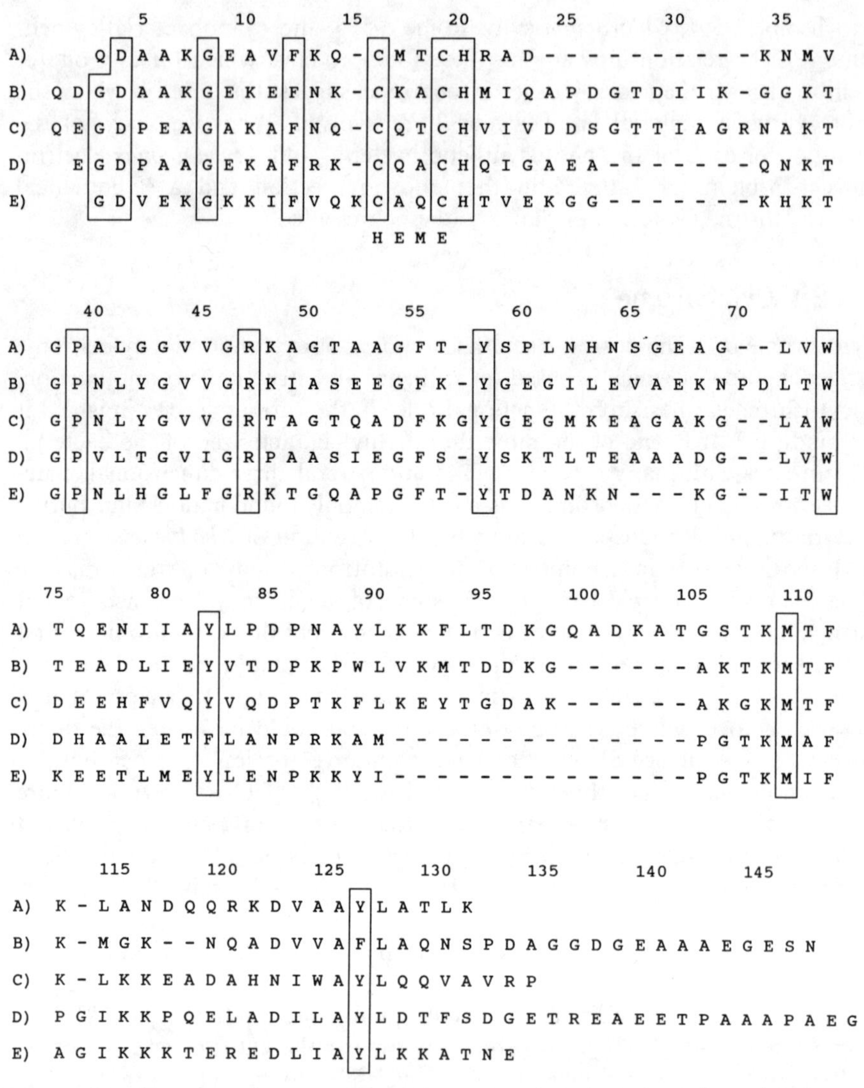

Figure 4.1 Sequence alignment of cytochromes c_2 based on three-dimensional structures. (A) *Rhodopseudomonas palustris* [24]; (B) *Paracoccus denitrificans* [80,137]; (C) *Rhodobacter sphaeroides* iso-1 [24]; (D) *Rhodobacter sphaeroides* iso-2 [25]; (E) horse.

deletions located by crystallography. Because there are several ways of using insertions and deletions to build trees, there should be some discussion of methods. One could build a matrix of the minimum numbers of internal gaps separating species and construct a tree in the normal fashion. However, this was not done because it does not take into account sequential or concerted events nor does it consider possible convergence or metabolic similarities and differences among species. All of these influences were considered in constructing the insertion–deletion trees herein.

Because there are several known three-dimensional structures for cytochrome c_2 and mitochondrial cytochrome c, insertions and deletions have been located with quite a high degree of confidence. Some of the insertions and deletions in cytochrome c_2 are shared among several species (residues 65–67, 91–98, 113, and 117–118 of Fig. 4.1), whereas others are unique (residues 57 and 70–71). Some apparently have occurred more than once along different lines of divergence (e.g., residue 15 in Fig. 4.1), which shows that even this type of evolutionary analysis is subject to some convergence and requires interpretation. Nevertheless, it is superior to the use of the matrix of amino acid substitutions for large differences. Once the shared gaps were identified as sequential (residues 65–67), concerted (residues 91–98 and 113), or convergent (residues 15 and possibly 65–67 in part), then tree construction could proceed.

An "insertion–deletion tree" for cytochrome c and c_2 is shown in Figure 4.2. Only internal gaps were used in its construction because terminal gaps may be too frequent and not as constrained by the structure–function relationship, and thus may be more subject to convergent mutation. The most conservative interpretation (minimal gaps) was assumed for comparison of sequences for which there are no three-dimensional structures. The base of the tree is assumed to be "A," which includes the species *Rhodopseudomonas viridis*, *Rhodopseudomonas acidophila*, and *Rhodomicrobium vannielii* [24]. Concerted 1, 3, and 8 residue gaps (residues 113, 65–67, and 91–98 of Fig. 4.1) separate the large cytochromes c_2 in the upper part of Figure 4.2 (M to V) from the small cytochromes c_2 and cytochromes c at the base of the figure (A to K). This is a significant and important division that is not apparent from examination of the matrix of sequence identities, presumably because it is lost in the noise resulting from convergence. It is highly unlikely that all three of these gaps could have happened more than once, although shared occurrence of any one of them could represent convergence (such as insertion 65–67 in part). Insertion 65–67 is shown in Figure 4.2 as the first step (L) in the sequential accumulation of the three gaps leading to the large cytochromes c_2. This is an interpretation that may or may not hold up to further scrutiny. The division into large and small cytochromes c_2 disagrees with the 16S rRNA α and β categories which are based on matrix data. The matrix of gap differences (not shown) is also inadequate to show this dichotomy. However, the division between large and small cytochrome c_2 may be correlated with other properties. For example, all of the species lacking the tetraheme reaction center cytochrome subunit have a large cytochrome c_2. All of the species in the upper half of the tree are facultatively aerobic, whereas all of the anaerobes plus some aerobes are in the lower half.

The *Rhodobacter* species are among the most divergent of the large cytochromes c_2 with four gaps separating them from the supposed root (Q and beyond). However, both *Rhodobacter sphaeroides* [25] and *Rhodobacter capsulatus* [26] have cytochrome c_2 isozymes that are at the base of the tree and

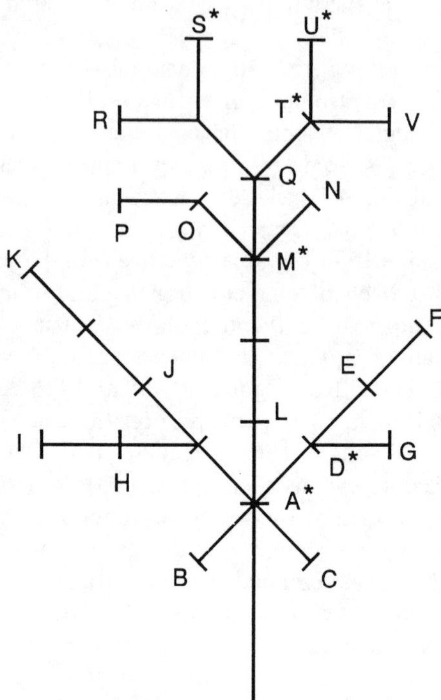

Figure 4.2 Evolutionary tree for cytochromes c_2 and c based on internal insertions and deletions located by comparison of three-dimensional structures as shown in the alignment of Figure 4.1. Each node or bar represents an insertion or deletion whether lettered or not. (A) *Rhodopseudomonas viridis, Rhodomicrobium vannielii, Rhodopseudomonas acidophila*; (B) *Bradyrhizobium japonicum* membrane-bound c_2; (C) *Methylobacterium extorquens*; (D) most eucaryotic mitochondrial cytochromes *c*; (E) *Rhodobacter capsulatus* membrane-bound *c*; (F) *Tetrahymena pyriformis*; (G) *Euglena gracilis* and *E. viridis*; (H) *Rhodospirillum molischianum* and *R. Fulvum* iso-1; (I) *R. molischianum* and *R. fulvum* iso-2; (J) *Rhodopila globiformis*; (K) *Rhodospirillum salexigens*; (L) *Agrobacterium tumefaciens, Rhodobacter sphaeroides* iso-2 and *Rhodopseudomonas marina*; (M) *Rhodospirillum rubrum* and *Aquaspirillum itersonii*; (N) *Rhodospirillum photometricum*; (O) *Rhodospirillum centenum*; (P) *Rhodopseuomonas palustris*; (Q) the genus *Rhodobacter*; (R) *Rhodobacter sulfidophilus*; (S) *Rhodobacter sphaeroides* iso-1; (T) *Rhodobacter capsulatus* and *Rb.* strain TJ12; (U) *Paracoccus denitrificans, Thiobacillus versutus*, and *Thiosphaera pantotropha*; (V) *Roseobacter denitrificans* (formerly *Erythrobacter* sp. strain OcH114). An asterisk represents species for which a three-dimensional structure has been determined.

much closer to the supposed root. Although one could argue which isozyme is representative of the species as a whole, the principal (large) isozyme normally functions in photosynthesis [13, 18] and should be representative of the evolution of that pathway. Thus, patterns of similarity of large cytochromes c_2 should correlate with those of the reaction center genes (for which there are currently too few sequences for comparison). Some of the cytochrome c_2 isozymes such as *Rb. sphaeroides* iso-2 presumably have other roles [25] and that may be why they have taken separate evolutionary paths. It will be interesting to compare their sequences and functional roles once additional examples are characterized.

The fine structure of the cytochrome c_2 family tree in the region of the genus *Rhodobacter* may be obtained from examination of the matrix of sequence identities, which is just barely significant at this level. That is, a matrix tree may be constructed for *Rhodobacter* by taking the insertion–deletion tree (branches Q and beyond) as a starting point (not shown). However, because of the very large divergence within this genus and others, the branch lengths can be only approximate. If we assume for the moment that cytochrome c_2 is representative of the species as a whole, then it is apparent that some rearrangement of genera and species would be required in the future, although more data from a greater number of genes and species are necessary before that would be possible. For example, *Rhodopseudomonas palustris* cytochrome c_2 is only distantly related to those of *Rhodopseudomonas viridis*, *Rhodomicrobium vannielii*, and *Rhodopseudomonas acidophila* [24] as shown in Figure 4.2, whereas these four species are usually clustered by 16S rRNA analyses [1, 2, 5]. *Rhodospirillum centenum* cytochrome c_2 (J. Van Beeumen, unpublished) is closer to that of *Rps. palustris* than to *Rhodospirillum rubrum* [24] as shown in Figure 4.2, but 16S rRNA places it adjacent to *R. rubrum* [2]. Furthermore, *R. centenum* and *R. rubrum* cytochromes c_2 are distant from those of *R. molischianum* and *R. salexigens* [24, 25a] as shown in Figure 4.2, but *R. molischianum* 16S rRNA has been reported to be adjacent to that of *R. rubrum* [2]. Thus, our results suggest that both traditional and nontraditional classifications may be in error.

In the past, we refrained from building evolutionary trees for the bacterial cytochromes c_2 because of insufficient evidence that such trees had any validity [22, 24]. This did not nor should it have prevented others from interpreting our data and arriving at their own conclusions [5, 27, 28]. Where they agreed, our data were used in support of 16S rRNA trees [28]. On the other hand, the cytochromes were rejected as poor chronometers because of possible changes in function where the trees disagreed [1]. Both conclusions are weak for the following reasons. Because the early cytochrome and 16S rRNA trees were based on matrix data, both are invalid except where species are much closer than average. Thus, actual agreement is extremely limited and confined to recognition of similarity between such closely related species as *Rb. sphaeroides* and *Rb. capsulatus* or *R. rubrum* and *R. photometricum* plus a few others. The generalized role of cytochromes in electron transfer is conserved, as is the role of 16S rRNA in protein synthesis. The particular reaction partners of cytochromes do change from one species to the next [13], although one common reaction partner of all cytochromes c_2 appears to be the cytochrome bc_1 complex. Nevertheless, it was shown that the functional role has had little influence on the evolution of mitochrondrial cytochrome c or bacterial cytochrome c_2 [22]. Even where the reaction partners of cytochrome c_2 are held constant, matrix trees do not agree with those for 16S rRNA. However, it is not necessarily the case that all cytochromes found in purple bacteria, such as the *Rhodocyclus gelatinosus* and *Rhodocyclus tenuis* cytochromes c_8 [29], function exactly the same as cytochromes c_2 or that the function has had no influence on evolution at all. The cytochromes c_8 may have different roles as well as markedly different sequences, but were erroneously included in cytochrome c_2 trees [23, 28]. Furthermore, there is no evidence that insertions and deletions in cytochrome c_2 have any influence on the functional properties. In the analogous situation with 16S rRNA, the details of protein synthesis are unknown; thus we cannot be certain that the proteins interacting with 16S rRNA are the same in all species. Such arguments cannot explain the differences in the cytochrome c_2 and 16S rRNA trees. The insertion–deletion tree for cytochrome c_2 in Figure 4.2 does not agree with the previous matrix trees for cytochrome c_2 nor does it agree with those for 16S rRNA. This is not because cytochrome c_2 is a poor chronometer as previously claimed, but because matrix trees are generally invalid for large evolutionary distances owing to convergence whether using cytochrome or 16S rRNA data.

The evolution of cytochrome c_2 is more thoroughly understood than that of any other bacterial macromolecule including 16S rRNA. That is not because cytochrome c_2 is inherently superior to any other macromolecule, but because we have a greater number of three-dimensional structures and a relatively large number of sequences and precisely located insertions and deletions. Nevertheless, it still falls short of our expectations. This is because convergence is not the only problem faced in bacterial tree construction, even though it is one of the most serious. Even for this data set, the information is incomplete; not all species that have a cytochrome c_2 are represented and in many cases only a single strain of each species has been examined. Insertions and deletions provide the framework, but fine structure must be determined by amino substitutions. Mutations (which form the basis of all trees) are roughly exponentially related to sequence differences; thus we are often looking at small differences between very large numbers of mutations [22]. The errors involved in translating sequence differences into mutations also increase with the extent of divergence and limit the resolution of methods based on sequence differences regardless of what macromolecule is used for analysis. That is, a 46% sequence difference in cytochrome c_2 may be equivalent to millions of mutations, whereas the difference between 45% and 46% sequence difference may represent either tens or thousands of mutations. Only precisely located insertions and deletions currently have the potential to extend bacterial trees beyond the species level to genera and higher taxa. However, because gaps are relatively rare events, such trees may suffer from errors owing to statistically small sample sizes [30], thus requiring data from several macromolecules to obtain consistent results. Some proteins and pathways may have been duplicated or acquired through gene transfer which also requires analysis of data from a number of proteins. Thus, Figure 4.2 will not be a perfect representation of the relationships among the species as a whole, but must be compared with a number of similarly constructed trees for other macromolecules such as the following.

4.2.2 HiPIP

HiPIP is a small [4Fe–4S] iron–sulfur protein that has a very high redox potential as compared with other [4Fe–4S] ferredoxins. It is not related to any other known protein and is found only in the purple bacteria [15] and in two nonphototrophs (a halophilic *Paracoccus* sp. and *Thiobacillus ferro-oxidans*). The alignment of the sequences based on comparison of three-dimensional structures [31] is shown in Figure 4.3. Important residues including the four cysteines that bind the [4Fe–4S] iron–sulfur cluster (residues 43, 46, 63, and 80 of Fig. 4.3) are boxed. There is an average of only 30% ± 15% sequence identity among the HiPIPs. Thus, they are also generally too divergent to use the matrix of sequence identities to quantify relationships, except for closely

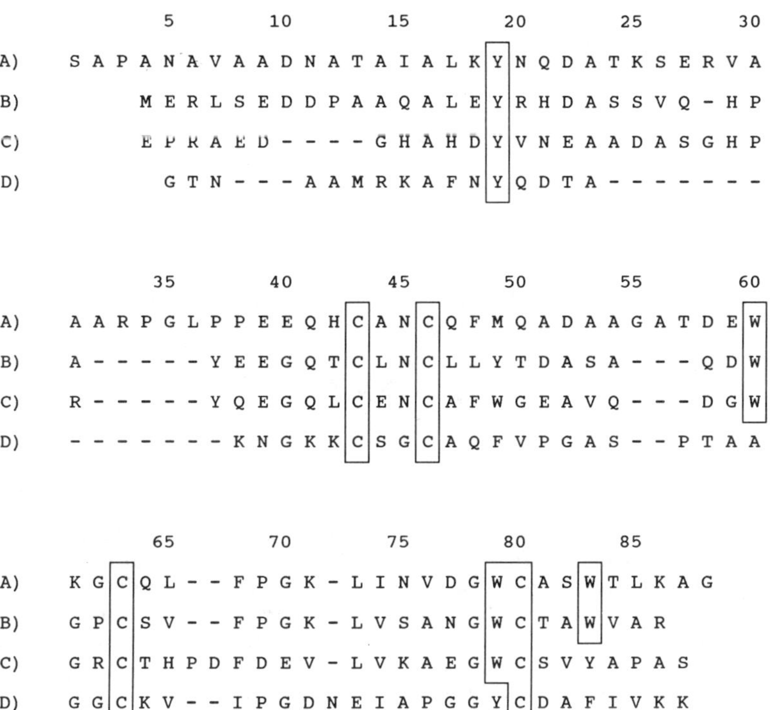

Figure 4.3 Sequence alignment of HiPIP based on three-dimensional structures. (A) *Chromatium vinosum* [32]; (B) *Ectothiorhodospira vacuolata* iso-2 [31, 138]; (C) *Ectothiorhodospira halophila* iso-1 [139, 140]; (D) *Rhodocyclus tenuis* [141, 142].

related species. As a generalization, this conclusion is valid for all of the gene products considered in this chapter. It has little to do with the particular molecule chosen for study, or with how fast it evolves, but rather is related to the large evolutionary distances between the genera and families of phototrophic and denitrifying bacteria and of even larger bacterial groups.

As an observation, insertions and deletions generally occur outside regions of secondary structure (also called surface loops). Because there is very little secondary structure in the HiPIPs, this may explain why insertions and deletions are common. Fortunately, there are four known three-dimensional structures for some of the most divergent HiPIPs that have established the locations of most of these insertions and deletions with a high degree of confidence as shown by Benning et al. [31]. Thus, HiPIP is second only to

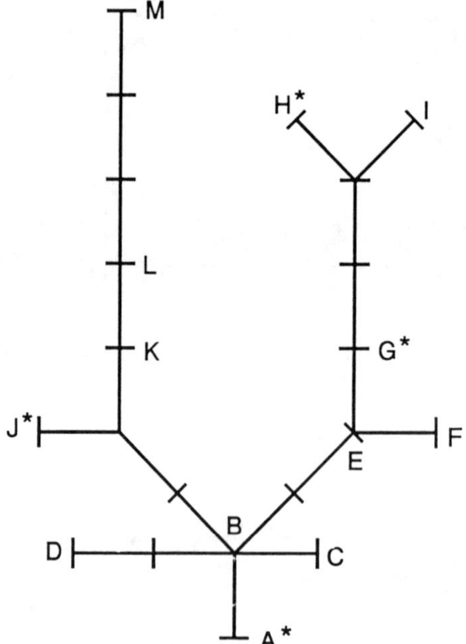

Figure 4.4 Evolutionary tree for HiPIP based on internal insertions and deletions located by comparison of three-dimensional structures as shown in the alignment of Figure 4.3. Each node or bar represents an insertion or deletion whether lettered or not. (A) *Chromatium vinosum* and *Thiocapsa roseopersicina*; (B) *Chromatium gracile* and *Chromatium tepidum*; (C) *Thiocapsa pfennigii*; (D) *Rhodocyclus gelatinosus*; (E) the genus *Ectothiorhodospira*; (F) halophilic *Paracoccus* sp. ATCC 12084; (G) *Ectothiorhodospira vacuolata*; (H) *Ectothiorhodospira halophila* iso-1; (I) *E. halophila* iso-2; (J) *Rhodocyclus tenuis*; (K) *Rhodospirillum salinarum* iso-1; (L) *Rhodopseudomonas marina, Rhodopila globiformis, Rhodomicrobium vannielii*, and *Thiobacillus ferrooxidans*; (M) *R. salinarum* iso-2. An asterisk designates species for which the three-dimensional structure has been determined.

cytochrome c_2 in its value as an evolutionary marker for purple photosynthetic bacteria.

The HiPIP data are more subject to interpretation than are those for c_2 because there are no obvious concerted events such as the three shared gaps in c_2. For example, some events that have been interpreted as sequential, alternatively may have occurred in parallel. An important consideration in construction of the insertion–deletion tree was to group metabolically similar bacteria as shown in Figure 4.4. The base of the tree is assumed to be "A," which includes *Chromatium vinosum* [32] and *Thiocapsa roseopersicina* [33], the largest of the HiPIPs. Other members of the family Chromatiaceae are clustered nearby and separated by one or two gaps. The family Ectothio-rhodospiraceae has diverged in one direction by three to seven gaps and the family Rhodospirillaceae has gone one to eight gaps in another direction. All of the Rhodospirillaceae that have a cytochrome c_2 as well as HiPIP: *Rhodopseudomonas marina* (R. P. Ambler and J. Van Beeumen, unpublished), *Rhodopila globiformis* [34], and *Rm. vannielii* (J. Van Beeumen, unpublished) group together at "L" in Figure 4.4 and are near the assumed base of the cytochrome c_2 tree. Thus, the cytochrome c_2 tree appears to begin approximately where the HiPIP tree ends. It is important to note that the *Rhodocyclus* species of HiPIP [35, 36] are closer to *C. vinosum* than are those of the other Rhodospirillaceae. *Rhodocyclus* species lack a cytochrome c_2, but instead have cytochromes similar to those of *C. vinosum* [15]. Matrix trees eventually may be constructed for fine tuning the evolution of some Chromatiaceae and Rhodospirillaccae within the larger insertion–deletion tree (not shown).

4.2.3 Cytochrome c'

The distribution of cytochrome c' is more widespread within purple bacteria than that of either HiPIP or cytochrome c_2 [15]. A number of sequences are known [37] and three-dimensional structures exist for three species [38]. An *Alcaligenes* sp. (now called *Achromobacter xyloseoxidans*) has a cytochrome c' obviously homologous to those of purple bacteria, yet has an azurin similar to those of *Pseudomonas* [20, 39]. More sequences are necessary to establish whether this species is more closely allied with purple bacteria or with *Pseudomonas*. The sequences are aligned based on the three-dimensional structures in Figure 4.5 and the most divergent average about 30% identity. However, cytochromes c' are primarily helical in structure and have fewer and smaller insertions and deletions than either cytochrome c_2 or HiPIP and the five gaps (residues 33, 54, 67–69, 81, and 108 of Fig. 4.5) are concentrated in only three interhelical regions of the structure, which make them less useful for evolutionary analyses. Nevertheless, *Rhodocyclus tenuis* cytochrome c' has no gaps relative to *C. vinosum* [37], which agrees with the results of the HiPIP analysis, which indicates that *Rhodocyclus* is closer to *Chromatium* than to other Rhodospirillaceae. The species of Rhodospirillaceae that group together in terms of cytochrome c_2 and HiPIP (*Rps. viridis*, *Rm. vannielii*, *Rps.*

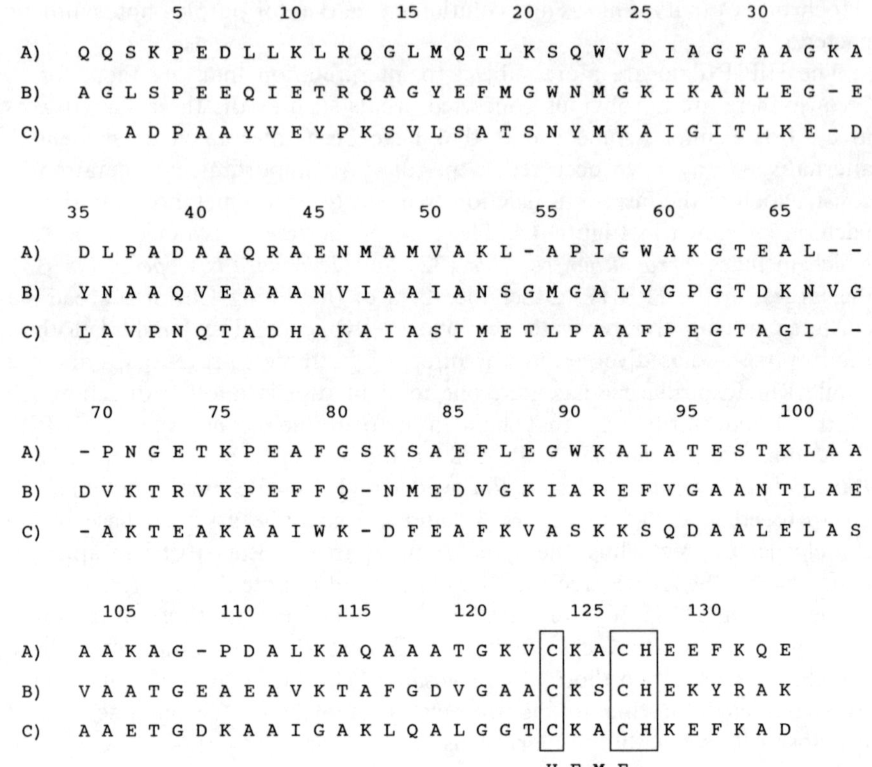

Figure 4.5 Sequence alignment of cytochrome c' based on three-dimensional structures. (A) *Rhodospirillum molischianum*; (B) *Chromatium vinosum*, (C) *Rhodospirillum rubrum* [37, 39].

acidophila, Rp. globiformis) generally do not have a cytochrome c' at all [15], which is again consistent with the other analyses. Matrix trees eventually may be constructed for *Rhodobacter* and *Chromatium* species, in agreement with the conclusions based on cytochrome c_2 and HiPIP insertion–deletion trees (not shown). Thus, in spite of the lack of a detailed insertion–deletion tree at this time, the cytochrome c' data are still valuable in a confirmatory sense and may eventually bridge the purple bacterial families more effectively than either cytochrome c_2 or HiPIP once a larger range of sequences and structures become available.

4.2.4 Cytochrome c_5

Cytochromes c_5 are generally membrane-bound proteins found in *Pseudomonas* and *Azotobacter* species. There are only a few sequences known [16] and one three-dimensional structure [40]. Although c_5 is homologous to mitochondrial cytochrome c, it is much more divergent than are the cytochromes c_2 and the sequences cannot be aligned with confidence even with the help of three-dimensional structures because they differ in helical content. That is, cytochrome c_5 has a helix in the midsection that should have been present in the larger cytochrome c_2 if they were related by a simple insertion–deletion event. This suggests that the differences in the midsection of cytochromes c_2 and the much smaller cytochromes c_5 are caused by several unspecified insertions and deletions rather than a single event. This conclusion differs from the interpretation of Dickerson [23], who reported that the structural differences between large and small cytochromes were due to a single mutation. Much closer (and soluble) homologs of cytochrome c_5 have been found in phototrophic bacteria as shown in Figure 4.6. Nevertheless, until more three-

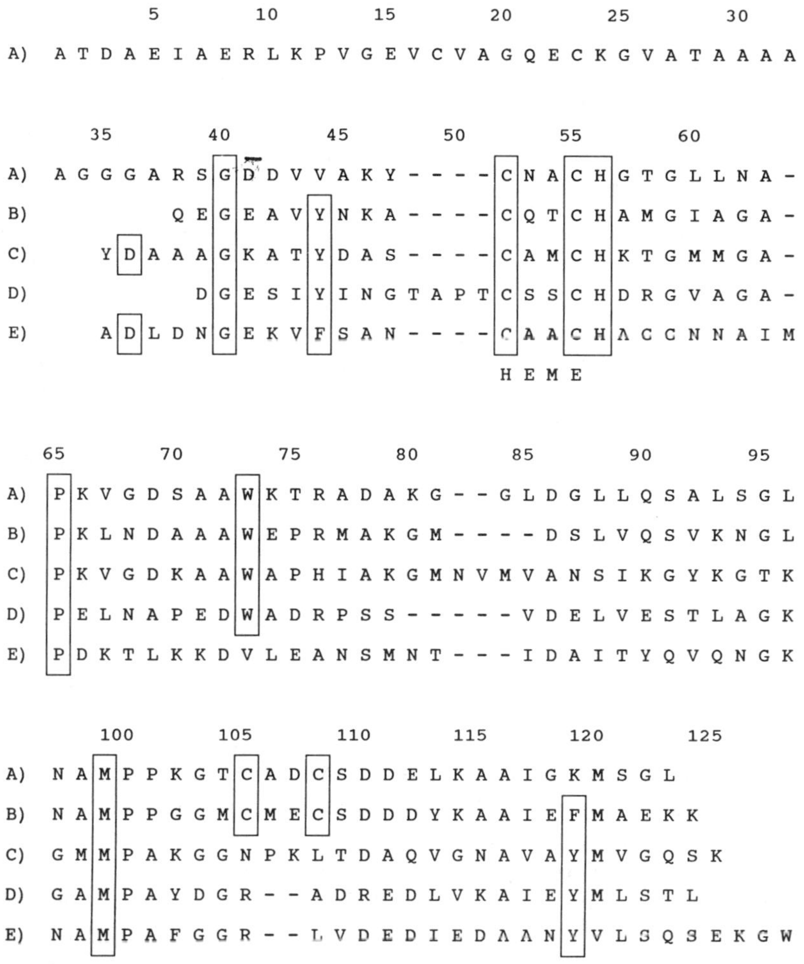

Figure 4.6 Sequence alignment of cytochrome c_5 homologs. (A) *Azobacter vinelandii* c_5 [64]; (B) purple bacterium strain HIR *c*-552 [16]; (C) *Chlorobium thiosulfatophilum* *c*-555 [41]; (D) *Ectothiorhodospira halophila c*-551 [42]; (E) *Porphyra tenera* c_6 [43].

dimensional structures are determined, insertions and deletions have limited utility within this group of proteins. The locations of gaps in the midsection (between positions 73 and 99 of Fig. 4.6) are particularly difficult to assign. Boxed residues (36, 40, 65, 99, and 119) are generally conserved in type I cytochromes, but are too few to unequivocally align the sequences. Rare and functionally important amino acid residues are currently more useful in this group for evolutionary purposes (such as W73 and the disulfide 105–108).

Cytochrome c-552 from the purple phototrophic bacterium strain H1R is the nearest homolog of cytochrome c_5 in that it has proline 65 (Fig. 4.6 numbering) that hydrogen bonds the histidine 56 heme ligand, it has tryptophan 73, and it has the cystine disulfide 105–108 after the methionine 99 heme ligand [16]. The cytochrome c-555 from *Chlorobium* species [41] and cytochrome c-551 from *Ectothiorhodospira* species [42] are somewhat more divergent in that they have the proline and tryptophan, but lack the cystine disulfide. Cyanobacterial cytochromes c_6 are the most divergent in that they generally lack the proline (*Porphyra* is an exception), the tryptophan, and the cysteine [43]. However, the cytochromes c_6 appear to share a two-residue deletion with *Ectothiorhodospira* c-551 just after the heme ligand methionine as shown in Figure 4.6. These data suggest the progression: $c_5 \rightarrow c$-552 $\rightarrow c$-555 $\rightarrow c$-551 $\rightarrow c_6$. An alternative interpretation is that c-555 and c-551 evolved in parallel following divergence from c_5. However, more sequences and structures for this group of proteins are necessary before we can further quantify the observed relationships. Matrix trees are currently invalid for this family of proteins, but may be useful once more closely related species are characterized. Cytochrome c_5 and homologs are currently the only macromolecules that tie together all of the major groups of phototrophs and allow us to extend the combined cytochrome c_2–HiPIP tree to all phototrophs. The only other macromolecules that could possibly serve the same purpose are nitrogenase subunits, bacterial ferredoxins, and possibly photosynthetic reaction centers as described in following sections.

4.2.5 Cytochromes c_8

Cytochromes c_8 were previously known as *Pseudomonas* cytochromes c-551 and were among the first bacterial proteins to be sequenced [44]. They have since been found in a variety of bacteria, including *Azotobacter* [16]. Unlike cytochromes c_5, the cytochromes c_8 are water soluble. There are a number of sequences known [16], but there is only one three-dimensional structure [45]. There are no insertions or deletions within the *Pseudomonas* species of cytochrome c_8 and the average percentage identity (55%) is much higher than for cytochromes c_2 or for other photosynthetic proteins. Although cytochromes c_8 are homologous to cytochromes c_2 and c_5, they are too divergent to align with confidence.

Near homologs of cytochrome c_8 occur in purple bacteria in the genus *Rhodocyclus* [29] as shown in Figure 4.7. Woese et al. [28] and Dickerson [23] erroneously combined them with the cytochromes c_2 although it should be apparent from Figure 4.7 that they are much closer to the *Pseudomonas* proteins. Some cytochromes c_8 such as *Rc. tenuis* may now function similarly to cytochrome c_2 in the purple bacteria, but they came from a very different background and cannot be compared on the same basis as c_2. Furthermore, it is doubtful whether the *Rc. gelatinosus* cytochrome c_8 has a role in photosynthesis because of its unusually low redox potential. A few rare and functionally important amino acid residues are conserved (such as W56, M66, and W85), which aids in alignment. The *Rhodocyclus tenuis* c_8 homolog has a five-residue insertion just before the sixth ligand methionine-66 and *Rhodocyclus gelatinosus* has a three-residue insertion just after the methionine. Furthermore, we have recently found that *C. vinosum* cytochrome *c*-551 [46] is a c_8 homolog [46a]. It has a single residue insertion just after the methionine. *C. vinosum* and *Rc. tenuis* c_8 homologs have tryptophans-56 and -85, but *Rc. gelatinosus* does not have W56. This is consistent with the results of HiPIP and cytochrome *c'* analyses, which show that *Rc. gelatinosus* is more divergent from *C. vinosum* than is *Rc. tenuis* and may not be on the same branch as the latter. One interpretation of the sequence data for cytochrome c_8 is: *Pseudomonas* → *C.*

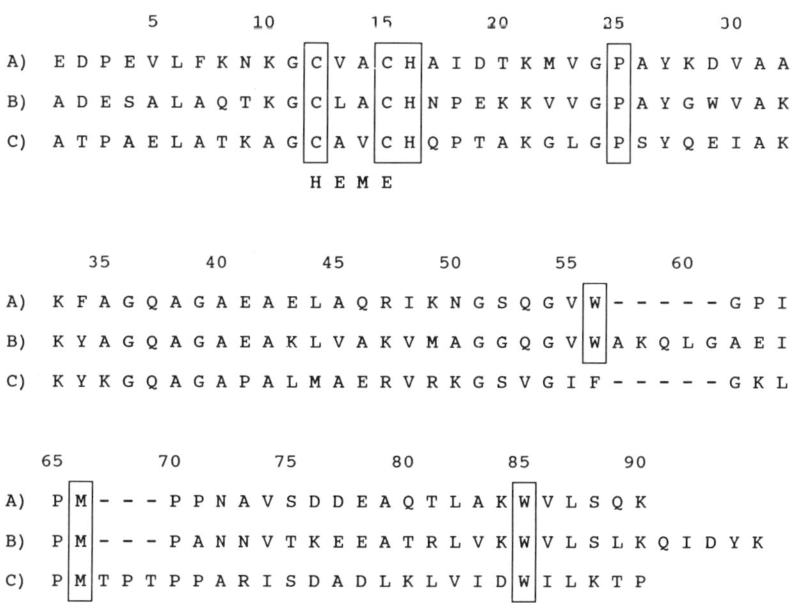

Figure 4.7 Sequence alignment of cytochrome c_8 homologs. (A) *Pseudomonas aeruginosa* [44]; (B) *Rhodocyclus tenuis* [29]; (C) *Rhodocyclus gelatinosus* [29].

vinosum → *Rc. tenuis* → *Rc. gelatinosus*. As noted above, the latter two species alternatively could have diverged from *Chromatium* in parallel. Cytochrome c_8 homologs have not yet been found in *Ectothiorhodospira* species nor have cytochrome c_5 homologs been found in *Chromatium* or *Rhodocyclus* species, but they are expected to be present and the evolutionary picture will be clarified greatly by such discoveries.

4.2.6 Copper Proteins

Azurin is a small blue copper protein present in most denitrifying pseudomonads, but not yet seen in *Ps. stutzeri* or in the strictly aerobic *Azotobacter*. It functions interchangeably in vitro with cytochrome c_8 as electron donor to nitrite reductase and is found in several more divergent denitrifying bacteria such as the *Alcaligenes* sp. mentioned previously [39]. There are a number of sequences available [39, 47] and three-dimensional structures are known [48–50]. As is the case with cytochrome c_8, there are no internal insertions or deletions in the pseudomonad azurins. However, the naturally occurring chimera known as *Chloroflexus aurantiacus* contains copper proteins called auracyanin A and B [51] which presumably mediate electron transfer between the cytochrome bc_1 complex and photosynthetic reaction center and are related to azurin as shown in Figure 4.8. There are three small gaps in auracyanins relative to azurin (approximately residues 80–81, 129–131, and 147 in Fig. 4.8) and the cysteine disulfide characteristic of azurin is absent. *Chloroflexus* uniquely contains a purple bacterial reaction center [52], including the tetraheme cytochrome subunit [53], but has green bacterial light harvesting bacteriochlorophyll proteins [54] and is facultatively aerobic. It thus appears to have evolved after the purple and green bacterial families via gene transfer in much the same manner as the cyanobacteria (as described below).

Plastocyanin is a related copper protein that, along with cytochrome c_6, interchangeably mediates electron transfer between the cytochrome $b_6 f$ complex and photosystem I in cyanobacteria and algae. It is much more divergent than is auracyanin, as shown in the alignment of Figure 4.8. The three-dimensional structure of plastocyanin [55] shows that it folds into an antiparallel β barrellike azurin and the gaps can be located with some confidence. There are at least eight insertions and deletions, one of which is fairly large (at positions 37–40, 55–58, 69–73, 85–105, 117–111, 137–141, 152, and 157 of Fig. 4.8). The azurin cystine disulfide (29–57 of Fig. 4.8) is absent in plastocyanin, but the copper ligands (H77, C150, H155, and M160 of Fig. 4.8) are conserved. A likely path of evolution of the copper proteins is: azurin → auracyanin → plastocyanin. Alternatively, the auracyanins may have diverged in parallel. An important point is that auracyanin is much closer to azurin than to plastocyanin, and thus may be more primitive than the latter. This path is

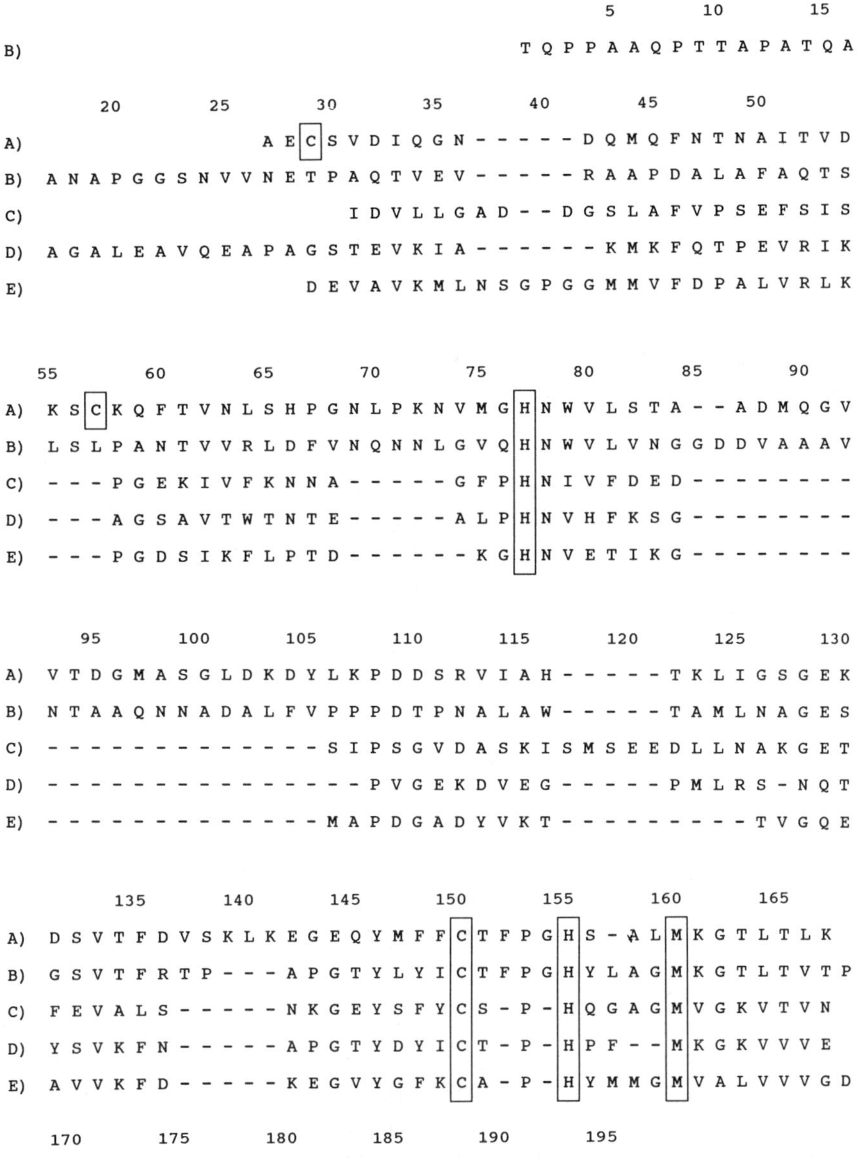

Figure 4.8 Sequence alignment of copper proteins based on three-dimensional structure. (A) *Pseudomonas aeruginosa* azurin [39, 143]; (B) *Chloroflexus aurantiacus* auracyanin B [51]; (C) poplar plastocyanin [144]; (D) *Methylobacterium extorquens* amicyanin [58]; (E) *Methylobacterium extorquens* pseudoazurin [58].

consistent with that of cytochrome c_5 and near homologs in photosynthetic bacteria, but would be strengthened if additional examples from purple bacteria could be found. In this regard, we have observed a blue copper protein in *Rps. palustris* strain TVC (T. E. Meyer, unpublished).

Amicyanin is a small copper protein found in a few methylotrophic bacteria including *Paracoccus denitrificans* and *Methylobacter extorquens* (formerly *Pseudomonas* AM1), both of which also contain a cytochrome c_2. It has a well-defined role as electron acceptor for methylamine dehydrogenase [56]. Thus, it is not unreasonable to expect amicyanin in purple bacteria which may utilize methylamine. The three-dimensional structure is known [57] and there are currently three sequences [58–60]. It is clearly much closer to plastocyanin than to azurin as shown in Figure 4.8 and appears to share six gaps, although they are not as easy to place as in the cytochromes c_2.

Pseudoazurin is a copper protein found in a small number of pseudomonads. The functional role is presumably similar to that of azurin. The three-dimensional structure [61, 62] indicates that it is more closely related to plastocyanin than to azurin but it has an extra C-terminal helix. It also shares six gaps with plastocyanin and amicyanin as shown in Figure 4.8. The impression is that these proteins may be equidistant from one another and reinforces the idea that plastocyanin originated with the pseudomonads. However, there are too few known sequences to provide much evolutionary data [58, 63]. Pseudoazurin has not yet been found in photosynthetic bacteria.

4.2.7 Other Electron Transfer Proteins

Cytochrome c_4 is a diheme membrane-bound cytochrome generally present in *Pseudomonas* and *Azotobacter* species and that has undergone gene doubling in the distant past. There is only one known complete sequence and several partial sequences [64]. The two halves of *Azotobacter* c_4 are aligned in Figure 4.9. Cytochrome c_4 is homologous to cytochromes c_2, c_5, and c_8, but is not close enough to any to align with confidence, especially in the absence of a three-dimensional structure. We have determined the sequence [17] of *Chromatium vinosum* cytochrome c-553(550) [65] and have found that it is the nearest photosynthetic homolog of c_4. The *Chromatium* cytochrome has distinctive size, heme content, low solubility, and spectral properties which make its recognition in other purple bacteria fairly certain without sequence data. Thus, *Thiocapsa pfennigii, Ectothiorhodospira vacuolata, Rc. tenuis,* and *Rc. gelatinosus* all appear to have a c_4 [15]. It therefore appears that cytochromes c_4, c_5, and c_8 generally may be present in the *Chromatiaceae, Ectothiorhodospiraceae,* and *Rhodocyclus* species. Once the sequences and three-dimensional structures are determined, the cytochromes c_4 may become more useful evolutionary markers.

Phototrophic bacterial flavocytochromes c are involved in sulfur metabolism and are generally found in *Chromatiaceae* and *Chlorobiaceae*, although not universally present [15]. They are composed of dissimilar subunits. The *C.*

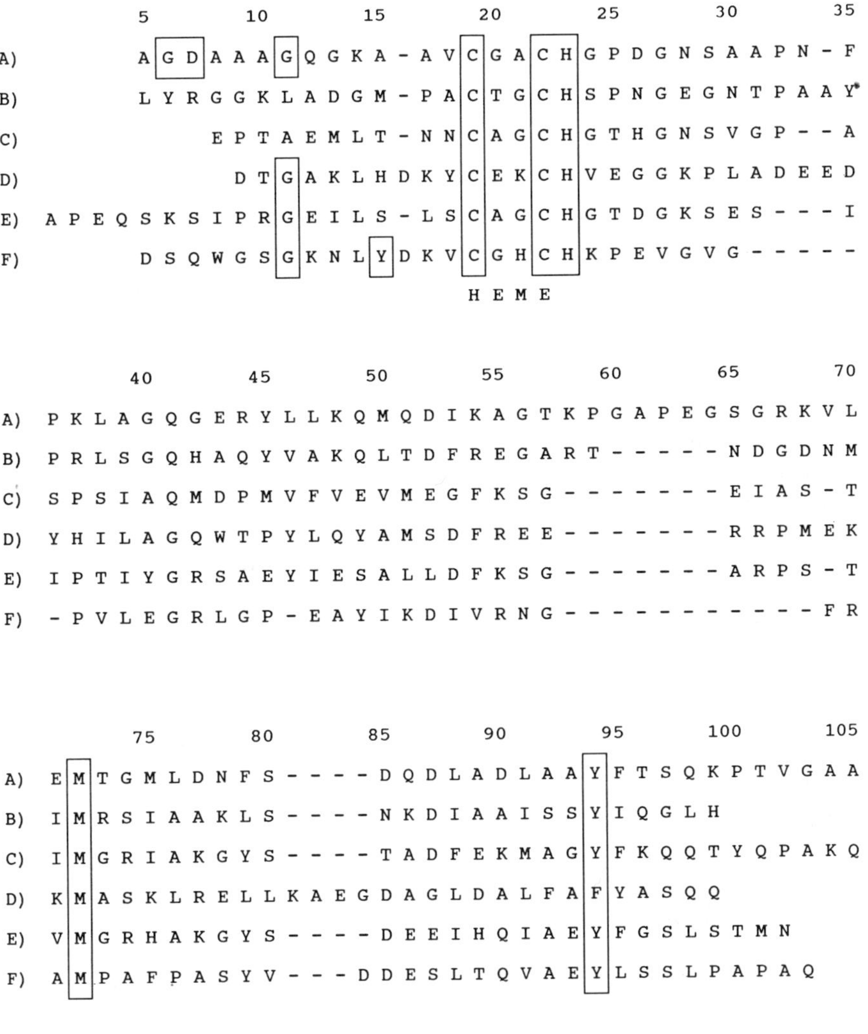

Figure 4.9 Sequence alignment of the two halves of (A, B) *A. vinelandii* cytochrome c_4 [64]; (C, D) *C. vinosum* flavocytochrome *c* diheme subunit [66, 67], with the complete (E) *Ch. thiosulfatophilum* flavocytochrome *c* monoheme subunit [68]; and (F) *Ps. putida* PCMH flavocytochrome *c* monoheme subunit [71].

vinosum cytochrome subunit is a diheme protein that shows evidence for gene doubling analogous to that of cytochrome c_4 [66,67]. The *Chlorobium* cytochrome subunit has a single heme [68] and corresponds to the first half of the *Chromatium* protein. The two halves of the *C. vinosum* cytochrome subunit and the *Chlorobium* cytochrome subunit are aligned with the two halves of *Azotobacter* cytochrome c_4 in Figure 4.9, although it is not certain that they are any more closely related to cytochrome c_4 than to other type I *c*-type cytochromes. The *Chromatium* and *Chlorobium* flavocytochromes *c* are undoubtedly structurally and functionally related to one another and provide an important link between the two families. The structure of *Chromatium* flavocytochrome *c* has recently been solved and shows that the two halves of the heme subunit do indeed fold in separate domains [69]. Once additional sequences are determined, they will provide even more valuable evolutionary data.

Pseudomonas putida also has a flavocytochrome *c*, which functions as a *p*-cresol methyl hydroxylase [70]. The cytochrome subunit has a single heme like that of *Chlorobium*, but the sequences [71] are not appreciably closer than to other type I cytochromes. The flavoprotein subunits of phototrophic and *Pseudomonas* flavocytochrome *c* show no similarity at all [66–69, 72]. *Pseudomonas* flavocytochrome *c* appears to be extrachromosomal and easily lost when cells are grown on nonselective media, suggesting that it originated via gene transfer.

Shewanella putrefaciens has a single subunit flavocytochrome *c* that functions as a fumarate reductase. The gene sequence [73] shows no similarity to the above proteins, although the tetraheme cytochrome domain is similar to a low-potential tetraheme cytochrome from the purple bacterium strain H1R [16] and to the tetraheme cytochromes c_3 from *Desulfovibrio* species [74,75]. It thus appears that the three types of flavocytochrome *c* arose independently and do not provide any evidence either for or against a relationship between pseudomonads and phototrophic bacteria.

A *Pseudomonas stutzeri* diheme cytochrome *c*-552 [76,77] has only been found in two species, the other being *Rhodocyclus tenuis* [17]. There is no evidence for gene doubling and there is no obvious similarity to any other cytochrome. It may never be as important an evolutionary marker as are other cytochromes because of its apparently limited distribution and lack of similarity to other proteins, but the knowledge that it is present in both *Pseudomonas* and purple bacteria is significant within the context of the distribution of *Pseudomonas* cytochromes in purple bacteria.

The *Ps. stutzeri* gene for a tetraheme cytochrome *c* that is an obligate component of the nitrite reductase pathway [77] has been found in only one other species, *C. vinosum* [67], in association with the genes for flavocytochrome *c*. Because this protein is apparently membrane bound, it may be more widespread than it presently seems. The two sequences are aligned in Figure 4.10. There is some evidence for gene doubling in that the potential sixth heme ligands M64, H77, M157, and H170 are conserved and the spacing between hemes 1 and 2 is similar to that between hemes 3 and 4. There is no obvious

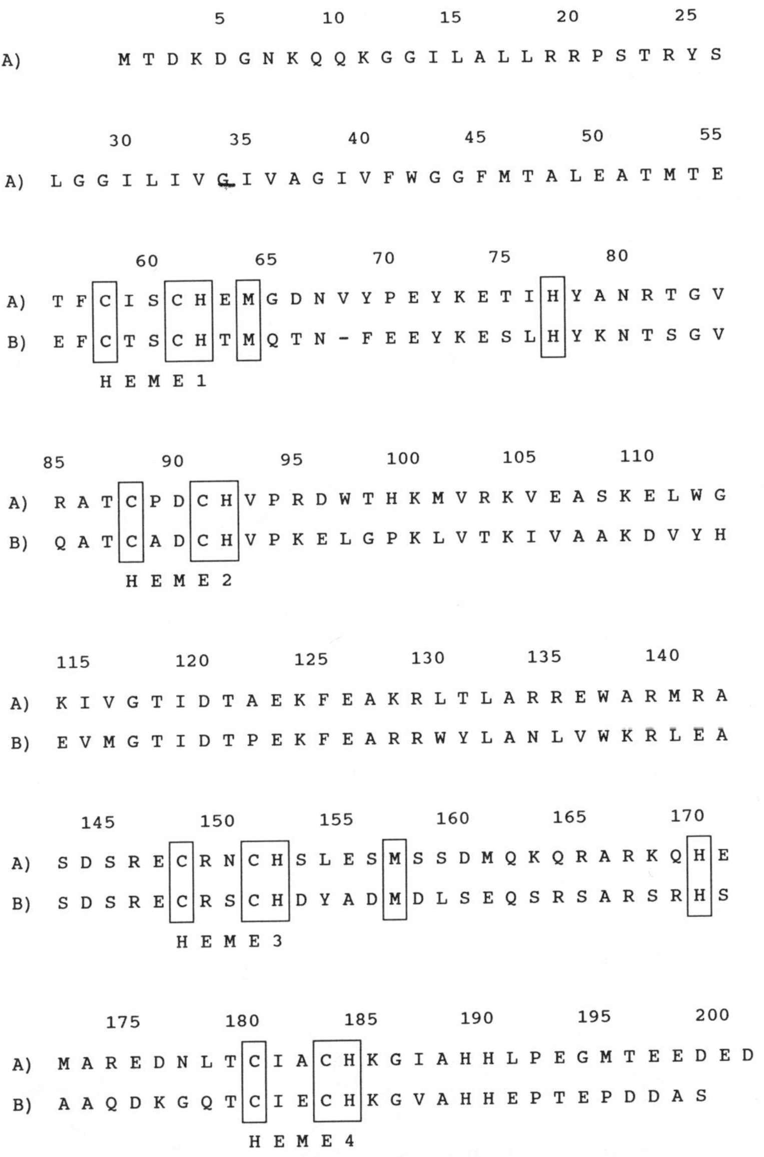

Figure 4.10 Sequence alignment of membrane-bound tetraheme cytochrome *c* from (A) *Pseudomonas stutzeri* [77], and (B) *Chromatium vinosum* [67].

similarity to any other known cytochrome. The tetraheme protein may be unique among cytochromes if two hemes have Met ligands and two have His ligands, suggesting two high-potential and two low-potential hemes.

Cytochrome *cd* occurs in a number of *Pseudomonas* species and functions as a nitrite reductase. It is also present in *Roseobacter denitrificans* [78] and

Paracoccus denitrificans [79], both of which may be close relatives of *Rhodobacter capsulatus* as indicated by cytochrome c_2 sequences [24, 80, 81]. Only the *Pseudomonas aeruginosa* [82] and *Pseudomonas stutzeri* [77] sequences are currently available although the three-dimensional structure has just been determined [82a]. The *c*-type heme domain is at the N-terminus of the protein and is related to type I *c*-type cytochromes, but is too divergent to align with any. Denitrification is unusual in purple bacteria, which suggests that the cytochrome *cd* gene was transferred from *Pseudomonas* to *Rhodobacter* relatives.

Other species of denitrifying bacteria including *Pseudomonas* and *Alcaligenes* utilize a copper-containing type of nitrite reductase [83–85], which has also been found in a denitrifying strain of *Rhodobacter sphaeroides* [86, 87]. The three-dimensional structure of *Achromobacter cycloclastes* nitrite reductase is known [88]. Thus, both nitrite reductases have the potential to aid in establishing the evolutionary relationship of *Pseudomonas* to phototrophs. Because this is the only known occurrence of the copper-containing nitrite reductase in purple bacteria, it could also represent gene transfer from pseudomonads.

The diheme bacterial cytochrome *c* peroxidase originally characterized in *Pseudomonas aeruginosa* [89] has also been found in *Rb. capsulatus* [17, 90] and *Paracoccus denitrificans* [17, 91, 92] as well as in other pseudomonads [13]. BCCP is related to type I cytochromes *c* and has undergone gene doubling analogous to that in cytochromes c_4 based upon the three-dimensional structure [92a]. Refinement of the structure is necessary to further characterize its evolutionary position. Additional sequences may then prove valuable in relating photosynthetic and aerobic bacteria, especially if eventually found in more species.

4.2.8 Bacterial Ferredoxin

The bacterial ferredoxins are a heterogeneous group of small, low redox potential ferredoxins containing [4Fe–4S] clusters that are found almost universally in bacteria including methanogens. There is one particular protein, the [7Fe–8S] ferredoxin from *Pseudomonas* and *Azotobacter*, which contains both [4Fe–4S] and [3Fe–4S] clusters [93]. Homologs of this protein appear to be widespread in the purple phototrophic bacteria, although no systematic surveys have been carried out. We have found it in *Rps. marina* and *Rps palustris* (J. Van Beeumen, unpublished) in addition to *Rb. capsulatus* [94]. The [7Fe–8S] ferredoxin appears to be constitutive [95, 96].

An apparently more abundant [8Fe–8S] ferredoxin in purple bacteria appears to be associated with nitrogen fixation and to be regulated by oxygen and ammonia consistent with this role [95, 96]. Both [7Fe–8S] and [8Fe–8S] ferredoxins have been sequenced from *Rb. capsulatus* [97–99] and *Azotobacter vinelandii* [100, 101]. The two proteins are most readily distinguished by amino acid sequence, yet few sequences have been determined for photosynthetic bacteria, or pseudomonads for that matter; thus the actual distribution is unknown. Most if not all purple and green photosynthetic bacteria fix nitrogen; thus all are expected to have the second type of ferredoxin (the [8Fe–8S] protein). The two ferredoxins from *A. vinelandii* and *Rb. capsulatus* are aligned with the *Chlorobium* nitrogenase-associated ferredoxin and with the fermentative ferredoxin from *Peptococcus* in Figure 4.11. Another ferredoxin is

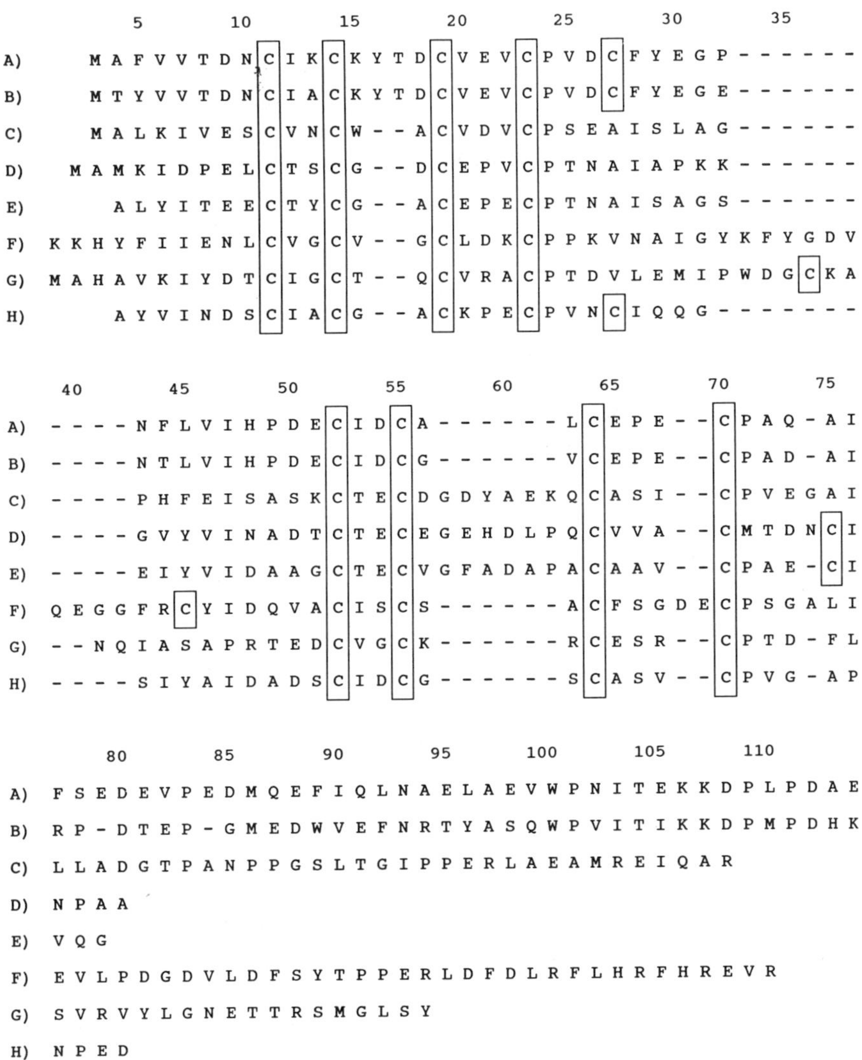

Figure 4.11 Sequence alignment of bacterial ferredoxins based on three-dimensional structures. (A) *Azotobacter vinelandii* [7Fe–8S] [100]; (B) *Rhodobacter capsulatus* [7Fe–8S] [97,98]; (C) *Azotobacter vinelandii* [8Fe–8S] [101]; (D) *Rhodobacter capsulatus* [8Fe–8S] [98,99]; (E) *Chlorobium thiosulfatophilum* [8Fe–8S] [145]; (F) *Chlorobium* F$_a$F$_b$ protein [126–146] [105]; (G) Liverwort *PsaC* gene product (F$_A$F$_B$) [107]; (H) *Peptococcus aerogenes* [146].

associated with nitrogenase in *Rb. capsulatus* although its precise role is unknown [102, 103]. The membrane-bound $F_A F_B$ protein, which is the immediate electron acceptor from the PSI reaction center in cyanobacteria such as *Anabaena* sp. [104], in the green bacteria such as *Chlorobium* [105, 106], and in plants [107], is another important homolog aligned in Figure 4.11.

Although there are not many three-dimensional structures within this group [93, 108], some insertions and deletions are easy to locate because of the large number of conserved chromophore-binding cysteines. The [7Fe–8S] ferredoxins have a two-residue insertion at position 15–16, they have an extra cysteine at position 27, and have a large C-terminal tail as shown in Figure 4.11. The nitrogenase-associated [8Fe–8S] ferredoxins have a six-residue insertion at 57–62 and have an extra cysteine at position 75. The membrane-bound PSI [8Fe–8S] ferredoxins have an 8–10 residue insertion at approximately position 33–42. Sequences of the nitrogenase-associated [8Fe–8S] ferredoxins from the purple bacteria *C. vinosum* [109], *Rps. palustris* [110], and *R. rubrum* [111] as well as *Rb. capsulatus* have been published and we have found related proteins in *C. salexigens*, *Rc. tenuis*, and *Rm. vannielii* (J. Van Beeumen and R. P. Ambler, unpublished). An [8Fe–8S] ferredoxin from the cyanobacterium *Anabaena* that is associated with nitrogenase also fits into this group [112]. Once additional sequences and structures are determined, these proteins may be as useful as the cytochrome c_5 homologs in tracing the path of evolution of photosynthesis, as they are found in all major families of phototrophs.

4.3 Photosynthetic Reaction Centers

All photosynthetic reaction centers (RCs) appear to be related to one another [113], although in the absence of three-dimensional structures this proposition is highly speculative. Thus, the D1 and D2 subunits of photosystem II from cyanobacteria and plants are homologous to purple bacterial reaction center L and M subunits [114]. The single subunit core of the reaction centers of the green bacteria *Heliobacterium* [115] and *Chlorobium* [105] are homologous to the two subunits of plant and cyanobacterial photosystem I. It is also conceivable that photosystem I and photosystem II are even more distantly related to one another, although completion of the three-dimensional structure of photosystem I [116] is necessary to firmly establish such a relationship. Even though it is highly speculative, one could also envision a relationship between reaction centers and the cytochrome b of the bc_1 or $b_6 f$ complexes [113]. The basis for this speculation is that both types of protein bind tetrapyrroles, they interact with quinones, and span the membrane. Furthermore, the simplest photosynthetic pathway includes reaction centers, bc_1 complexes, and soluble mediators. This suggests that the first photosynthetic pathway might have originated by substitution of chlorophyll for heme in a duplicated cytochrome b subunit of a preexisting electron transfer pathway such as the bc_1 complex. A possible order in which reaction centers may have evolved is: cyt b → purple RC → green RC → cyanobacterial photosystem I and photosystem II. Because cyanobacteria and plants have both purple and green RC homologs, they are likely to have arisen by gene transfer of one of the two reaction centers. The alternative possibility, that the reaction centers first evolved in cyanobacteria and then one or the other was deleted in purple and green bacteria, does not agree with the evolution of cytochrome c_5 homologs as described above.

Structures are available for two species of purple bacterial reaction centers from *Rps. viridis* [117] and *Rb. sphaeroides* [118,119] and sequences have been determined for L and M subunits of *Rps. viridis* [120], *Rb. sphaeroides* [121,122], *Rb. capsulatus* [123], *R. rubrum* [124], *Chloroflexus* [52], *Roseobacter* [125], and *Rc. gelatinosus* [126]. There are not many insertions and deletions in comparison of the L or of the M subunits with one another, or in the cross-comparison of L with M. Thus, these sequences are currently not very useful for tree construction, but have the potential to clarify the evolution of purple bacteria and cyanobacteria when more sequences and structures are determined, in particular those of photosystem I and photosystem II.

The tetraheme cytochrome subunit of the purple bacterial reaction center is more interesting with respect to evolution in that it is apparent that there are going to be a number of insertions and deletions once a sufficient data base is constructed. Currently, there are three complete sequences from *Rps. viridis* [127], *Chloroflexus* [53], and a *Rc. gelatinosus* strain [126]. The partial sequence of *Roseobacter denitrificans* [125] is also available. There is evidence for two gene doublings both from the sequences and the three-dimensional structure [117]. The two halves of the *Rps. viridis* and *Chloroflexus* tetraheme reaction center cytochromes are aligned in Figure 4.12. Each heme is preceded by a helix that contains the sixth ligand for that heme. Presumably as a result of the most recent gene doubling, one heme changed sixth ligands from Met to His in *Rps. viridis*. There may still be a cytochrome that retains all four Met ligands. It may also become apparent which cytochrome is ancestral by comparison of the two halves such as shown in Figure 4.12. That protein in which the two halves are most similar presumably would be the oldest. Neither of the illustrated proteins meets that criterion, but the *Chromatium* tetraheme cytochrome will be interesting in that regard once the sequence is determined.

The green bacterial reaction centers also have a cytochrome subunit, but it contains only a single heme [128]. There is no evidence that it is ancestral to the purple bacterial RC cytochrome subunit and in fact it appears that it could be a type I c-type cytochrome in which the sixth heme ligand follows the heme binding site rather than precedes it as in the tetraheme RC cytochrome.

One step in the biosynthesis of chlorophyll and bacteriochlorophyll is reduction of the tetrapyrrole. Protochlorophyllide and chlorin reductases that accomplish this task are homologous to the nitrogenase H subunits [129]. This is consistent with the observation that all photosynthetic bacterial families are

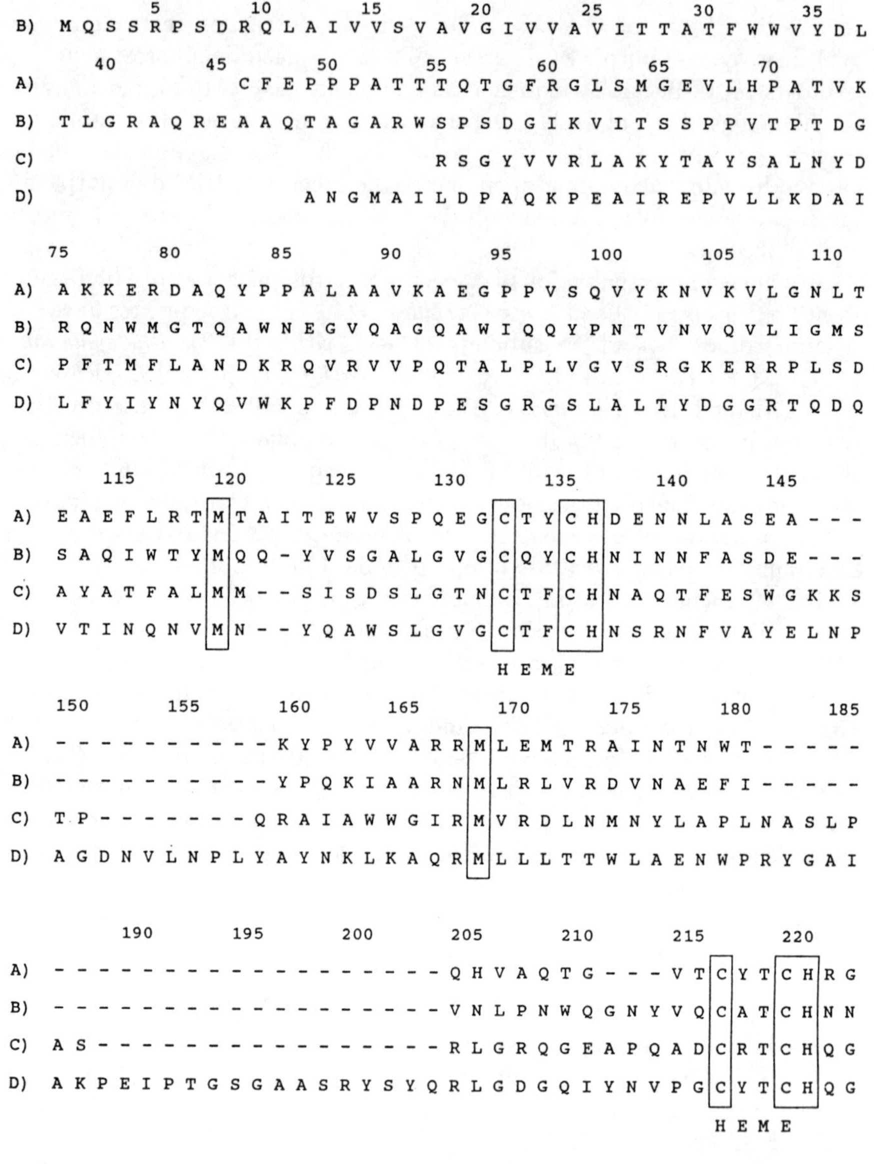

Figure 4.12 Sequence alignment of the two halves of the tetraheme reaction center cytochromes *c*. (A) *Rhodopseudomonas viridis* [126] first half; (B) *Chloroflexus aurantiacus* [53] first half; (C) *Rps. viridis* second half; (D) *Ch. aurantiacus* second half.

capable of nitrogen fixation and that the earliest photosynthetic bacteria may have been related to *Pseudomonas* and *Azotobacter*. Comparison of nitrogenases and chlorophyll biosynthetic enzymes may thus be a useful exercise in the future. The [2Fe–2S] ferredoxin of cyanobacteria and plants, which is involved in reduction of pyridine nucleotides, is also homologous to a nitrogenase-associated ferredoxin discovered in *Rb. capsulatus* [95]. Eventually, all of the pathways and proteins discussed above may clarify the evolution of photosynthesis, but much more sequence and structural data are needed before that will be possible.

4.4 16S rRNA Evolution

Ribosomal RNA is not directly relevant to a discussion of bacterial electron transfer protein evolution and photosynthesis. However, it has been reported to be superior to cytochromes for evolutionary analyses [1, 28]. Consequently, some discussion of their relative merits is warranted. We have already pointed out some differences in the two methods in the section on cytochrome c_2, but we believe that 16S rRNA as presently analyzed may provide significantly less information than claimed [1, 4]. Ribosomal 16S rRNA is found in all bacteria and therefore has the potential to contribute as much or more to the study of the evolution of photosynthesis as any of the above described proteins. However, although there are a large number of partial 16S rRNA sequences covering the breadth of the bacterial kingdom, there are fewer complete sequences known, especially among the phototrophs and pseudomonads. There is no three-dimensional structure for the ribosome and there are often multiple genes for rRNA molecules [130]. Furthermore, there is no reason to believe that convergent mutations are any less of a problem with 16S rRNA than they are with proteins and there is some evidence that they are more so because there are only four bases as opposed to 20 amino acid residues. Complementary base pairing in secondary structural regions further increases the likelihood of convergent mutations. In fact, the maximum divergence of 16S rRNA across the whole bacterial kingdom including the so-called archaebacteria appears to be about 40%. For the purple bacteria, it is about 20%.

If the very large evolutionary distances among photosynthetic bacteria limit the use of matrices of sequence differences for homologous proteins because of convergence [22], then the very same limitations should apply to use of 16S rRNA for the same species. We estimate that convergence in 16S rRNA may limit use of matrix data for identities less than about 85%.

It is likely that 16S rRNA can safely be used in the same way that the matrix of sequence differences for cytochrome c_2 is used in comparison of closely related species such as *Rb. capsulatus* and *Rb. sphaeroides*. One can also establish whether two strains are identical within experimental error, but >99% identity in 16S rRNA is insufficient to prove that they are the same species [131]. The branching patterns for published 16S rRNA trees also appear to be highly unreliable when bacteria with dissimilar CG content are compared [132, 133].

If three-dimensional structures were available for ribosomes, then insertions and deletions could be analyzed in the same way as for proteins. This might alleviate some of the problems with convergence. Thus, in the future, 16S rRNA could possibly be as useful in bacterial evolutionary studies as proteins or even more so because of the wider range of occurrence. These conclusions are more or less heretical because the majority of microbiologists now accept as dogma evolutionary speculations based on matrices of sequence differences in 16S rRNA no matter how divergent. If the simple divergent 16S rRNA tree for bacteria were correct and representative of the whole genome, it would require that all the genes for homologous functions that occur on more than one branch such as photosynthesis, nitrogen fixation, denitrification, and aerobic respiration be present in the ancestral organism. Bacterial evolution would then involve more loss than gain over billions of years of time.

Although the advocates of 16S rRNA trees now accept gene transfer on a small scale, they believe that the ribosome represents a nontransferable core and only small genes such as cytochromes can be transferred. Although single small genes and gene fragments are more likely to be transferred and incorporated into the genome than are larger pieces of DNA [134, 135], there is some evidence to suggest that relatively large fragments have been transferred such as the nitrogenase gene cluster (to *Klebsiella*), the denitrification pathway (to *Roseobacter* and *Rhodobacter*), the chlorosome light-harvesting proteins (to *Chloroflexus*), and the photosynthetic reaction center (to cyanobacteria). It has been estimated that 11% of the genes in *E. coli* were acquired by gene transfer [6]. Because unfavorable codons in transferred genes will eventually be replaced, evidence for such gene transfer will gradually disappear with time. Thus, the 11% gene transfer may be an underestimate. In light of the magnitude of the above events, why should the ribosome or a portion thereof be exempt? In fact there is some evidence for genetic crossing over of ribosomal DNA in *Aeromonas* [136]. However, there is no evidence to suggest that gene transfer of electron transfer pathways including cytochromes or protein biosynthetic pathways including 16S rRNA is common in bacteria over the short term. Gene transfers, duplications, and chromosomal rearrangements are likely to occur less often than insertions and deletions and may even provide another means of reconstructing evolution once actual events are documented.

4.5. Evolution of Photosynthesis

Based on the distribution and evolution of homologs of Pseudomonad electron transfer proteins, particularly in purple bacteria, but also in other photosynthetic bacteria, we believe that photosynthesis originated in a Pseudomonad-like anaerobic bacterium. This is a scenario that fits the available data, but is not the only possible interpretation. We believe that this is the first time that sequence evidence has been assembled in favor of a monophyletic origin of photosynthesis and for pseudomonad precursors. As such, it provides a unique basis for discussion. We do not believe that the first bacteria were photosynthetic as suggested by 16S rRNA analyses, but instead were fermentative and gram-positive. Electron transfer proteins evolved to form anaerobic respiratory pathways coupled to proton gradient formation and this eventually led to photosynthesis. Formation of the gram-negative cell and the accompanying periplasmic space was a major event in this process. Aerobic respiration was the last major electron transfer pathway to evolve in bacteria. All such pathways are likely to have been transferred among species at a low frequency, resulting in the composite species we know today.

The photosynthetic bacteria that have the greatest number of ancient characteristics would be in the family Chromatiaceae, which are anaerobic and utilize reduced sulfur compounds for CO_2 fixation and contain more pseudomonad-like proteins than any other. The families Heliobacteriaceae and Chlorobiaceae appear to have diverged before Ectothiorhodospiraceae and Rhodospirillaceae. The green bacteria combined precursors of light harvesting and reaction center proteins into a larger reaction center and changed the electron acceptor from quinone to a lower redox potential [4Fe–4S] cluster. This allowed them to increase energy conservation by adding the NADH dehydrogenase complex to the pathway in addition to the bc_1 complex. The cyanobacteria represent a chimera made up of both purple and green bacterial reaction centers that further evolved to add a water splitting apparatus and the light harvesting phycocyanin family of pigments.

A schematic evolutionary tree is shown in Figure 4.13, which qualitatively summarizes the information presented above. The cytochrome c_2 and HiPIP branches of the tree are the only sections which are reasonably quantitative as shown in Figures 4.2 and 4.4 because that is where most of the available data are located. *Chloroflexus* and cyanobacteria are both represented as chimeras bridging green and purple bacteria, although much more work is necessary to determine when, how much, and in which directions the gene transfers may have taken place. It presently appears that *Chloroflexus* has more purple bacterial characteristics than green, but that may only be a consequence of the small data base. The presumably oldest purple and green bacteria are autotrophic sulfur bacteria which, along with the cyanobacteria, have a limited capacity to utilize organic compounds. *Thiobacillus denitrificans* is an interesting bacterium in this regard in that it combines these characteristics with the presence of cytochromes cd and c_8, proteins that are widespread in pseudomonads and also present in some purple bacteria. It thus has more of the characteristics of the presumed ancestor of the photosynthetic bacteria than any other. Proof for the above speculation requires much more structural work both with the molecules described herein and for many others. Because gene transfer is an important element of bacterial evolution, we may only be able to trace the development of particular genes and pathways until we are able to

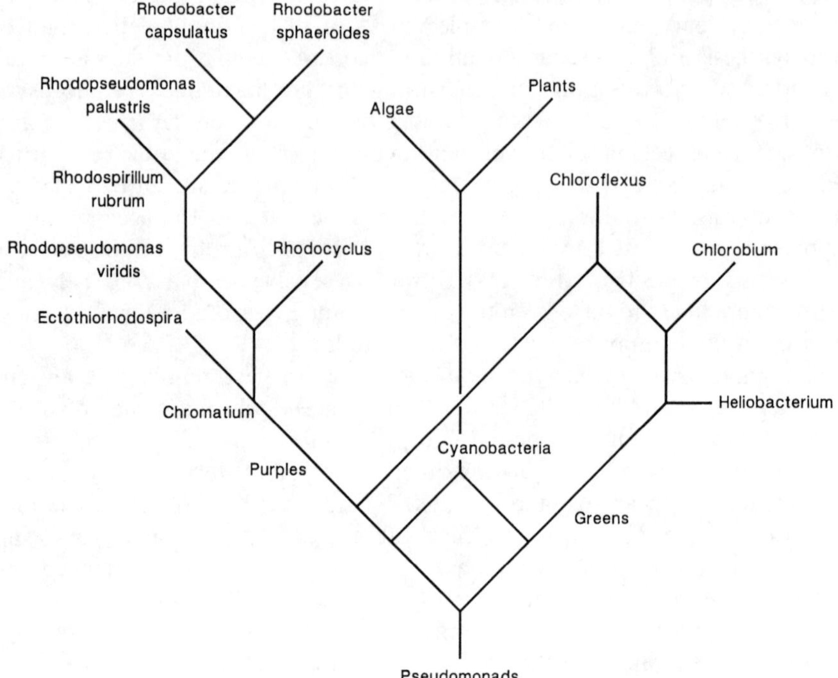

Figure 4.13 Summary evolutionary tree for photosynthetic bacteria. Branch lengths are arbitrary.

obtain complete genome sequences. We can be confident that the more data we obtain in the near future, the more incongruent will be our results until the data base is sufficiently enlarged. Only to the extent that we can recognize instances of gene transfer and determine their extent and direction will we eventually be able to outline the evolution of bacteria. At present, it is too early to tell whether we ultimately will be successful in that endeavor.

Acknowledgment

This work was supported in part by a grant from the National Institutes of Health, GM 21277.

References

1. Woese, C. R. *Microbiol. Rev.* 1987; *51*, 221–271.

2. Kawasaki, H., Hoshino, Y., Yamasato, K. *FEMS Microbiol. Lett.* 1993; *112*, 61–66.

3. Kawasaki, H., Hoshino, Y., Hirata, A., Yamasato, K. *Arch. Microbiol.* 1993; *160*, 358–362.

4. Olsen, G. J., Woese, C. R., Overbeek, R. *J. Bacteriol.* 1994; *176*, 1–6.

5. Gibson, J., Stackebrandt, E., Zablen, L. B., Gupta, R., Woese, C. R. *Curr. Microbiol.* 1979; *3*, 59–64.

6. Medigue, C., Rouxel, T., Vigier, P., Henaut, A., Danchin, A. *J. Mol. Biol.* 1991; *222*, 851–856.

7. Krawiec, S., Riley, M. *Microbiol. Rev.* 1990; *54*, 502–539.

8. Pfeifer, F., Griffig, J., Oesterhelt, D. *Mol. Gen. Genet.* 1993; *239*, 66–71.

9. Walsby, A. E., Haynes, P. K. *Biochem. J.* 1989; *264*, 313–322.

10. Takao, M., Kobayashi, T., Oikawa, A., Yasui, A. *J. Bacteriol.* 1989; *171*, 6323–6329.

11. Tindal, B. J., Truper, H. G. *System. Appl. Microbiol.* 1986; *7*, 202–212.

12. Margulis, L., Sagan, D. *Microcosmos. Four Billion Years of Evolution from Our Microbial Ancestors*, Summit Books, New York, 1986.

13. Meyer, T. E., Kamen, M. D. *Adv. Protein Chem.* 1982; *35*, 105–212.

14. Bartsch, R. G. Cytochromes. In *The Photosynthetic Bacteria*, Clayton, R. K., Sistrom, W. R. (eds.), Plenum Press, New York, 1978; pp. 249–279.

15. Bartsch, R. G. *Biochim. Biophys. Acta* 1991; *1058*, 28–30.

16. Ambler, R. P. *Biochim. Biophys. Acta* 1991; *1058*, 42–47.

17. Van Beeumen, J. *Biochim. Biophys. Acta* 1991; *1058*, 56–60.

18. Pettigrew, G. W., Moore, G. R. *Cytochromes c. Biological Aspects*, Springer-Verlag, New York, 1987.

19. Moore, G. R., Pettigrew, G. W. *Cytochromes c. Evolutionary, Structural and Physicochemical Aspects,* Springer-Verlag, New York, 1990.

20. Ambler, R. P. *Biochem. J.* 1973; *135*, 751–758.

21. Benning, M. M., Wesenberg, G., Caffrey, M. S., Bartsch, R. G., Meyer, T. E., Cusanovich, M. A., Rayment, I., Holden, H. M. *J. Mol. Biol.* 1991; *220*, 673–685.

22. Meyer, T. E., Cusanovich, M. A., Kamen, M. D. *Proc. Natl. Acad. Sci. USA* 1986; *83*, 217–220.

23. Dickerson, R. E. *Nature* 1980; *283*, 210–212.

24. Ambler, R. P., Daniel, M., Hermoso, J., Meyer, T. E., Bartsch, R. G., Kamen, M. D. *Nature* 1979; *278*, 659–661.

25. Rott, M. A., Witthuhn, V. C., Schilke, B. A., Soranno, M., Ali, A., Donohue, T. J. *J. Bacteriol.* 1993; *175*, 358–366.

25. a. Ambler, R. P., Daniel, M., Meyer, T. E., Kamen, M. D. *Biochimie* 1994; *76*, 583–591.

26. Jenney, F. E., Jr., Daldal, F. *EMBO J.* 1993; *12*, 1283–1292.

27. Dayhoff, M. O. *Precambrian Res.* 1983; *20*, 299–318.

28. Woese, C. R., Gibson, J., Fox, G. E. *Nature* 1980; *283*, 212–214.

29. Ambler, R. P., Meyer, T. E., Kamen, M. D. *Nature* 1979; *278*, 661–662.

30. Sneath, P. H. A. *Syst. Appl. Microbiol.* 1989; *12*, 15–31.

31. Benning, M. M., Meyer, T. E., Rayment, I., Holden, H. M. *Biochemistry* 1994 *33*, 2476–2483.

32. Dus, K., Tedro, S. M., Bartsch, R. G. *J. Biol. Chem.* 1973; *248*, 7318–7331.

33. Tedro, S. M., Meyer, T. E., Bartsch, R. G., Kamen, M. D. *J. Biol. Chem.* 1981; *256*, 731–735.

34. Ambler, R. P., Meyer, T. E., Kamen, M. D. *Arch. Biochem. Biophys.* 1993; *306*, 215–222.

35. Tedro, S. M., Meyer, T. E., Kamen, M. D. *J. Biol. Chem.* 1976; *251*, 129–136.

36. Tedro, S. M., Meyer, T. E., Kamen, M. D. *J. Biol. Chem.* 1979; *254*, 1495–1500.

37. Ambler, R. P., Bartsch, R. G., Daniel, M., Kamen, M. D., McLellan, L., Meyer, T. E., Van Beeumen, J. *Proc. Natl. Acad. Sci. USA* 1981; *78*, 6854–6857.

38. Ren, Z., Meyer, T. E., McRee, D. E. *J. Mol. Biol.* 1993; *234*, 433–445.

39. Ambler, R. P. Sequence Data Acquisition For The Study of Phylogeny. In *Developpements Recents dans l'Etude Chimique de la Structure des Proteine*, Previero, A., Pechere, J. F., Coletti-Previero, M. A. (eds.), Inserm, Paris, 1971; pp. 289–305.

40. Carter, D. C., Melis, K. A., O'Donnell, S. E., Burgess, B. K., Furey, W. F. Jr., Wang, B. C., Stout, C. D. *J. Mol. Biol.* 1985; *185*, 279–295.

41. Van Beeumen, J., Ambler, R. P., Meyer, T. E., Kamen, M. D., Olson, J. M., Shaw, E. K. *Biochem. J.* 1976; *159*, 757–774.

42. Ambler, R. P., Meyer, T. E., Kamen, M. D. *Arch. Biochem. Biophys.* 1993; *306*, 83–93.

43. Ambler, R. P., Bartsch, R. G. *Nature* 1975; *253*, 285–288.

44. Ambler, R. P. *Biochem. J.* 1963; *89*, 349–378.

45. Matsuura, Y., Takano, T., Dickerson, R. E. *J. Mol. Biol.* 1982; *156*, 389–409.

46. Van Grondelle, R., Duysens, L. N. M., Van der Wel, J. A., Van der Wal, H. N. *Biochim. Biophys. Acta* 1977; *461*, 188–201.

46. a. Samyn, B., De Smet, L., Van Driessche, G., Meyer, T.E., Bartsch, R. G., Cusanovich, M. A., Van Beeumen, J. J. *Eur. J. Biochem.* 1996; *236*, 689–696.

47. Barber, M. J., Trimboli, A. J., McIntire, W. S. *Arch. Biochem. Biophys.* 1993; *303*, 22–26.

48. Norris, G. E., Anderson, B. F., Baker, E. N. *J. Mol. Biochem.* 1983; *165*, 501–521.

49. Korszun, Z. R. *J. Mol. Biol.* 1987; *196*, 413–419.

50. Nar, H., Messerschmidt, A., Huber, R., Van de Kamp, M., Canters, G. W. *J. Mol. Biol.* 1991; *221*, 765–772.

51. McManus, J. D., Brune, D. C., Han, J., Sanders-Loehr, J., Meyer, T. E., Cusanovich, M. A., Tollin, G., Blankenship, R. E. *J. Biol. Chem.* 1992; *267*, 6531–6540.

52. Shiozawa, J. A., Lottspeich, F., Oesterhelt, D., Feick, R. *Eur. J. Biochem.* 1989; *180*, 75–84.

53. Dracheva, S., Williams, J. C., Van Driessche, G., Van Beeumen, J. J., Blankenship, R. E. *Biochemistry* 1991; *30*, 11451–11458.

54. Zuber, H. Structural Studies on the Antenna Complexes and Polypeptides of *Chloroflexus aurantiacus* and Other Green Photosynthetic Bacteria. In *Green Photosynthetic Bacteria*; Olson, J. M., Ormerod, J. G., Amesz, J., Stackenbrandt, E., Truper, H. G. (eds.), Plenum Press, New York, 1988; pp. 53–55.

55. Colman, P. M., Freeman, H. C., Guss, J. M., Murata, M., Norris, V. A., Ramshaw, J. A. M., Venkatappa, M. P. *Nature* 1978; *272*, 319–324.

56. Anthony, C. Quinoproteins and Energy Transduction. In *Bacterial Energy Transduction*, Anthony, C. (ed.), Academic Press, New York, 1988; pp. 293–316.

57. Durley, R., Chen, L., Lim, L. W., Mathews, F. S., Davidson, V. L. *Prot. Sci.* 1993; *2*, 739–752.

58. Ambler, R. P., Tobari, J. *Biochem. J.* 1985; *232*, 451–457.

59. Van Beeumen, J. J., Van Bun, S., Canters, G. W., Lommen, A., Chothia, C. *J. Biol. Chem.* 1991; *266*, 4869–4877.

60. Ubbink, M., Van Kleef, M. A. G., Kleinjan, D. J., Hoitink, C. W. G., Huitema, F., Beintema, J. J., Duine, J. A., Canters, G. W. *Eur. J. Biochem.* 1991; *202*, 1003–1012.

61. Adman, E. T., Turley, S., Bramson, R., Petratos, K., Banner, D., Tsernoglou, D., Beppu, T., Watanabe, H. *J. Biol. Chem.* 1989; *264*, 87–99.

62. Petratos, K., Banner, D. W., Beppu, T., Wilson, K. S., Tsernoglou, D. *FEBS Lett.* 1987; *218*, 209–214.

63. Hormel, S., Adman, E., Walsh, K. A., Beppu, T., Titani, K. *FEBS Lett.* 1986; *197*, 301–304.

64. Ambler, R. P., Daniel, M., Melis, K., Stout, C. D. *Biochem. J.* 1984; *222*, 217–227.

65. Cusanovich, M. A., Bartsch, R. G. *Biochim. Biophys. Acta* 1969; *189*, 245–255.

66. Van Beeumen, J. J., Demol, H., Samyn, B., Bartsch, R. G., Meyer, T. E., Dolata, M. M., Cusanovich, M. A. *J. Biol. Chem.* 1991; *266*, 12921–12931.

67. Dolata, M. M., Van Beeumen, J. J., Ambler, R. P., Meyer, T. E., Cusanovich, M. A. *J. Biol. Chem.* 1993; *268*, 14426–14431.

68. Van Beeumen, J., Van Bun, S., Meyer, T. E., Bartsch, R. G., Cusanovich, M. A. *J. Biol. Chem.* 1990; *265*, 9793–9799.

69. Chen, Z. W., Koh, M., Van Driessche, G., Van Beeumen, J. J., Bartsch, R. G., Meyer, T. E., Cusanovich, M. A., Mathews, F. S. *Science* 1994; *266*, 430–432.

70. Hopper, D. J., Bossert, I. D., Rhodes-Roberts, M. E. *J. Bacteriol.* 1991; *173*, 1298–1301.

71. McIntire, W., Singer, T. P., Smith, A. J., Mathews, F. S. *Biochemistry* 1987; *25*, 5975–5981.

72. Mathews, F. S., Chen, Z., Bellamy, H. D., McIntire, W. S. *Biochemistry* 1991; *30*, 238–247.

73. Pealing, S. L., Black, A. C., Manson, F. D. C., Ward, F. B. Chapman, S. K., Reid, G. A., *Biochemistry* 1992; *31*, 12132–12140.

74. Ambler, R. P. *Biochem. J.* 1968; *109*, 47p.

75. Voordouw, G., Brenner, S. *Eur. J. Biochem.* 1986; *159*, 347–351.

76. Denariaz, C. M., Liu, M. Y., Payne, W. J., LeGall, J., Marquez, L., Dunford, H. B., Van Beeumen, J. *Arch. Biochem. Biophys.* 1989; *270*, 114–125.

77. Jungst, A., Wakabayashi, S., Matsubara, H., Zumft, W. G. *FEBS Lett.* 1991; *279*, 205–209.

78. Doi, M., Shioi, Y., Morita, M., Takamiya, K. *Eur. J. Biochem.* 1989; *184*, 521–527.

79. Timkovich, R., Dhesi, R., Martinkus, K. J., Robinson, M. K., Rea, T. M. *Arch. Biochem. Biophys.* 1982; *215*, 47–58.

80. Van Spanning, R. J. M., Wansell, C., Harms, N., Oltmann, L. F., Stouthamer, A. H. *J. Bacteriol.* 1990; *172*, 986–996.

81. Okamura, K., Miyata, T., Iwanaga, S., Takamiya, K., Nishimura, M. *J. Biochem.* 1987; *101*, 957–966.

82. Silvestrini, M. C., Galeotti, C. L., Gervais, M., Schinina, E., Barra, D., Bossa, F., Brunori, M. *FEBS Lett.* 1989; *254*, 33–38.

82. a. Fulop, V., Moir, J. W. B., Ferguson, S. J., Hajdu, J. *Cell* 1995; *81*, 369–377.

83. Fenderson, F. F., Kumar, S., Adman, E. T., Liu, M. Y., Payne, W. J., LeGall, J. *Biochemistry* 1991; *30*, 7180–7185.

84. Nishiyama, M., Suzuki, J., Kukimoto, M., Ohnuki, T., Horinouchi, S., Beppu, T. *J. Gen. Microbiol.* 1993; *139*, 725–733.

85. Ye, R. W., Fries, M. R., Bezborodnikov, S. G., Averill, B. A., Tiedje, J. M. *Appl. Environ. Microbiol.* 1993; *59*, 250–254.

86. Sawada, E., Satoh, T., Kitamura, H. *Plant Cell Physiol.* 1978; *19*, 1339–1351.

87. Michalski, W. P., Nicholas, D. J. D. *Biochim. Biophys. Acta* 1985; *828*, 130–137.

88. Godden, J. W., Turley, S., Teller, D. C., Adman, E. T., Liu, M. Y., Payne, W. J., LeGall, J. *Science* 1991; *253*, 438–442.

89. Ronnberg, M., Kalkkinen, N., Ellfolk, N. *FEBS Lett.* 1989; *250*, 175–178.

90. Hanlon, S. P., Holt, R. A., McEwan, A. G. *FEMS Microbiol. Lett.* 1992; *97*, 283–288.

91. Goodhew, C. F., Wilson, I. B. H., Hunter, D. J. B., Pettigrew, G. W. *Biochem. J.* 1990; *271*, 707–712.

92. Pettigrew, G. W. *Biochim. Biophys. Acta* 1991; *1058*, 25–27.

92. a. Fulop, V., Ridout, C. J., Greenwood, C., Hajdu, J. *Structure* 1995; *3*, 1225–1233.

93. Stout, C. D. *J. Mol. Biol.* 1989; *205*, 545–555.

94. Jouanneau, Y., Meyer, C., Gaillard, J., Vignais, P. M. *Biochem. Biophys. Res. Commun.* 1990; *171*, 273–279.

95. Saeki, K., Suetsugu, Y., Tokuda, K., Miyatake, Y., Young, D. A., Marrs, B. L., Matsubara, H. *J. Biol. Chem.* 1991; *266*, 12889–12895.

96. Suetsugu, Y., Saeki, K., Matsubara, H. *FEBS Lett.* 1991; *292*, 13–16.

97. Duport, C., Jouanneau, Y., Vignais, P. M. *Nucleic Acids Res.* 1990; *18*, 4618.

98. Saeki, K., Suetsugu, Y., Yao, Y., Horio, T., Marrs, B. L., Matsubara, H. *J. Biochem.* 1990; *108*, 475–482.

99. Schatt, E., Jouanneau, Y., Vignais, P. M. *J. Bacteriol.* 1989; *171*, 6218–6226.

100. Morgan, T. V., Lundell, D. J., Burgess, B. K. *J. Biol. Chem.* 1988; *263*, 1370–1375.

101. Joerger, R. D., Bishop, P. E. *J. Bacteriol.* 1988; *170*, 1475–1487.

102. Moreno-Vivian, C., Hennecke, S., Puhler, A., Klipp, W. *J. Bacteriol.* 1989; *171*, 2591–2598.

103. Jouanneau, Y., Meyer, C., Gaillard, J., Forest, E., Gagnon, J. *J. Biol. Chem.* 1993; *268*, 10636–10644.

104. Mulligan, M. E., Jackman, D. M. *Plant Mol. Biol.* 1992; *18*, 803–808.

105. Buettner, M., Xie, D. L., Nelson, H., Pinther, W., Hauska, G., Nelson, N. *Proc. Natl. Acad. Sci. USA* 1992; *89*, 8135–8139.

106. Illinger, N., Xie, D. L., Hauska, G., Nelson, N. *Photosynth. Res.* 1993; *38*, 111–114.

107. Oh-oka, H., Takahashi, Y., Kuriyama, K., Saeki, K., Matsubara, H. *J. Biochem.* 1988; *103*, 962–968.

108. Adman, E. T., Siecker, L. C., Jensen, L. H. *J. Biol. Chem.* 1973; *248*, 3987–3996.

109. Hase, T., Matsubara, H., Evans, M. C. W. *J. Biochem.* 1977; *81*, 1745–1749.

110. Minami, Y., Wakabayashi, S., Yamada, F., Wada, K., Zumft, W. G., Matsubara, H. *J. Biochem.* 1984; *96*, 585–592.

111. Von Sternberg, R., Yoch, D. C. *Biochim. Biophys. Acta* 1993; *1144*, 435–438.

112. Mulligan, M. E., Buikema, W. J., Haselkorn, R. *J. Bacteriol.* 1988; *170*, 4406–4410.

113. Meyer, T. *BioSystems* 1994; *33*, 167–175.

114. Michel, H., Deisenhofer, J. (1988) *Biochemistry 27*, 1–7.

115. Liebl, U., Mockensturm-Wilson, M., Trost, J. T., Brune, D. C., Blankenship, R. E., Vermas, W. *Proc. Natl. Acad. Sci. USA* 1993; *90*, 7124–7128.

116. Krauss, N., Hinrichs, W., Witt, I., Fromme, P., Pritzkow, W., Dauter, Z., Betzel, C., Wilson, K. S., Witt, H. T., Saenger, W. *Nature* 1993; *361*, 326–331.

117. Deisenhofer, J., Epp, O., Miki, K., Huber, R., Michel, H. *Nature* 1985; *318*, 618–624.

118. Allen, J. P., Feher, G., Yeates, T. O., Komiya, H., Rees, D. C. *Proc. Natl. Acad. Sci. USA* 1987; *84*, 6162–6166.

119. Chang, C. H., El-Kabbani, O., Tiede, D., Norris, J., Schiffer, M. *Biochemistry* 1991; *30*, 5352–5360.

120. Michel, H., Weyer, K. A., Gruenberg, H., Dunger, I., Oesterhelt, D., Lottspeich, F. *EMBO J.* 1986; *5*, 1149–1158.

121. Williams, J. C., Steiner, L. A., Ogden, R. C., Simon, M. I., Feher, G. *Proc. Natl. Acad. Sci. USA* 1983; *80*, 6505–6509.

122. Williams, J. C., Steiner, L. A., Feher, G., Simon, M. I. *Proc. Natl. Acad. Sci. USA* 1984; *81*, 7303–7307.

123. Daldal. F., Bylina, E. J., Alberti, M., Begusch, H., Hearst, J. E. *Cell* 1984; *37*, 949–957.

124. Belanger, G., Berard, J., Corriveau, P., Gingras, G. *J. Biol. Chem.* 1988; *263*, 7632–7638.

125. Liebetanz, R., Hornberger, U., Drews, G. *Mol. Microbiol.* 1991; *5*, 1459–1468.

126. Nagashima, K. V. P., Shimada, K., Matsuura, K. *Photosynth. Res.* 1993; *36*, 185–191.

127. Weyer, K. A., Lottspeich, F., Gruenberg, H., Lang, F., Oesterhelt, D., Michel, H. *EMBO J.* 1987; *6*, 2197 2202.

128. Okkels, J. S., Kjaer, B., Hansson, O., Svendsen, I., Moller, B. L., Scheller, H. V. *J. Biol. Chem.* 1992; *267*, 21139–21145.

129. Burke, D., Hearst, J. E., Sidow, A. *Proc. Natl. Acad. Sci. USA* 1993; *90*, 7134–7138.

130. Menke, M. A. O. H., Liesack, W., Stackebrandt, E. *Arch. Microbiol.* 1991; *155*, 263–271.

131. Fox, G. E., Wisotzkey, J. D., Jurtshuk, P., Jr. *Int. J. Syst. Bacteriol.* 1992; *42*, 166–170.

132. Lockhart, P. J., Penny, D., Hendy, M. D., Larkum, A. D. W. *Photosynth. Res.* 1993; *37*, 61–68.

133. Woese, C. R., Achenbach, L., Rouviere, P., Mandelco, L. *Syst. Appl. Microbiol.* 1991; *14*, 364–371.

134. Arber, W. *Gene* 1993; *135*, 49–56.

135. Smith, J. M., Dawson, C. G., Spratt, B. G. *Nature* 1991; *349*, 29–31.

136. Sneath, P. H. A. *Int. J. Syst. Bacteriol.* 1993; *43*, 626–629.

137. Ambler, R. P., Meyer, T. E., Kamen, M. D., Schichman, S. A., Sawyer, L. *J. Mol. Biol.* 1981; *147*, 351–356.

138. Ambler, R. P., Meyer, T. E., Kamen, M. D. (1994) *Arch. Biochem. Biochem.* 1994; *308*, 78–81.

139. Breiter, D. R., Meyer, T. E., Rayment, I., Holden, H. M. *J. Biol. Chem.* 1991; *266*, 18660–18667.

140. Tedro, S. M., Meyer, T. E., Kamen, M. D. *Arch. Biochem. Biophys.* 1985; *241*, 656–664.

141. Tedro, S. M., Meyer, T. E., Kamen, M. D. *Arch. Biochem. Biophys.* 1985, *239*, 94–101.

142. Rayment, I., Wesenberg, G., Meyer, T. E., Cusanovich, M. A., Holden, H. M. *J. Mol. Biol.* 1992; *228*, 672–686.

143. Hoitink, C. W. G., Woudt, L. P., Turenhout, J. C. M., Van de Kamp, M., Canters, G. W. *Gene* 1990; *90*, 15–20.

144. Collyer, C. A., Guss, J. M., Sugimura, Y., Yoshizaki, F., Freeman, H. C. *J. Mol. Biol.* 1990; *211*, 617–632.

145. Hase, T., Wakabayashi, S., Matsubara, H., Evans, M. C. W., Jennings, J. V. *J. Biochem.* 1978; *83*, 1321–1325.

146. Backes, G., Mino, Y., Loehr, T. M., Meyer, T. E., Cusanovich, M. A., Sweeney, W. V., Adman, E. T., Sanders-Loehr, J. *J. Am. Chem. Soc.* 1991; *113*, 2055–2064.

From Dihydrogen to Dioxygen: Evolution of Biological Oxidation–Reduction

Robert J. P. Williams

5.1 Introduction

This chapter considers the development of biological redox reactions from their early beginnings to the present. I shall assume that the early Earth provided a milieu that contained available hydrogen and carbon in a variety of forms equivalent to HCHO, formaldehyde, as well as CO, CO_2, CH_4, and H_2 itself [1]. One polymeric equivalent of formaldehyde is any sugar $\{CH_2O\}_n$. Formaldehyde or sugars could have come about from the reaction

$$CO + H_2 \rightarrow HCHO$$

and acetyl groups could have arisen from the reaction

$$2\,CH_4 + 2\,CO \rightarrow 2\,CH_3CO- + H_2$$

We shall assume that many such forms of reduced carbon were present. Reactions such as these are common on metal surfaces or low-valence sulfides (and today in organo-metallic chemistry in solution). The early sea was constantly exposed to such inorganic materials at high temperatures and especially close to the edges of ocean vents. Therefore an early supply for life of reducing equivalents of low potential was found in many carbon compounds [2]. The H_2/H^+ potential in water at pH 7 is close to $-0.5\,V$. A free energy diagram for C/H/O compounds is shown in Figure 5.1, and given plenty of CO_2 and reducing equivalents, many prebiotic organic componds could be made.

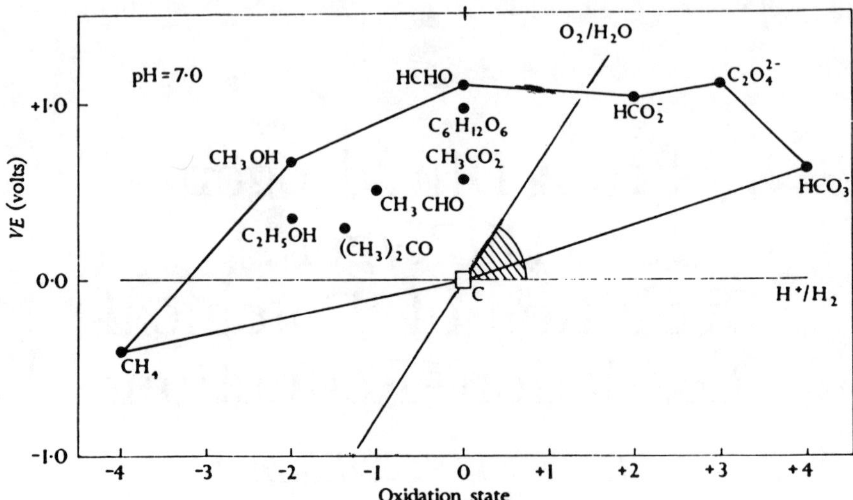

Figure 5.1 The relative stability of the standard states of carbon compounds plotted as the volt equivalent (VE) of each compound against its oxidation state. The VE is the free energy change in volts of reactions of the kind

$$C + 2H_2 \rightarrow CH_4$$

$$C + O_2 \rightarrow CO_2$$

and standard states are at unit pressure for gases or unit molarity for solutions at 298 K. The values are given for pH 7 so as to be useful in consideration of biological reactions or (roughly) of reactions in the sea. The H^+/H_2 potential shown is the standard potential at pH 0 (the potential for H^+/H_2 at pH 7 is -0.43 V). The oxidation state of any compound is conventionally the sum of twice the number of O atoms less the number of H atoms and negative charges that is, that of CO_2 is $+4$, of CH_4 is -4, and of HCO_2^- is $+2$. All points above 0.0 V are unstable to reduction by unit pressure of H_2 since stability increases downwards. Similarly O_2 is very unstable with respect to H_2O and all carbon compounds are unstable to unit pressure of O_2. Thus we write in standard states

$$HCHO + O_2 \rightarrow CO_2 + H_2O(HCO_3^- + H^+ \text{ at pH 7})$$

with a release of energy of $0.6 + 1.6$ VE (1 V = 13.8 kcal). In evolution from a reducing to an oxidizing atmosphere carbon compounds are always unstable except for CH_4 and CO_2. For further explanation see Ref. [5].

To gain the energy necessary for all biological syntheses [3] from hydrogen sources in redox reactions there had to be a sink for reducing equivalents. Again the early Earth provided various sources of energized sulfur-containing materials that may have acted as such a sink for electrons. An example is pyrites (FeS_2), which can react as follows:

$$H_2 + FeS_2 \rightarrow FeS + H_2S$$

The redox potential of sulfur to sulfides could be as high as $+0.0$ V. In other words, redox states of the prebiotic Earth were far from equilibrium, so that energy could be obtained via some combination of components [2, 4].

5.2 Redox Potentials of Nonmetals and Metals

Before seeing how biological systems could make use of this energy difference, we must look at the redox state of a variety of other nonmetals and metals. We can thus see the readily possible chemistry in the redox range -0.5 (H_2/H^+) to 0.0 V (S_n/H_2S) early in evolution [5]. Figure 5.2 shows that few nonmetal oxides have potentials in this range, so that initially we exclude from redox reaction considerations the oxides of N, O, F, Cl, Br, I, and Se as having too high a redox potential and reduced oxides of B, Si, P, Ge, and As as having too low a redox potential. The only nonmetal oxides of interest are those of C and H and possibly S. The hydrides of interest are really only those of C, N, O, S, and Se. Thus very few nonmetals were of value in redox reactions in early biology. We shall see too that even some carbon compounds have such high redox states that they could be utilized only later in evolution. Effectively the low redox potential excluded incorporation of covalently bound F, Cl, Br, and I. Although B, Si, and P could be incorporated (by condensation), none could take part in redox chemistry.

The low redox potential also limited the supply of metals especially since H_2S was plentiful and precipitated many sulfides (Fig. 5.3) [6]. The metals readily available included nonredox metals such as Na, K, Mg, and Ca but available levels of Zn and Cd would have been very low. The metal elements of Group 3 and 4 such as Al, Sc, and Ti are very insoluble as oxides and are still today of low availability. The redox active metals available would have included therefore V, Mn, Fe, Co, Ni, and perhaps Mo but not Cu. Copper has a very insoluble sulfide. Some of the not so "available" metals were often present as multimetal species or even colloids of sulfides, for example, Fe, Co, and Ni whereas others such as V and Mo were present as complex sulfido anionic species, for example, MoS_4^{2-}. In this early chemistry sulfide dominated redox metal chemistry and perhaps few redox metals could escape from association with it at least in part. An exception is Mn (see Fig. 5.3).

The redox potentials of the metals under consideration and in the range -0.5 V to $+0.0$ V are seen in Figure 5.4, taking their chemical form to be that in water at pH 7 [5]. Immediately we see that in aqueous solution it is complex formation or precipitation only that brings into play the redox chemistry of any metals except V and Mo, as the redox potential must be below 0.0 V. (Here, that is, in water (Fig. 5.4) complex formation or precipitation means just the formation of oxides or hydroxides.) In fact even if we consider hydroxide formation only Fe has the redox couple Fe^{2+}/Fe^{3+} in the right range of the initial environment of Earth (0.0 to -0.5 V). It is a general truth that the M^{2+}

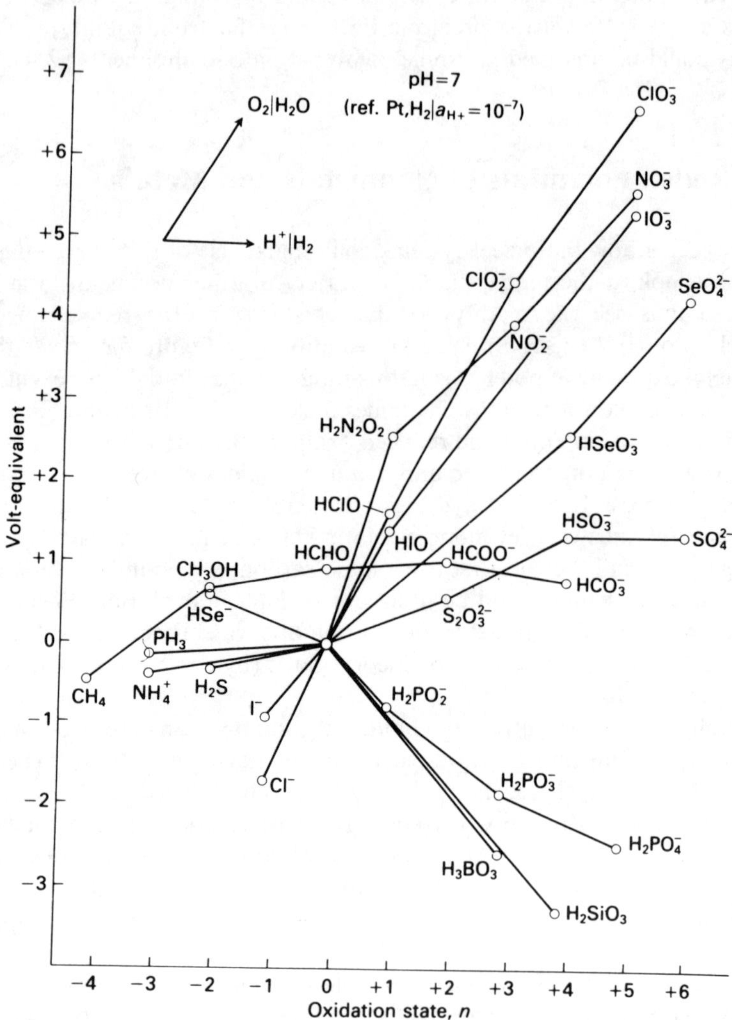

Figure 5.2 The relative free energies in volts of the oxidation states of nonmetals at pH 7.0 (See legend to Fig. 5.1. for explanation.)

state at pH 7 and in water is very stable relative to all other oxidation states for Mn, Fe, Co, and Ni. Thus the value of these metals as biological catalysts may well have started from their sulfide complexes and clusters which we know in the case of Fe and Ni can have redox potentials below $+0.0$ V especially in protein complexed form [7]. The early iron systems were then ferredoxins and those of nickel were the sulfide complexes in hydrogenases (H_2/H^+). Mn and Co could not have been used in this form. However, there are other ways apart

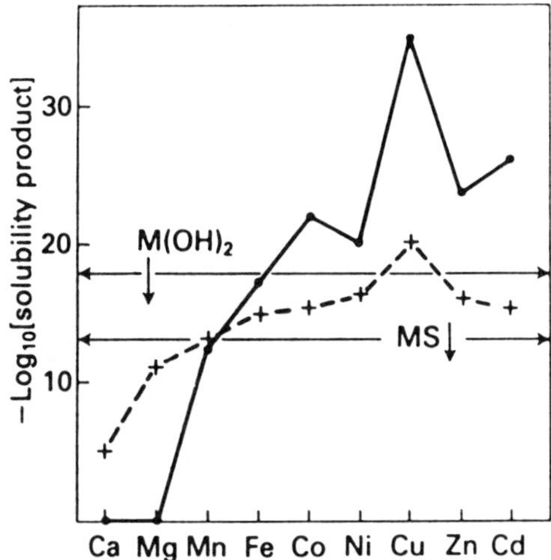

Figure 5.3 The solubility products of divalent metal sulfides (full line) and hydroxides (broken line). The horizontal lines divide the elements at $10^{-4} M$ precipitated at pH 7 as hydroxides or as sulfides in the presence of $10^{-3} M H_2S$. Note: MoS_2 is rather insoluble too. For further explanation see Refs. [5] and [6].

from sulfide binding of moving the potentials into the useful range, for example, by the synthesis of and complex formation with tetrapyrroles. It is not known when these organic chelates first appeared but it could have been very early indeed, as Eschenmoser [8] has shown that they could have been formed in the prebiotic era. All three metals — Fe(heme), $Co(B_{12})$, and Ni(F-430) — when complexed in their specific tetrapyrrole derivatives, have potentials in the required range of 0.0 to -0.5 V. Note too that these molecules can be soluble in lipid membrane phases, without binding to proteins. Of the redox metals this leaves only Mn not present in sulfide or porphyrin complexes. The significance of this observation may well be that it allowed Mn to become the central metal ion in the catalysis of the $H_2O \rightarrow O_2$ reaction (see below). Mn^{2+} forms only relatively weak complexes and a soluble sulfide whereas Mn^{3+} and Mn^{4+} form much stronger hydroxy-/oxy- complexes but not with sulfides.

Before we turn to the functional value of these redox systems we must guess at the elements present in early organisms from the above chemical information and our analytical knowledge of anaerobic bacteria (prokaryotes) and archaebacteria today. They are known to contain (Mn), Fe, Co, Ni, Mo, W, and (V). Of the nonmetals other than H, C, N, O, and S, which are redox active, only Se has been found. Knowing the composition of the early sea all these elements as well as Na, K, Mg, Ca, Si, and P would have been of potential value. Note Group III elements, eg. Al, are missing to this day. No form of life is known that does not have at least 15 elements from the above list (Fig. 5.5).

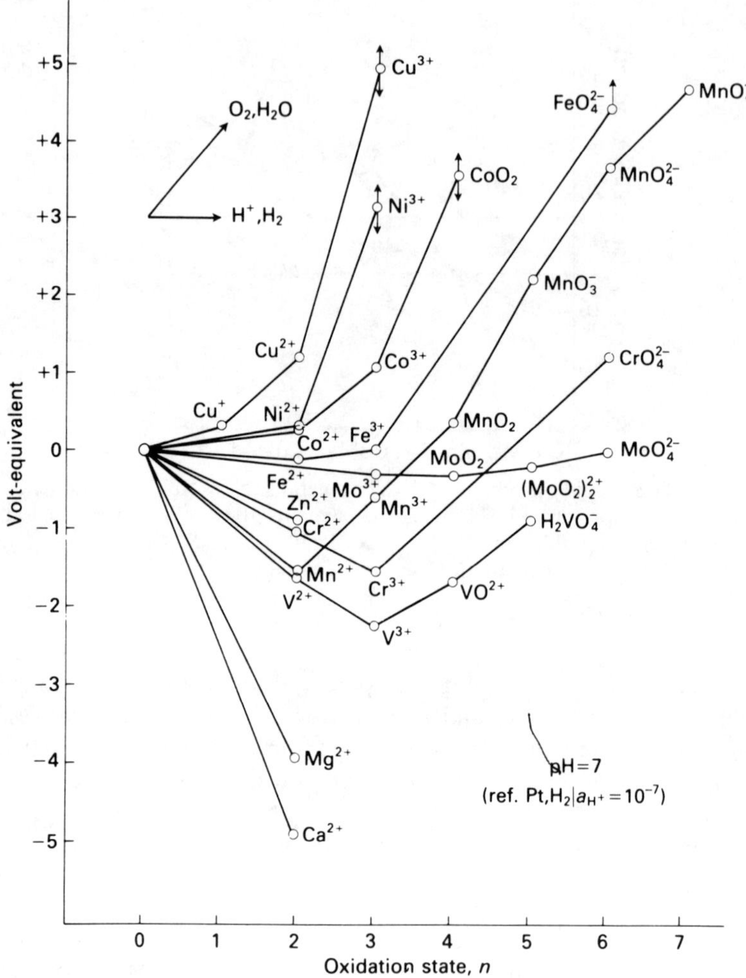

Figure 5.4   The relative free energies in volts of the oxidation states of metals at pH 7.0. (See legend to Fig. 5.1 for explanation.)

5.2.1 Functional Significance of Elements

In simple early prokaryotes there was but one inner compartment, the cytoplasm. To this day the cytoplasm of all cells has a redox potential range controlled partly by the pH of around 7.0 and partly by the reactions based on $-S-S-/(-S^-)_2$ at around 0.0 V and H^+/H_2 (NADH) at around -0.5 V. The buffering of pH, important for redox activity, is managed by a combination of nonmetal ionizations of protein side chains, many phosphates, and bicarbonate, all of pK_a around 6.5. The redox potential control of all cells

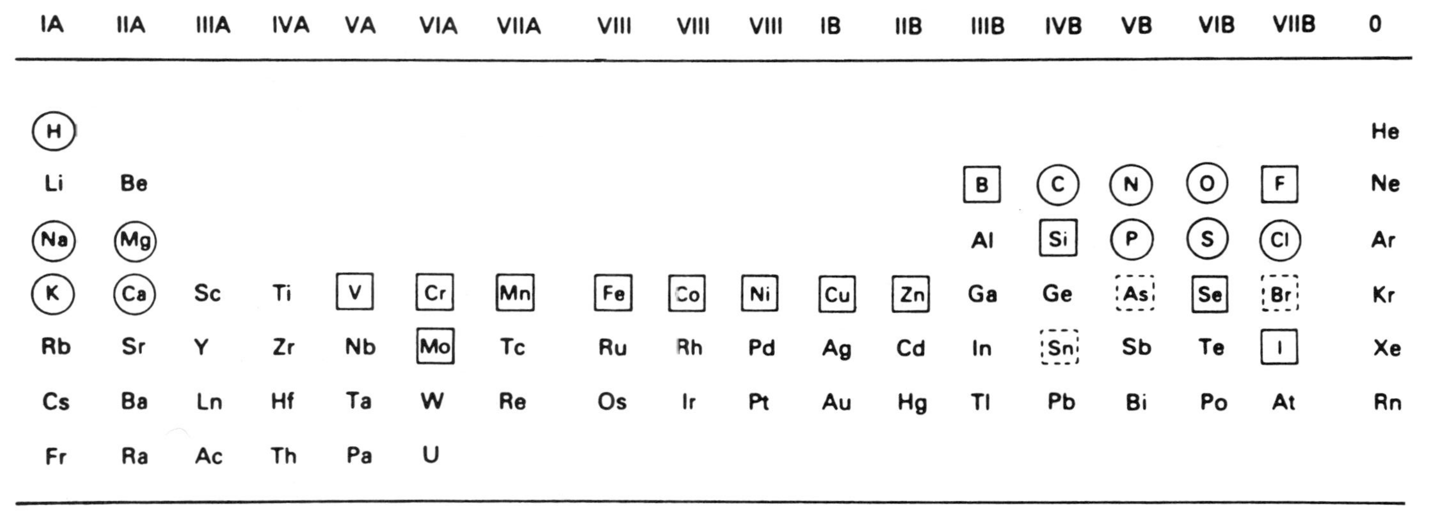

Figure 5.5 The elements of the periodic table and their relevance in life systems.

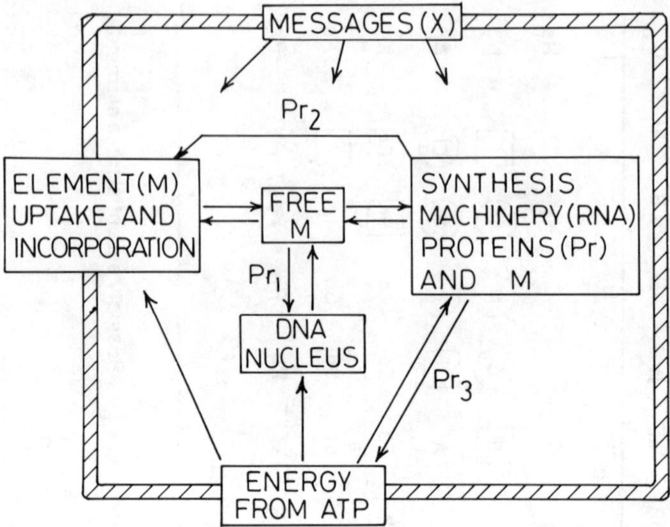

Figure 5.6   A schematic diagram of the many ways in which elements (M) are involved, with proteins (Pr) especially, in *all* cellular activity. There is no life with fewer than 15 elements, M.

known to us is partly managed by mobile cofactors linked to H_2/H^+ such as NAD/NADH (-0.45 V) and S_n/S^- ($+0.0$ V) within glutathione. The essential character of the interior of every primitive cell was (and is) that it could (and can) metabolize by pairs of redox reactions in this range [9] very simple and basic inorganic compounds, for example, H_2, CO, CH_4, CO_2, NH_3, and N_2, using energy, perhaps originally from the ocean trenches, to make polymers (Fig. 5.6). The functional need for other elements has to be related to the fundamental kinetic stability of these nonmetal elements in the biological polymers. It is not possible to activate the nonmetal small molecules without the aid of the above redox active metals. Some of the enzymes known to be

Table 5.1   Some Primitive Metal-Requiring Enzymes

Enzymes	Metals
Hydrogenases	Ni/S, Fe/S
CO handling	Ni[a]
Citric acid cycle	Fe/S, Fe[a]
Nitrogenase	Fe/S, V/S, Mo/S
Ribonucleotide reductase	Fe/S, Co[a], Fe

[a]In prophyrin-like ligands.

involved are given in Table 5.1 and incorporate the expected metals, very frequently using sulfur and/or porphyrin donors.

Now apart from simple enzymic reactions for synthesis and degradation there was the need to capture energy, especially for condensation reactions in order to make all the polymers of biology. We assume that cells were membrane limited by lipid material and that sugars were available (see the introduction). It is very likely that a major early route to obtain energy used the available sulfides (or possibly porphyrins) as catalysts in membrane-bound metal complexes much as we see them today in Complex I and III of the mitochondrial electron transfer chain. A very early event would then be the establishment of a pH or electrostatic potential gradient, and effectively a redox gradient across the membrane particle of these complexes (Fig. 5.7) [10, 11]. The energy source is then the positioning of catalysts so as to allow a gradient of metabolizable hydrogen, from sugars as H^-, to produce OH^- and H^+ on either side of the membrane utilizing diffusible reactants. The positioning of redox catalysts in space must have been a fundamental advance early in evolution, generating isolated zones in cells at special redox potentials. The complexity of the system could then grow because it had a built-in energy source, the proton gradient, for the pumping of ions and molecules and for the synthesis of ATP (pyrophosphate) [10, 11]. A plausible description of these events is beyond the scope of this chapter.

It is worth noting the types and related value of the different organic coenzymes that are known to have been present in these very early organisms. As mentioned, at the lowest potential around -0.45 V there was the NAD$^+$/

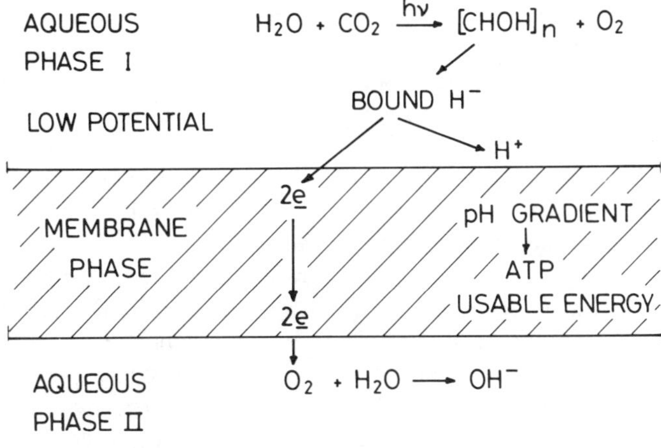

Figure 5.7 The way in which redox reactions of carbon compounds are linked via redox reactions to pH gradients and then to ATP formation. S_n, SO_4^{2-}, NO_3^- and so on could be used in place of O_2.

NADH couple. This coenzyme is water soluble and acted, as it does to this day, to distribute reducing equivalents as H^- between metabolic pathways. Next is a very different coenzyme with a potential nearly as low, namely flavin. The flavins are *bound* cofactors of proteins and do not exchange. Further on in the hierarchy of redox potential are the quinones, at close to 0.0 V, which given their long fatty chains, function in lipid layers to distribute H^- much as NADH does in water. These are three of the major organic redox active cofactors together with glutathione, which is also in the aqueous phase, but is now at the relatively high redox potential of $\sim 0.0\,V$. Also present, even in the very earliest organisms of which we are aware, was a variety of organic cofactor means of distributing fragments such as of C_1 (CH_3-, $-CHO$); of $C_2(CH_3CO-)$; of $N(-NH_2)$; and of $P(ATP)$. We shall not describe this distribution of fragments of carbon compounds of different redox potential here [3]. Thus sometimes metal complexes were used to activate or carry redox-active fragments for chemical synthesis and degradation while sometimes organic moieties were used. We must see the earliest redox systems especially as ways of distributing H^+ so as to gain energy, to pump required molecules and ions, and as a source of synthetic construction.

A feature of these early cells, as far as we know them today, is that as well as developing degradative and synthetic machinery, distribution devices, and energy capture devices using redox-active metals they also developed control and regulatory systems over redox as well as acid–base reactions. In the evolution of the "best" it is obvious that such control devices will appear, given time; otherwise the cells will waste much energy in uncontrolled activities. The controls and regulation must work between uptake mechanisms of, say, metals, the production of the protein components of the cell's machinery, and the activity of this degradative/synthetic machinery (Fig. 5.6). The components of the redox machinery are both metal ions and proteins in stoichiometric complexes so both must be controlled *in proportion*, selectively yet cooperatively. It is necessary to have a certain amount of a given protein to go with metal M_1 and a certain amount of another protein to go with metal M_2 but it is also obvious that M_1 and M_2 must be present approximately in well-controlled amounts. This requires feedback control. Let us consider first the controls and regulation of iron metabolism because iron was and is a dominant element in biology as far as redox reactions are concerned.

(In passing, note that given the nature of this chapter we have made no mention of the further essential requirements of very primitive cells to manage approximately osmotic regulation and charge balance. Both are under the control of the movement of simple cations and anions across cell membranes but there is little or no redox chemistry involved. The major simple ions concerned were and are Na^+, K^+, and Cl^- while many organic anions were retained by the cell membrane. We shall not describe the controls over these gradients but will analyze the need to control the concentrations of the redox-active elements).

5.3 Early Cellular Metabolism

5.3.1 Early Redox Reactions, Uptake, Control, and Regulation

The scheme of feedback control over iron and probably manganese uptake today (aerobic life) using siderophores in bacteria is now well documented [12, 13]. An example is shown in Figure 5.8. The level of free Fe^{2+} (and Mn^{2+}?) in a cell is monitored by a metal binding (FUR) protein [12]. The protein also binds to DNA and when uncomplexed by iron (manganese) shuts off part of the synthesis of enzymes for metabolism and synthesis, for example, the citric acid cycle. Thus in the absence of excess $Fe^{2+}(Mn^{2+})$, FUR is an apoprotein, and then allows the synthesis of siderophores for Fe uptake. This is one homeostatic device. [Note it is coupled to oxidative O_2-dependent metabolism and this regulation could not have been present in the early history of cells. It is given here as the best documented example of regulation by redox metal ions even though the citrate cycle may have operated in reverse in primitive organisms, when Fe^{2+} may have been in plentiful supply].

An alternative scenario for the early control of iron levels without the production of siderophores relies on the *equilibria* between iron and sulfide in the environment [13]:

$$Fe^{2+} + HS^- \rightleftharpoons FeS + H^+$$

and

$$n\,Fe^{3+}S^{2-} + \frac{n}{2}\,S^{2-} \rightleftharpoons n\,Fe^{2+}S^{2-} + \frac{n}{2}\,S$$

There are multiple equilibria of this kind including many Fe_nS_n polymers that have been incorporated into protein polymers and precipitates from the earliest times. To this day the standing free Fe^{2+} in cells may well be related to the controlled synthesis of protein incorporating Fe_nS_n clusters and precipitates in some cases [14]. There is also the possibility that the Fe_nS_n units or precipitates dissociate and equilibrate in free solution. If this system is to regulate as well as to control, then one of these systems of free ions or free clusters must bind to a DNA binding protein and turn on or shut off protein synthesis. It is highly probable that such systems are operative today, in both prokaryote and eukaryote cells, for example, in the FNR gene system, and the association–dissociaton of Fe or Fe/S units may well be redox state dependent [14].

Such major internal signalling systems of prokaryotes are related to the acid–base metabolism too through the phosphate signalling pool via ATP synthesis as well as to the redox metabolism through iron or sulfur signalling. The phosphate signals are further related to the Na, K, Ca, and Mg levels through pumps, to polymer synthesis, and to the metabolic pathways of

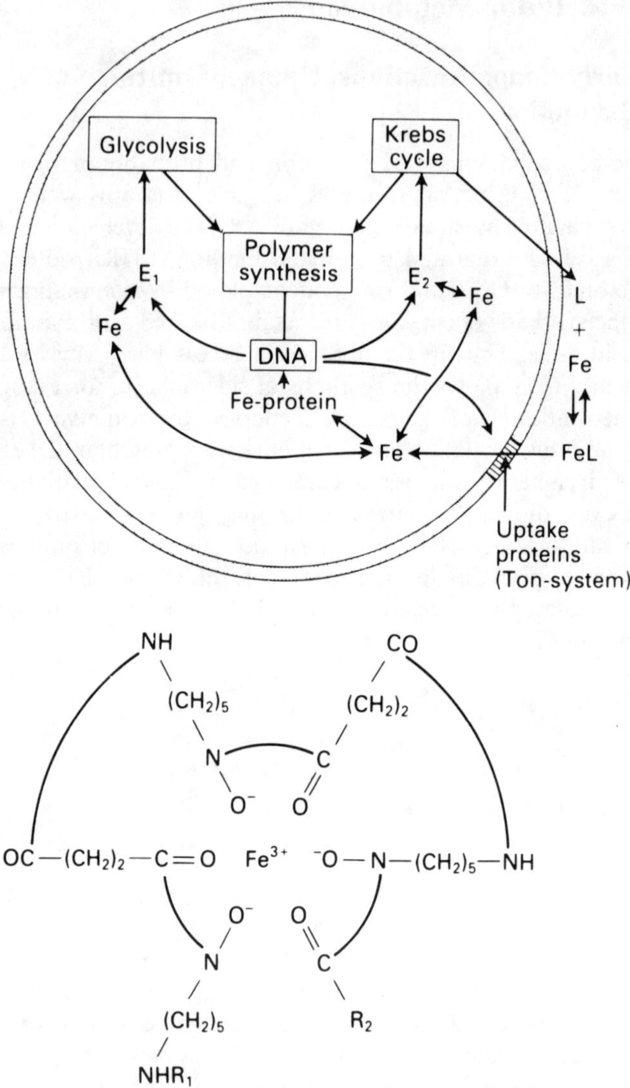

Figure 5.8   An example of feedback control over uptake, here of iron. The Fe-protein (FUR) as an apoprotein switches on scavenger, L, synthesis and the synthesis of uptake (Ton) systems while switching off Krebs cycle reactions. The iron protein does the reverse. Below, a typical siderophore scavenger.

C/H/N/O such as the glycolytic pathway and citrate cycle. Iron was always also related to the citrate cycle and fatty acid synthesis. The generalization follows that redox and acid/base element levels in cells are closely controlled and *mutually* correlated. The mutual interaction of the phosphate and the iron signals today, for example, is known to be as follows [12]:

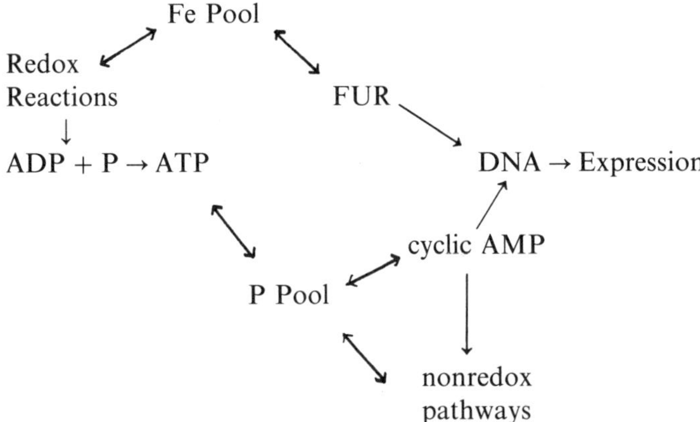

but even the use of Fe–S systems may have been controlled by phosphorylation. In this case iron and phosphate homeostasis would have been extremely interactive controls, especially in primitive cells.

The intention of this section is to show that the primitive redox systems were not just part of metabolism but were totally integrated in every aspect of cellular activity.

5.3.2 Heme Feedback Control

When the porphyrin-bound forms of metal ions were synthesized the slow exchange of iron effectively made free iron and heme (Fe porphyrin) into two different metals just as NADH and quinol are two different forms of bound H. A separate feedback relationship developed between $Fe^{2+}(Fe_nS_n)$ and DNA and between heme and DNA. The heme feedback protein system is called CYT and it regulates the production of many cytochromes. Once again this regulation requires control over porphyrin synthesis as well as iron uptake and protein synthesis. The value of two very different iron pools is that they can be linked to quite independent redox active zones and compartments In prokaryotes the compartments are cytoplasmic, membrane, and periplasmic (Fig. 5.9). As an example of this compartmentalization of iron proteins, cytochromes c and c' became prevalent in the periplasm. (Other metal ions may also have two well separated functional pools that evolved differentially in time e.g., Ni and Mo).

5.3.3 Early Sensing and Signalling

The periplasmic space, which developed in prokaryotes, is a zone that is partially linked to the cytoplasm and partially to the environment via two membrane systems (Fig. 5.9). Its redox potential and pH are poorly controlled. It follows that this zone can contain sensing equipment unrelated to metabolism and that messages can be passed from the environment via the periplas-

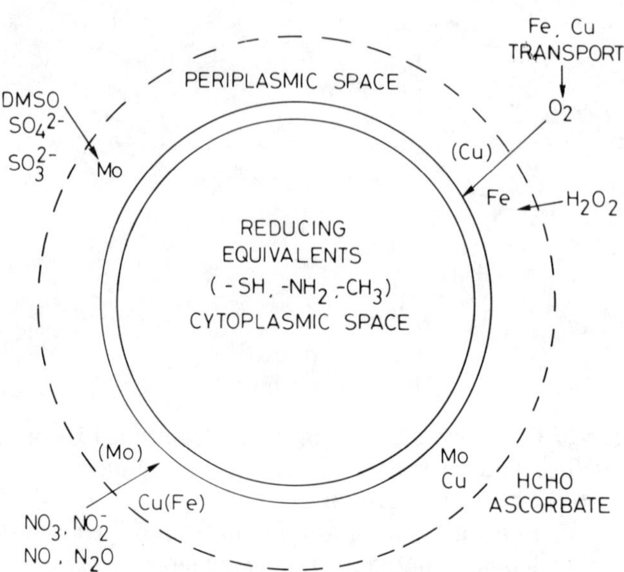

Figure 5.9   The distribution of some Cu and Mo enzymes in modern prokaryotes illustrating the separation of element functions between cytoplasm and periplasm.

mic space to the interior of the cell. There are some proteins in the periplasmic space that appear to behave in this way. Notably cytochrome c' in this space is an allosteric dimer that responds to CO (and NO) but not to O_2. Sensing of gradients of chemicals in the environment is a useful refinement for a cell provided that it can then swim along the gradient to food or to a safer place. Note that all cytochromes c belong in the periplasmic space and came very early in evolution. The NO and CO sensors of today's signals maybe related to the cytochrome c' systems [15, 16].

Of course physical sensing can be extended further into the environment by filaments and by a network of appendages linked to them. Single cells developed a variety of proteins and polysaccharides that gave them the ability to swim freely, to attach themselves to surfaces, and to provide protection. The design of these extracellular systems was such that they interacted with calcium ions which could then act as crosslinks strengthening the outermost regions of the cell or could even be used to link the organic matrix to inorganic solids such as calcium salts or silica to make rough shells out of composites. We mention this situation of rough and ready building outside the early cells so as to contrast it with later systematic extracellular building, often dependent on dioxygen reactions making crosslinks of several kinds (see below). In particular, $-S-S-$ links could not have been stable outside early cells.

These extracytoplasmic constructs require the ability of special polymers, proteins especially, to be directed into the appropriate part of space. Redox systems on their own are useful but spatial positioning gives dimension to their value.

All the above features are mentioned because the initial build-up of dioxygen affected them all very greatly.

5.4 Summary of Anaerobic Redox Systems

In this section I must make it clear that the redox chemistry of C, H, N, and S as well as the absence of such redox chemistry for many elements such as O, P, Cl, Na, K, Mg, and Ca plus the catalytic activity of the redox states of Fe, Co, Ni, W, V, and Mo made it possible to generate all the pathways and products of primary metabolism including synthesis of proteins, RNA, DNA, and polysaccharides. However, if we omit almost any one of the above elements we cannot describe even a primitive living cell no matter what the progression of evolutionary steps among polymers. It may well be that other elements, especially Mn and Zn, played minor but essential roles too, in RNA and DNA synthesis for example. Zinc, unlike manganese, would have been of very limited availability because of the presence of sulfide but it is a very useful acid–base catalyst. It is against this background that the influence of a dioxygen atmosphere has to be seen. It is clear that relatively speaking dioxygen affected this primary (cytoplasmic) metabolism of cells very little.

Many other changes took place in cells even before or perhaps at about the same time as the initial build-up of dioxygen, especially the development of filaments and of internal compartments that led to the eukaryotes. As far as one can know, this evolution of the eukaryote cell, linked to changed use of calcium, brought about isolation of many functions away from the cytoplasm and separated from the nucleus. The development of compartments meant that new redox and pH compartments and membrane zones could evolve and with them new dispositions of redox-active metals and cofactors. The development of filaments inside cells does not appear to required a change in redox potential and it may have been this change that allowed cells to develop internalized vesicles. Certainly the filaments changed the stability of the membrane by supplying an underlying network of polymers rather than an external coat. However, many eukaryotic cells differ in other ways from prokaryotes with regard to these membranes. Thus the development of cholesterol, which some of them have, had to await a special pathway of synthesis using dioxygen.

Small prokaryotic cells, such as anaerobic bacteria, operating at low internal redox potential are the progenitors of present cells but the peculiarities of their membranes and compartments appear to have precluded much change of shape of cells from that of a cylinder. There were, however, numerous later larger eukaryote anaerobic cells showing a high degree of structural development but apparently there were no true multicellular organisms. Thus we can be reasonably confident that the evolution of multicellular life depended on the advent of dioxygen.

5.5 The Advent of Dioxygen

5.5.1 Energy, Light Capture, and Dioxygen Production

If we believe that life started in the dark trenches of the oceans, one possibility if the origin of life did not require light, we must next introduce the switch to light capture [3]. Again this was a major event that eventually altered the redox scale available to biological systems from $-0.5/0.0$ V to $-0.5/+0.8$ V. It is this event that led to dioxygen production. It occurred in the prokaryotes and seemingly was dependent on development of pigments inside membranes. We do not know the succession but today there are a variety of means based on rhodopsins, chlorophylls (i.e., tetrapyrroles), and perhaps on carotenoids. The scale of light energy is greater than 2.0 V and is seen to be large enough to overlap H_2/H^+ and O_2/H_2O redox potentials, a gap of 1.3 V (Fig. 5.2).

We know that at some period around 2 billion years ago dioxygen began to accumulate considerably in the atmosphere (Fig. 5.10). We know of several possible mechanisms but they all depend on photochemistry [17]. In other words, the escape from the redox conditions of the early Earth depended on the excitation of photoreactive centers to higher redox states. In place of the charge separation due to the $H_2/2H^+$ and S_n/S^{2-} there evolved (or there was initially present in the sea on a very small scale) systems for absorption of light

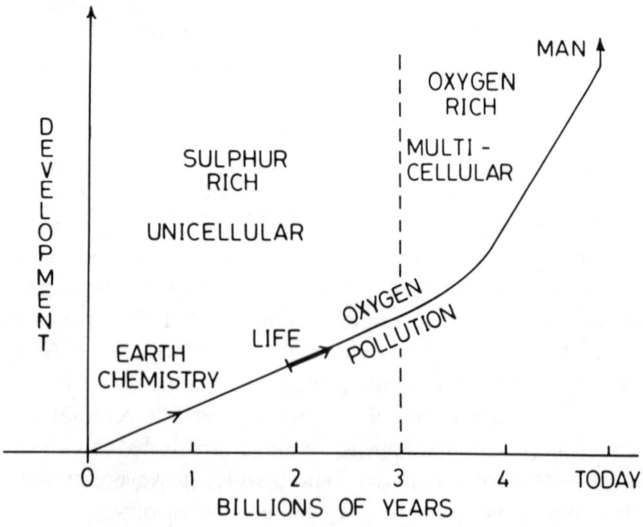

Figure 5.10   The change in the Earth's environment with time and its effect on biological evolution.

of the kind [17]:

$$Fe^{3+}X \xrightarrow{h\nu} Fe^{2+} + X^{\cdot +}$$

which would generate dioxygen. The simplest would be

$$Fe^{3+}OH^- \xrightarrow{h\nu} Fe^{2+} + OH^{\cdot}$$

followed by

$$4\,OH^{\cdot} \rightarrow 2\,H_2O + O_2\uparrow$$

Fe^{2+} could react with sulfur to return free Fe^{3+} and extra sulfide. This is not a biological reaction path but could have produced both H_2O_2 and O_2, which were highly poisonous for all early life. There is an alternative very similar possibility that there evolved a special organic molecule dimer that rested in a biological membrane. On exposure to light it was able to give a diradical:

$$XX \xrightarrow{h\nu} {}^{+\cdot}XX^{\cdot -}$$

Such molecular dimers, in *internally* disproportionated high and low redox states, could fluoresce, when the energy would be lost, but in the asymmetric membrane described previously electron transfer could and did sooner or later generate a simple charge separation. The system that has developed today is of course X = chlorophyll, a derivative of porphyrin and soluble in lipids.

$$M_1 \qquad XX \qquad M_2$$
$$\downarrow h\nu$$
$$M_1 \qquad {}^{+\cdot}XX^{\cdot -} \qquad M_2$$
$$\downarrow$$
$$M_1^+ \qquad XX \qquad M_2^-$$

The scheme is very close to that of photoabsorption by chloroplasts today. If this history is correct, energy capture from light replaced energy capture from preexisting unstable chemical gradients. The above reaction scheme is that of photosystem I. If we couple this photo charge separation with a supply of reducing equivalents from S^{2-} and supply of oxidising agents H^+ then we have

$$M_1^+ \qquad XX \qquad M_2^-$$

$$e \qquad\qquad\qquad\qquad e$$

$$S_n\downarrow \leftarrow S^{2-} \qquad\qquad H^+ \rightarrow \text{bound H or } H^-$$
$$NAD^+/NADH$$

with a return of the photo-centre to its initial state. This could have been the first photosystem but it wastes much of the photoenergy. If we replace S^{2-} by OH^- then the OH^- gives $OH^{\cdot}$ and as before H_2O_2 and O_2 are subsequently produced, then much more of each light quantum is used. This is photosystem II of thylakoids today. The M_1 most suitable for O_2 production on Earth is Mn because it was not bound to sulfur centers; moreover Mn^{2+} does not bind O_2 and $2\,MnO^{2+}$ can give O_2 (see Fig. 5.3). (A third possibility involves light-induced charge separation through conformational switches as in bacteriorhodopsin but this does not involve redox steps and is not described further here).

Now at first the new products H_2O_2, $O_2^{\cdot-}$, and O_2 were highly toxic and it could be that the next step in evolution was detoxification rather than use of the products. In fact we find to this day simple detoxifying enzymes for H_2O_2 and $O_2^{\cdot-}$ in the cytoplasm of prokaryotes and organelles that do not use O_2 to a large extent [18]. They are based on simple Fe and Mn centers. Several of these superoxide dismutases and catalases interestingly are now under the gene regulation of FUR, the iron uptake gene system [12], and hence linked to Fe(Mn) regulation. These developments may well have preceded the use of dioxygen as a *valuable* chemical [18]. Note the connection between Mn and O_2 production and very early detoxification from $O_2^{\cdot-}$ and H_2O_2.

5.5.2 Metabolism and Dioxygen

It is quite likely that prokaryotes subsequently evolved a variety of redox mechanisms to make organic poisons from O_2 and to counter the poisons made by other organisms. We can include all of the production of these organic moeities under secondary metabolism and they range from the simplest oxides such as NO, ClO^-, $O_2^{\cdot-}$, and H_2O_2 to the most complex products such as penicillins, halogenated organic compounds, epoxides, hydroxysterols, alkaloids, and so on [19]. Chemical warfare undoubtedly broke out on a very large scale with the presence of dioxgen. It is a general feature of evolution that a new product is of its very nature a danger to existing organisms, frequently increasing the mutation rate and leading slowly first to species able to detoxify their surroundings before valuable use comes into play. Certainly the prokaryotes have developed the ability to produce O_2, to detoxify its products, and to use it in many ways for synthesis and in energy capture.

5.5.3 Changes in Inorganic Chemistry with O_2

We need to see that the advent of O_2 and such reagents as H_2O_2 increased the possible range of redox potentials of organic compounds but it also *gradually* raised that of all chemical elements toward nearly $+1.0\,V$ (Figs. 5.1, 5.2, and 5.4). Not only did the atmosphere of dioxygen produce higher oxidation states of many metals and nonmetals but it also removed from regions exposed to dioxygen the strongly reducing elements in the form of gases

H_2S, H_2Se, H_2, CH_4 and so on or in the form of ions, for example, $HS^- \rightarrow SO_4^{2-}$ ions. The removal of these gases and chemicals occurred and occurs in surface waters and on land but not in deep water or other extremely reducing zones of stagnant water. As a consequence, the evolution of life opposite a dioxygen atmosphere was localized and there remain to this day the developed forms of anaerobic life in other zones. It is these developed forms of anaerobes upon which we depend for the description of early life given above.

The relatively abundant elements that increased in availability with the coming of dioxygen were (cobalt?), nickel, copper, zinc, and cadmium (Fig. 5.3). Elements of reduced availability included nitrogen, which became available only as N_2 and not NH_3 or HCN, and iron in particular, because it became Fe^{3+} in an oxidizing atmosphere and was precipitated as an hydroxide, whereas previously it had been available as Fe^{2+}. Elements for which the available states changed included W, Mo, and V, from sulfido- to oxo- anions, and Se and S, from X^{2-} to XO_4^{2-}. These changes allowed a change in the elemental composition of living systems. However, we note that at first the coming of dioxygen introduced considerable new metal hazards, poisons (as well as poisonous organic compounds) for all life in that elements such as Cu and Cd are poisonous. Nitrogen and even carbon became available with great difficulty. Early life forms must have struggled hard to cope with the environmental effects of dioxygen [20].

Before we consider the high-potential systems of today we must note than an O_2 atmosphere had to evolve in stages. At first the O_2 would be removed by NH_3, giving N_2, and by H_2S (sulfides), giving sulfate. The redox potentials available would only swing slowly upward following the sequence from H_2S/S_n (0.0 V) to $O_2(+0.8\text{ V})$ in Figures 5.2 and 5.4. We see then that there had to be a succession of new possibilities for living chemistry based largely on C/H/O/N but involving first sulfur as SO_4^{2-} [18], then selenium as SeO compounds, then iodine, and much more recently other halogens and nitrogen oxides (see Figs. 5.2 and 5.4). The changes occur with the changes in metals, which would be in the sequence molybdenum (Mo–O systems), iron precipitation as $Fe(OH)_3$, introduction of extra Zn, and then introduction of Cu (Fig. 5.4).

5.5.4 Changing Ligands for Metals

In an atmosphere of H_2S most transition metals cannot easily be bound successfully by N/O donor ligands. An exception is Mn, as previously mentioned, and also the tetrapyrroles which, using a cavity, bind with quite exceptional stability. (In effect their synthesis often uses ring closure based on prior less stable binding of a metal ion). The appearance of dioxygen opened up a great variety of new possibilities while restraining H_2S or RSH as competing donors inside cells but especially outside cells. The examples of the change extend from bacteria to higher organisms. It is unlikely that there were many new developments of thiolate ligand systems. What we must expect and

Table 5.2   Switches in Metal Coordination in Evolution [20]

Ligand	Original, M	Present-day, M	Example
1. Thiolates	Fe^{3+}, Ni^{2+}	$Zn^{2+}/Cu^{2+}/Cd^{2+}$	Rubredoxin
			Alcohol dehydrogenase
			Zinc fingers
			Metallothionein
2. $-CO_2^-/N$	Mn^{2+}/Mn^{3+}	$Fe^{3+}, Zn^{2+}, Ni^{2+}$	SOD
			Fur
3. Porphyrin	Fe, Ni, Co	Mg, Fe, Co	Chlorophylls
			Porphyrins
4. O^{2-}	$\overset{O}{Mn\diagdown\diagup Mn}$	$\overset{O}{Fe\diagdown\diagup Fe}$	Ribonucleotide
			Reductase
5. Tyrosine	-?	Fe	Catalase
6. Dithiolate	W(?)	Mo	Aldehyde reductase

find is new metals in old coordination sites, these thiolates, and old metals in new coordination sites based on N/O donors (Table 5.2) [20].

5.5.5 Changes in Membranes

Whereas archaebacteria appear to use a special class of long-chain ethers to form membranes and prokaryotes use long-chain alcohols in esters, the eukaryotes use, along with the long-chain alcohols, a high percentage of cholesterol, which is a product of oxidative dioxygen metabolism but possibly was developed at low dioxygen tension. It may have been the ensuing reduction in membrane fluidity which allowed the capture of organelles and the preparation of internal vesicles. Instead of the fixed shape of bacteria, many of the cells of eukaryotes have elastic but strong shapes, for example, amoeba. Thus by changing membrane flexibility together with changing internal filaments what appeared was a vast new range of shapes and sizes of cells with a completely new internal structure. The localization of chemical events in cells changed drastically though the primary metabolism itself remained fixed. Thus we find the citrate cycle in mitochondria and photosynthesis in chloroplasts, and neither simply placed in the cytoplasm [3]. Hence much redox chemistry was made distant from DNA and it was further removed from risk by a nuclear membrane and detoxifying systems in the cytoplasm.

5.5.6 The Development of Compartments

As far as is known, simultaneously or just before the appearance of dioxygen enormous changes occurred in cell structures. As stated earlier, first a cell became composed of many somewhat separated zones usually protected by membranes and stabilized by filaments. Of course the outside of prokaryotes had probably developed a periplasmic space even earlier. It therefore is important to ask in this article if different regions of cells are of different chemical composition and redox activity. We consider first what is known about aerobic prokaryotes, because here the compartments are so simple.

It would appear that the exposure of such prokaryotes to dioxygen did not alter the redox balance in the cytoplasm greatly. There were a few new sets of intracellular enzymes for utilizing or metabolizing with the aid of dioxygen but note the detoxifying enzymes. The enzymes are iron, manganese, or heme iron proteins. Of course we must remember that the redox potentials in different compartments were now quite different, (Fig. 5.9).

The first example is that in the periplasmic space the prokaryotes generated quite new enzyme systems able to handle the new oxidized states of nonmetals. As Figure 5.9 shows, copper and molybdenum were utilized in particular. The new metabolites were thus met by evolution with new metallo and other enzymes with new redox potential ranges. We have already mentioned the additional possible use of heme in cytochromes c' for signalling the presence of CO and NO [15, 16].

Upgrading of energy capture was now also possible because the redox potential difference of 1.25 V from -0.45 to $+0.80$ allowed the synthesis of three ATP per two electrons from reductants such as NADH. The essential copper components of these enzymes, for example, cytochrome oxidase, were and are on the outer face of the inner membrane. The iron enzymes remain facing the interior. Figure 5.11 gives the more intricate pattern of compartment of different redox potential in eukaryote cells.

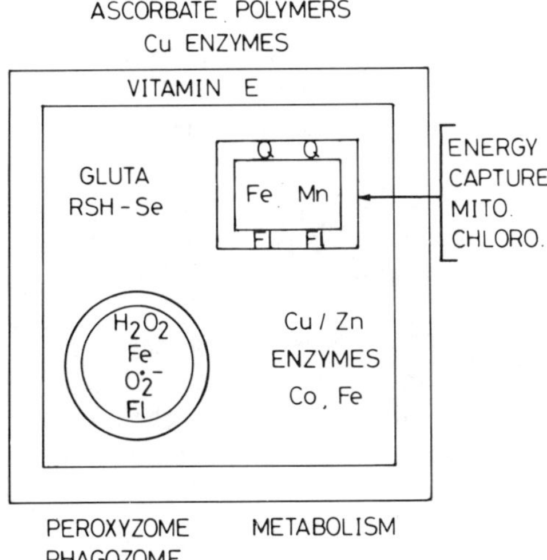

Figure 5.11 The distribution of elements in an aerobic cell within vesicles and organelles, for example, a peroxyzome and a mitochondrion. Q, quinone; Fl, flavin; Gluta, glutathione peroxidase. Some systems are for (secondary) metabolism, some for protection, and some for energy capture.

In the discussion of these changing redox potentials we must remember again the effect of complex formation. There is a curious feature here [5]. While binding to most ligands brings the redox potentials of Mn, Fe, Co, and Ni down so that they can be used in reducing media, <0.3 V in the M^{2+}/M^{3+} states, and can even bring in the M^{3+}/MO^{2+} states of Mn and Fe to such useful lowered potentials as <1.0 V, the effect of binding on molybdate and copper potentials is very different [5]. The redox potentials of molybdenum from Mo(IV) to Mo(VI) remain in the range of around -0.2 to $+0.4$ V no matter what the ligands. The copper potential (the Cu^+/Cu^{2+} aq. ion potential is $+0.23$ V) is actually *raised* even as high as $+0.8$ V in many complexes. Thus, after the hazards created by oxidation of sulfide (and iron) were eliminated, a redox-active poisonous metal, copper, of a quite novel character, was generated in the environment of cells. It turns out that this appearance of high redox potential copper with dioxygen, and the maintaining of the copper within *external* oxidases, was a major step in evolution as we shall show.

5.6 Advanced Aerobic Cells

5.6.1 The Eukaryotic Cell and Dioxygen

It would appear today that the aerobic eukaryotic cell developed a number of elementary chemical features that either depended on the metabolic reactions of dioxygen or were later refined in their use by that metabolism. It is useful to have a list of these:

1. Secondary metabolism, usually relying on iron catalysts in the cell, generating, for example, the hydroxylated sterols.
2. Extracellular metabolism, often dependent on copper or molybdenum catalysts.
3. Calcium-dependent signalling.
4. Protein modifications, especially in the extracellular zones, often dependent on copper catalysts, allowing special mineralization.
5. Zinc-dependent regulation of expression.
6. Antioxidant activity of Cu, Fe(heme), and Se.
7. Increased energy capability using Fe(heme) and Cu catalysts. Certain features of early metabolism are absent in the presence of dioxygen.
8. No direct response to H_2, CH_4, CO owing to Ni, Co, Fe catalyzed reactions since these gases effectively disappeared. Nickel biological chemistry, changed utterly, almost disappearing.

A typical example is a yeast cell, but separated cells from multicellular organisms are very similar.

The features of the cell that changed are not often of primary metabolism, except for the use of O_2 in fuel cell reactions, but are usually in the extracytoplasmic matrix with which we include the vesicle compartments especially in signalling, and in protective systems (Fig. 5.11). Just as an anaerobic prokaryotic cell had to keep a homeostatic chemistry in a rather varied environment, so did an eukaryotic cell. This must mean that the newly utilized elements or elements performing altered functions had to form two new networks of feedback interactions, one inside and the other outside the cell, only one of which was connected to the previous cytoplasmic network that controlled the primary metabolism in prokaryotes (see earlier, Fig. 5.6). The new networks had to link between all the new compartments (see Figs. 5.11 and 5.12). Of course the final

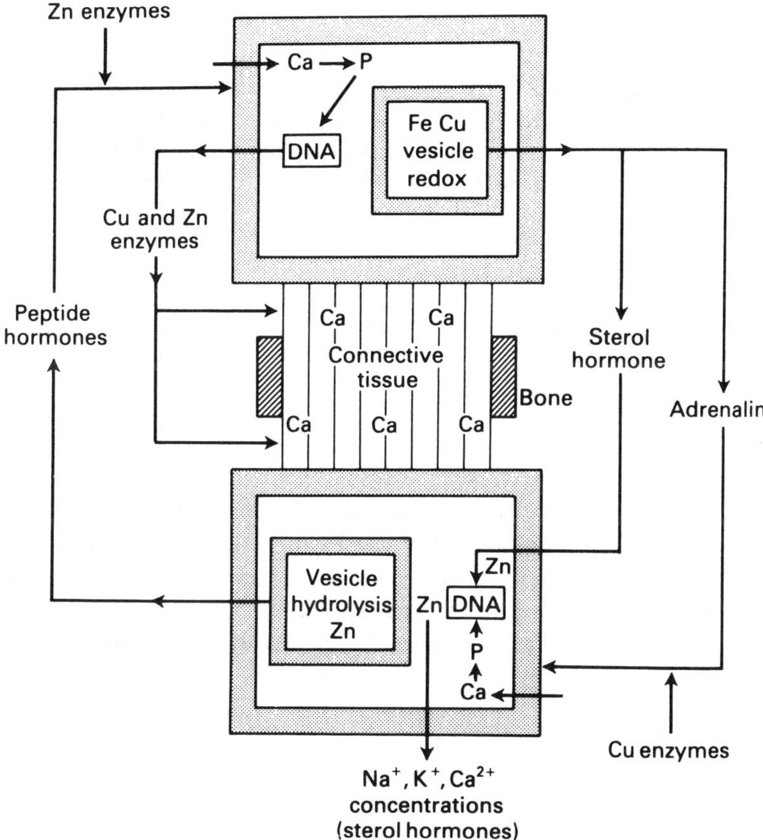

Figure 5.12 There are a number of new compartments in advanced organisms including extracellular zones, vesicles for hydrolysis and oxidation and the delivery of hormones. All the systems must be linked together. Skince Zn, Cu, and Fe are newly involved all three need to be linked back to the simple metabolism inside (P) and outside (Ca) cells. The exact nature of the intricate homeostasis is not understood.

link has to be to quite strict regulation at the DNA level as well as controls at the enzyme activity level, so there has to be a connection to phosphate metabolism. Clearly the more complex the feedback network the less its flexibility whence temperature and pressure controls as well as chemical controls evolved in sympathetic relationship.

5.6.2 The Extracellular Systems of Multicellular Organisms

In the development of multicellular organisms there are four primary needs:

1. A fixed positional relationship of cells, which requires a developed and stable extracellular matrix.
2. New means of tissue repair and protection, and of hydrolysis of extracellular structure to gain space for growth.
3. Communication messengers between cells, which must not be confused with the preexisting intracellular messengers that evolved in primitive prokaryotes; we have mentioned NADH, glutathione, Fe, and HPO_4 especially as interactive message-related chemicals inside all cells (Fig. 5.6).
4. Receptors in cells to communicate to DNA the new conditions outside the cell, which again must be different from those in the primitive varieties.

5.6.3 The Extracellular Matrix

To position cells in space the extracellular matrix has to be relatively rigid. Yet for cells to multiply, that is, for organisms to grow not just in a colony but with shape, the matrix must open up. The only way to achieve both is via a scissors and paste procedure. The paste is a crosslinking of polymers caused by the modification of proteins or polysaccharides by oxidative reaction. Different conventional oxidative crosslinks are possible including $-S-S-$ bridges which are rare in the cytoplasm. It is not possible to crosslink through condensation because there is no reagent such as ATP outside the cell that can be used to remove water. The further known modifications and crosslinkings are given in Table 5.3. They are oxidative and depend largely on copper or iron and heme iron [20]. In fact the iron is used almost invariably (in animals) in the initial

Table 5.3   Crosslinking Processes

Modification	Catalyst
Hydroxylation of proline	Iron proline hydroxylase
Hydroxylation of lysine	Iron lysine hydroxylase
Oxidative coupling of collagen	Copper lysine oxidase
Crosslinking of chitin	Copper tyrosinase

[a]$-S-S-$bridges.

Table 5.4 Breakdown of Crosslinked Matrix

Matrix	Catalyst
Collagen	Collagenase, Zn
Elastin	Elastase, Zn
Chitin	Chitinase, Fe·heme (H_2O_2)

modification of the crosslinking material inside cells in or on vesicle membranes, while copper is used outside the cell to do the final oxidative, free-radical, crosslinking or inside vesicles. It should be noted that the presence of copper with oxygen generated new oxidative *radical* metabolism, that is, new external organic chemistry, for example, the use of oxidized phenols and indoles in chitins, and the hydroxy lysines of collagen. The cutting of the extracellular proteins for growth, also novel, is managed by new zinc enzymes (Table 5.4). The release of copper and zinc from mineral sulfides by oxidation due to O_2 thus allowed development of a set of catalysts essential for the construction of connective tissue but obviously these metals had to be controlled and regulated.

5.6.4 New Internal Signal in the Cytoplasm for Cu and Zn

As stated, a cell now needed a new communication network to look after the building of extracellular organization, that is, copper and zinc systems. It appears as if this is managed in part through a new series of sulfur-rich proteins in the cytoplasm — the metallothioneins, which bind both Cu(I) and Zn(II) (also Cd(II)) [21]. In one sense the metallothionein acts as a protective metal-ion buffer but it is quite likely that it or a related protein can transport copper out of cells and that it, or another closely related protein, for example, ACE-1, binds to DNA and regulates the production of copper and zinc binding proteins [22] (compare FUR for Fe regulation in prokaryotes). The joint control of Cu and Zn is made attractive by the need to use both to control connective tissue. [This system of zinc signalling seems to be unrelated to zinc fingers, which connect to new secondary metabolism and signalling also associated with new functions of iron; see below]. Thus it appears that with the appearance of dioxygen there developed a new Zn, Cu regulatory system connected to DNA in addition to the old internal ones based on H^+, S (glutathione), phosphates, Mg^{2+}, Fe^{2+}, and various carbon compounds.

What appears to have happened in chemical terms is that while Fe–S systems that use S^{2-} have been retained in the Fe_nS_n clusters, the major binding by RS^- of proteins has become that of copper and zinc. Even the iron rubredoxin center of prokaryotes [18] seems to be absent and replaced by zinc in eukaryotes. The main displacement of iron in cells is by zinc. We note that iron now behaves in quite new ways.

5.6.5 New Functions for Iron with Dioxygen in Cells

The new iron enzymes in eukaryotes were responsible for a large part of secondary metabolism using dioxygen and the new production of a special kind of cell-to-cell messengers, for example, (see below). They were also responsible for the oxidative removal of many secondary metabolite poisons and excess messengers. Many of the enzymes belonged in the heme-containing P-450 cytochrome class. In the plant world the related hormones were made oxidatively and destroyed by other heme enzymes, peroxidases, outside as well as inside vesicles in cells. In fact even the extracellular matrix of plants is also broken down oxidatively by heme/Mn chitinases. A further set of iron oxidases made the bacterial poisons, for example, penicillins [19], and they were related to the enzymes for the initial steps of collagen oxidation. They contain iron bound to N, O donors, not thiolate. We next examine how these hormones came to be used as a result of oxidation by iron.

5.6.6 Secondary Metabolism: Sterol Hormones

Earlier we have pointed out that the appearance of dioxygen produced a great diversity of new organic chemicals, for example, alkaloids, sterols, and so on [19]. These products had to be rejected from cells because initially they could have been but little better than poisons. It is reasonable, though by no means clear, that the overlap in chemistry between poisons that had to be detoxified and poisons that became messengers between cells was considerable. To this day it appears that new messengers are closely related chemically to poisons, as are their modes of production and destruction. In Table 5.5 we list some of the messengers and enzymes involved. (Note especially that halogenation is closely associated with poisons and with the hormone thyroxine). Both must be produced in vesicles or on the outer membrane so that they can be ejected but this also keeps the oxidative activity away from the cytoplasm and DNA.

Now the production of the associated receptor proteins, providing links to general protein synthesis, had again to be regulated in a new way that did not distress primary metabolic events and controls in cells for fear of confusion between the two and so the cells required a new regulatory set of DNA-binding proteins. To a very large degree the new system of regulation was based on new zinc finger [23] proteins despite the fact that much of the oxidation is

Table 5.5   Fe/Zn-Dependent Hormones

Hydroxylated sterols
Sex hormones
Vitamin D
Mineral corticoids
Thyroxines
Ecdysones

based on iron. [Because of the loss of iron from the environment and also perhaps because it carries the risk of oxidative attack on DNA if used in multicellular regulatory systems, the new regulation had to be related to another new (acid–base) system of signals. Presumably the increased availability of zinc made it a natural replacement]. In fact a very complex relationship between iron-based sterol hormone messengers and zinc-based receptors, zinc fingers, has evolved. These messenger systems work in tandem with the original regulation systems of the primitive cells, for example, phosphate regulation, as we shall see.

5.6.7 Further Development of Vesicular Systems and Signals

It is difficult to categorize vesicle systems as other than of closely extracellular character (see Fig. 5.11). Most of the transport pumps that move chemicals out of the cell also act to fill vesicles, for example, calcium, sodium, and chloride pumps. Protons are often pumped strongly into vesicles, making them acidic. The difference in the thermodynamic conditions changes the redox balances because some redox potentials have very different pH dependence from others. Some interesting developments in the redox chemistry made possible by the appearance of dioxygen and vesicles are illustrated by adrenaline synthesis in the chromaffin granule [24]. Hydroxylations are catalyzed by iron enzymes on the outer face (cytoplasmic) of the vesicle while the copper hydroxylases are inside. The final oxidation then takes place inside to give stored adrenaline. Other hormones are made and stored in this way, for example, 5-OH tryptamine, but we must also note the vast number of amidinated peptide hormones made oxidatively using copper enzymes [24]. Many of these systems are the basis of new signalling but these signals do not go directly to the nucleus. They use secondary messengers of two kinds. One is inositol phosphate which is not used in prokaryotes. This system developed together with the intensive building of calcium-containing *internal* vesicles for localized internal signalling due to connections from extracellular receptors for these hormones. The hormone binding leads to intracellular cascades based on released inositol phosphates and lipids followed by calcium release from vesicles and phosphate reactions in the cytoplasm. This brings us to the internal network of cellular signals that already existed earlier in prokaryotes, that is, the cyclic-nucleotide and phosphorylation network described above (Section 5.3), and the question of how it became linked to the internal signals that act at the membrane. Before we analyse this situation we must describe a further set of signals that also activate calcium entry directly *through the outer membrane.*

5.6.8 Further Development of Zinc Biochemistry and Signals

It may well be that zinc was used even by primitive prokaryotes in various digestive hydrolytic and synthetic condensation reactions, for example, carbonic anhydrase, peptidases, and RNA synthetases [25, 26]. These activities would

Table 5.6   Cu/Zn-Dependent Signals Metal Dependence

Hormone	Production	Receptor	Destruction
Adrenaline	Cu/Fe oxidation	G-Protein, Ca, P	Cu oxidation
Amidated peptide	Cu oxidation	G-Protein, Ca, P	Zn hydrolysis
Other peptides	Zn hydrolysis	G-Protein, Ca, P	Zn hydrolysis
Sterols, thyroxine	Fe oxidation	Zn-fingers	Fe oxidation
"Calcium"	Na, K, depolarization	Ca- Calmodulin	Ca-pump

P is the phosphate-dependent set of signals in a cell.

have allowed it to produce peptides and, when they could be rejected, the peptides could have been used as poisons, for example, to form channels allowing entry of Na^+ or Ca^{2+} into enemy cells. Just as in the case of the secondary metabolites, such as sterols, this chemistry could have been converted into the modern messenger system between cells, which we see today, once multicellular organisms evolved. Now, however, these peptide messengers and those produced by copper oxidation often only act on a temporary channel opening for say, Ca^{2+} and not directly on DNA. It would then be the calcium ion that would carry the message to metabolism or to DNA regulation via phosphate reactons inside the cytoplasm. This calcium receptor connection was again completely new in eukaryotes and is due to calmodulins. Both the enzymes for production and destruction of the peptides and hormones such as adrenaline are novel zinc or copper enzymes [24]. The problem now is that there are a multiplicity of new and old signals to be reconciled and this must be done through phosphate metabolism eventually, since phosphate is the major regulator of prokaryotes and has remained so throughout evolution (Table 5.6).

5.6.9 Signals Between Cells: Calcium and Phosphates and the New Reactions

The signalling between cells is extremely complex but we can simplify the process to understand principles [27]. In essence a signal can enter the cell and interact with it at two levels: (1) metabolic paths, enzymes and (2) DNA/RNA protein production. The signals can be generated by diffusion of a molecule through the cell membrane or through recognition of the signal at the membrane by a receptor coupled to an internal signal (Fig. 5.12) [28, 29]. The major internal (original) signal system is the phosphate pool (P-pool) which became related to the Ca^{2+} pool in eukaryotes in two ways as described previously. The Ca^{2+} pool and the P-pool communicate to metabolism directly whereas the P-pool connects also to DNA/RNA. Thus very many messages could be made to depend on triggering changes in calcium or phosphorus inside cells by direct invasion of Ca^{2+} (or even cyclic AMP in slime molds) and then by synthesis or reaction of cyclic AMP or cyclic GMP, ATP, or GTP etc. These may be due to cell surface acid–base binding reactions, for example, of certain organic compounds, such as peptides, where no complication from oxidative reactions arises.

We now see that to develop new signalling to enzyme systems with the evolution of multicellular organisms after the appearance of dioxygen, it was essential for a messenger system to connect the external to the internal cellular solutions to generate a series of responses. Essential features of these responses were and are that in different situations:

1. They acted quickly because this gives a competitive advantage; no protein synthesis occurs but changes in metabolism take place to harmonize cell–cell activities.
2. They connected to preexisting internal controls, for example, especially to phosphorylation, which could lead on to DNA regulation and some differentiation.
3. They were made in tune with changes in other new message systems on a long time-scale basis to assist growth.

The beauty of the extended *use* of the rejected Ca^{2+} ions (not seen in prokaryotes) differentially in the first two cases is then clear. Since we know that previously in early prokaryotes this element in high concentration was a poison it could act (like drugs) as a pulsed stimulant or as a lethal agent. However, through the action of newly synthesised proteins after the appearance of dioxygen, calmodulins in eukaryotes, calcium entry was given multiple stimulant connections to metabolism, cell tension and shape, vesicle discharge, kinases and phosphatases. By control of the length of duration of the pulse, by signals, cases (1) and (2) could be separated. It is only the second that connects to the phosphate regulation of DNA transcription. Feedback was achieved by connection to pumps [28]. (With the developments following the advent of dioxygen, did the phosphate calcium system then extend to the IP_3 (inositol 1,4,5-trisphosphate) connections? Is IP_3 a recent messenger?) (Meanwhile the Na^+ ion became the primary ion current carrier between cells so that it too had to be controlled outside cells; see below).

The role of calcium in messages must not allow us to overlook the essential development of its compounds in the extracellular matrix that organizes cells. Here calcium acts often with phosphate as a crosslinking and strengthening agent of the extracellular matrix [30]. A series of new proteins were also needed to control extracellular calcium and some of these, such as the proteins containing Gla (γ-carboxyglutamate), like some of the hormonal peptides containing hydroxyproline, need oxidative modification. Again the calcium was incorporated into biominerals to further stabilize the extracellular systems. It is obvious that if this system of cell controls and these syntheses were to work in harmony then series of different hormones had to be correlated and/or connected in their activities. This would not apply so much to short pulses of action, for example, nerve and adrenaline activities of Na^+ and Ca^{2+}, but had to apply to growth. Thus there had to be connections between the new activities of Ca and P and those of Cu, Zn, and Fe especially with regard to (3) above (see Fig. 5.12).

If we remain briefly with the calcium (and the sodium) system then this connection back to the other messenger systems, hormones, for long-term change of function is managed through regulation of proteins for calcium and sodium uptake and for connective tissue synthesis which is now known to be by sterol hormones linked to both iron and zinc systems (see above). Just as FUR is the link for iron feedback in many prokaryotes so are sterol-dependent proteins the link to ionic balances of Na^+ and Ca^{2+} but they themselves depend on Fe and Zn. Figure 5.12 shows some of these links. However, we can then go on to look at the production and effect of peptide hormones and see how they too depend on balanced interactive feedback with crosstalk. (A parallel development is the NO signal system to the G-protein receptors which uses a heme-binding site and an iron-dependent oxidative creation of a signal and again connects to phosphorylation systems). In this way a network of crossinteracting feedbacks of the 20 inorganic elements was achieved. It is not perfect and undoubtedly its failure is seen in conditions in old age, for example, osteoporosis.

5.6.10 Protective Devices and Detoxification

While all the matrix, vesicle, repair, and signalling systems were upgraded by the use of both new inorganic and organic chemistry in the high-redox potential range there became possible two new protection systems apart from those related to hormones listed in Table 5.6 (see Fig. 5.4). Here we have space to only name the new superoxide dismutase (Cu/Zn) and the new glutathione peroxidase using selenium, both of which do not affect DNA while protecting the cytoplasm. At another level various poisons were generated and mechanisms of protection from these very poisons evolved.

5.7 Feedback Systems

Although the appearance of dioxygen introduced new complexity into biological chemistry it had to be controlled and regulated with respect to the primitive primary metabolism, and related to the environment. This requires an intensive feedback mechanism of communication from cell to cell and organ to organ. We cannot tackle this problem here except in a diagram, (Fig. 5.13). In essence we can imagine a sequence of events that has two arms—biological and environmental.

Down one arm a cooperative organized chemistry develops within biology whereas down the other an "inorganic" disorganized chemistry develops in the

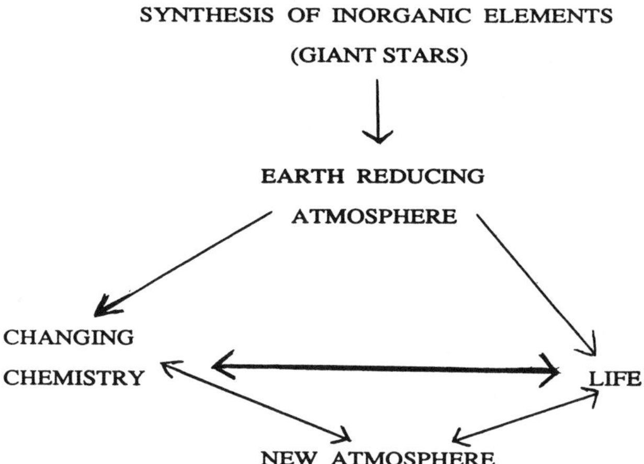

Figure 5.13 The interactive evolution of inorganic and biological chemistry.

environment. The two enter into a feedback relationship, so that as in the Gaia hypothesis the two are part of an either advancing or retarding unity. The system does not have the stability claimed for Gaia if the biological system or the environment generates or removes new chemical species and unfortunately perhaps this occurs all the time, aided by man. The development of feedback mechanisms is extremely slow so that the speed at which new chemistry is introduced into the environment has to be in terms of millions of years. Man must learn from the appearance of dioxygen, the greatest of all environmental impacts, that he too can change evolution by accident but he has to accommodate to geological and biological evolutionary time scales, which by their nature are interlocked (Fig. 5.13).

5.8 Conclusion

The purpose of this chapter is to show that the evolution of a variety of redox states of elements on the developing surface of the Earth has been and remains a major force in the evolution of life. The two exist in harmony — the so-called "inorganic" and the so-called "organic" — so that an ecological symbiosis is what we observe and we are one of its products. Niches at various redox potentials remain, making possible a huge diversity of living organisms. There are inputs to the environment other than dioxygen that are detrimental to its stability; among these are products of natural events, for example, volcanoes, as well as products of man's activities. These have to be seen for their possible outcome on the basis of what happened when dioxygen appeared in quantity.

Evolution is biased by pollution but we do not know in what direction. Increasingly we need to reflect before we act, but unfortunately short-term gain is sometimes man's present priority.

A more detailed account of the subject matter of this article is to be found in ref. 31.

References

1. Newsom, H. E., Jones, J. H. *Origin of Earth*, (eds.), Oxford University Press, Oxford, 1990.

2. Wächtershäuser, G. *Microbiol. Rev.* 1988; *52*, 452–486.

3. Stryer, L. *Biochemistry*, Freeman, New York, 1989.

4. Woese, C. R. *Microbiol. Rev.* 1987; *51*, 221–271.

5. Phillips, C. S. G., Williams, R. J. P. *Inorganic Chemistry*, Vol. I, Oxford University Press, Oxford, 1966; pp. 314–320 and 639–642.

6. Frausto da Silva, J. R. R., and Williams, R. J. P. *The Biological Chemistry of The Elements*, Oxford University Press, Oxford, 1991; pp. 18–20.

7. Walsh, C. T., Orme-Johnson, W. H. *Biochemistry* 1987; *26*, 4901–4906.

8. Eschenmoser, A. *Nova Acta Leopold* 1992; *67*, 201–233.

9. de Duve, C. *Proc. Natl. Acad. Sci. USA* 1987; *84*, 8253–8256.

10. Williams, R. J. P. *J. Theor. Biol.* 1961; *1*, 1–13.

11. Mitchell, P. D. *Nature* 1961; *191*, 144–148.

12. Lorenzo V., de., Wee, S., Herrero, M., Nielands J. B. *J. Bacteriol.* 1987; *169*, 2624–2630.

13. Williams, R. J. P. *FEBS Lett.* 1982; *140*, 3–10.

14. Green, J., Trargeser, M., Six, S., Unden, G., Guest J. R. *Proc. Roy. Soc. Biol. Sci.* (*Lond.*) 1991; *244*, 137–144.

15. The properties of cytochromes *c'* are fully reviewed in *Biochim. Biophys. Acta* 1991; *1058*, 1–84.

16. Henry, Y., Ducrocq, C., Drapier, J. C., Servent, D., Pellat, C., Guissani, A. *Eur. Biophys. J.* 1991; *20*, 1–15.

17. Braterman, P. S., Cairns-Smith, A. G. *Origins Life Evol. Biosp.* 1987; *17*, 221–228.

18. Chen, L., Liu, M-Y., Legall, J., Fareleira, P., Santos, H., Xavier A. V. *Eur. J. Biochem.* 1993; *216*, 443–448.

19. Mann, J. *Secondary Metabolism*, Oxford University Press, Oxford 1987.

20. Frausto da Silva, J. R. R., Williams, R. J. P. *The Biological Chemistry of The Elements*, Oxford University Press, Oxford, 1991; Chap. 21.

21. Kägi, J. H. R., Kojima, Y (eds.) *Metallothionein II. Experientia*, 1988; Suppl. 52.

22. Richards, M. P. *J. Nutr.* 1989, *11*, 1062–70.

23. Klug, A., Rhodes, D. *Trends Biochem. Sci.* 1987; *12*, 464–469.

24. Njus, D., Kelly, P. M. *Biochim. Biophys. Acta* 1993; *1144*, 235–248.

25. Mills, C. F. (ed.) *Zinc in Human Biology*, Springer-Verlag, Berlin, 1988.

26. Vallee, B. L., Auld, D. S. *Biochemistry* 1990; *29*, 5647–5659.

27. Williams, R. J. P. *Cell Calcium* 1992; *13*, 273–275.

28. Calcium flow and its interaction with phosphate signals is described in *Cell Calcium* 1992; *13*, 353–472,

29. João, H. C., Williams, R. J. P. *Eur. J. Biochem.* 1993; *216*, 1–18.

30. Frausto da Silva, J. J. R., Williams, R. J. P. *The Biological Chemistry of The Elements*, Oxford University Press, Oxford, 1991; Chap. 10.

31. Williams, R. J. P., Frausto da Silva, J. R. R. *The Natural Selection of the Chemical Elements*, Oxford University Press, Oxford, 1996.

Photosystem II and the Quinone–Iron-Containing Reaction Centers: Comparisons and Evolutionary Perspectives

A. W. Rutherford and W. Nitschke

The use of light energy to drive biological reactions is thought to be almost as old as life itself [1–3]. The most important of the existing proteins that catalyze the conversion of solar energy into chemical energy are called photosynthetic reaction centers. These are integral membrane proteins in which are located a range of electron transfer components. In particular, chlorophyll-type pigments are present that are able to absorb visible light and then undergo charge separation.

Four kinds of reaction center are known. In plants, algae, and cyanobacteria, the photosynthetic membranes contain two kinds of reaction center, phososystems I and II (PSI and PSII) (Fig. 6.1). The reaction center of PSI is very similar to those found in green sulfur bacteria and in heliobacteria (as a group these reaction centers are known as RCI) [4], whereas the PSII reaction center bears many similarities to those found in purple bacteria and in green filamentous bacteria (i.e., *Chloroflexus*), known collectively as RCII [5,6]. The phylogenic relationships between the various photosynthetic species are described in Chapter 7. From a knowledge of the characteristics of all the existing reaction centers, it is possible to speculate on the evolution of these proteins. At the same time, the evolutionary viewpoint is often useful in understanding the function (or lack thereof) of certain existing characteristics and in some cases can allow the comparative approach to be extended so that structural and functional insights can be obtained.

In Chapter 7, we describe the characteristics of RCI-type reaction centers and their relationship to the better characterized purple bacterial reaction center. In the present chapter, we describe the characteristics of PSII in

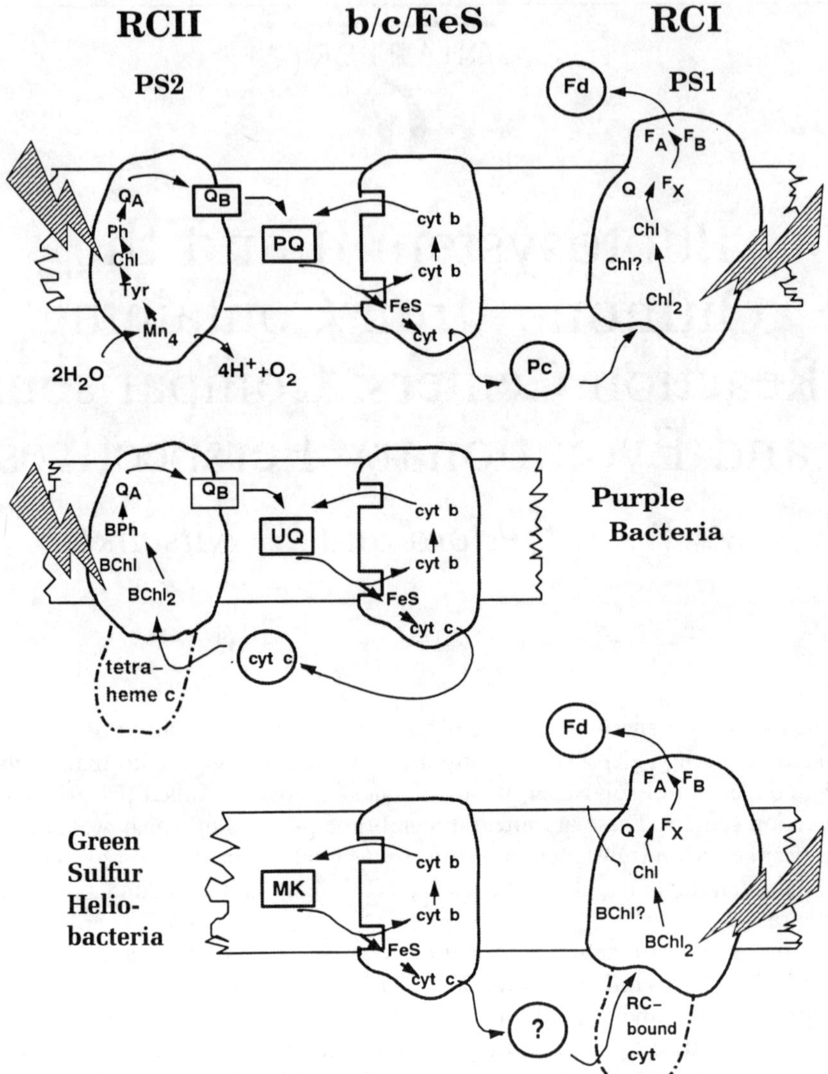

Figure 6.1 Photosynthetic electron transfer associated with the four main types of photosynthetic reaction center. The uppermost drawing represents the photosynthetic membrane found in plants, algae, and cyanobacteria and shows the three main membrane-bound electron transfer proteins: PSII, the cytochrome b–c–Rieske iron–sulfur complex, and PSI. The purple bacterial reaction center is drawn beneath PSII to stress the similarities between these two kinds of reaction center and the same approach is used for PSI and the green sulfur/heliobacterial reaction centers. The components involved in electron transfer within and between these complexes are shown as abbreviations. Tyr, tyrosine; Chl, chlorophyll, Ph, pheophytin, BChl, bacteriochlorophyll; Q, quinone; PQ, UQ, and MK, plasto-, ubi-, and mena-quinones; Pc, plastocyanin; Fd, ferredoxin; F_X, F_A, F_B, FeS, iron–sulfur clusters. The reaction center bound tetraheme cytochrome in the purple bacteria is present in most but not all known species; its presence in green sulfur and heliobacteria is discussed in Chapter 7. The striped woodpecker symbol is supposed to represent a photon of visible light which initiates charge separation as the first step in the electron transfer reactions. The arrows represent the electron transfer pathways.

comparison with the purple bacterial reaction center and we try to draw some insights on the evolutionary origins of these characteristics. There are very many review articles covering the extensive literature in this area. In this chapter, whenever possible, the citations given refer to review articles.

6.1 Background

6.1.1 Solar Energy to Chemical Energy: The Heart of Photosynthesis

The fundamental processes of excitation, charge separation, and charge stabilization are common to all of the photosynthetic reaction centers. The initial step in the energy conversion process of photosynthesis is the absorption of a photon of light by chlorophyll (or bacteriochlorophyll in bacteria). This results in the formation of its first excited singlet state. When this occurs in chlorophyll that has been extracted from the protein, the energy stored as the first excited singlet state is rapidly wasted by fluorescence or heat-generating back reactions. In vivo, however, the chlorophyll molecules are arranged in a specific protein environment in which such wasteful back reactions are minimized.

The majority of chlorophylls play a so-called light-harvesting, or antenna, role in which the excited chlorophyll is arranged near other chlorophylls in an environment in which the excitation can be rapidly and efficiently passed from one chlorophyll to another. In this way, the excited state is able to reach the reaction center chlorophylls so that photochemistry can take place. In many cases the excitation is funnelled, by small energy losses, toward longer wavelength, lower energy chlorophylls of the reaction center [7].

The reaction center chlorophylls are arranged in the protein so that the excited chlorophyll actually transfers an electron to a nearby chlorophyll or chlorophyll-like molecule. In this way, a charge pair is formed and thus solar energy is used to initiate a chemical reaction. The charge pair formed is stabilized by a sequence of electron transfer events in which the electron and the positive charge move away from each other with a loss of energy occurring at each step. The initial charge stabilization events occur in the reaction center protein (see Figs. 6.1 and 6.2) [8,9].

6.2 Comparisons of PSII and Purple Bacterial Reaction Centers

6.2.1 Introduction

In the 1970s, the biochemistry and spectroscopic characterization of the bacterial reaction center were very well developed while knowledge of PSII was relatively primitive and the closeness of the relationship between these two

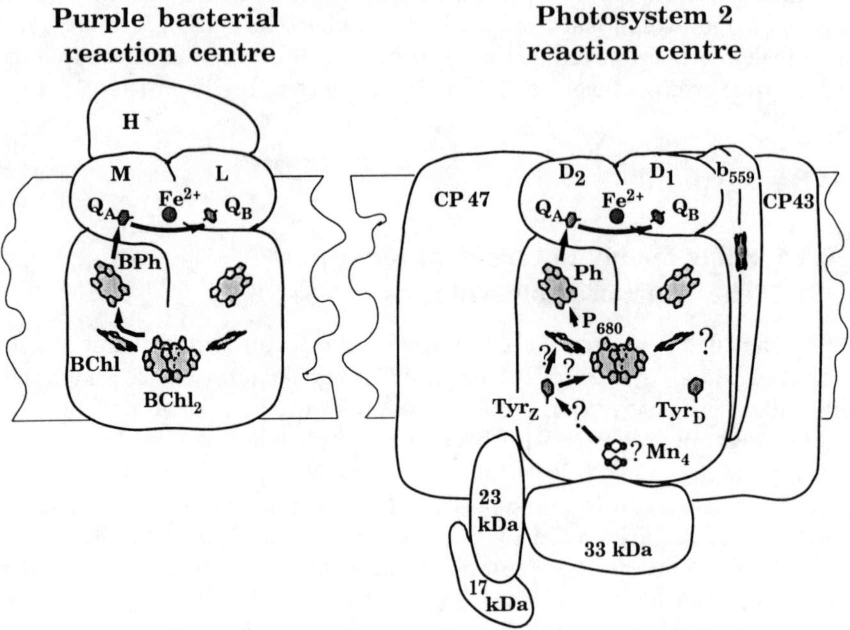

Figure 6.2 A comparative scheme of the structure of the purple bacterial reaction center and PSII. Both structures are based on the published crystal structure of the bacterial reaction center (see text). D1 and D2 are two core subunits of PSII which are equivalent to L and M in purple bacteria. The H-subunit in the bacterial reaction center has no counterpart clearly defined in PSII. The CP43 and CP47 are chlorophyll-containing proteins that play a light harvesting role but seem to be important in some aspects of O_2 evolution. These proteins are suggested to be evolutionarily related to the peripheral, antenna part of the RCI-type reaction center (see text). Other polypeptides are described in the text. The structure of P-680 is unclear; a number of models are discussed in the text but that favored is that P-680 is the monomeric chlorophyll. It should be emphasized that the positions of the chlorophyll in PSII is uncertain as is the position and structure of the Mn complex; details are given in the text. Other abbreviations are as in Figure 6.1 and as in the text. The positions of the chromophores are taken from Brookhaven Protein Data Bank, entry 1 PRC; see Chapter 7 for citation. The arrows indicate the electron transfer pathway. In the purple bacterial reaction center, the involvement of the monomeric BChl in electron transfer is likely but still debated (see text). The electron transfer pathways from TyrZ to the chlorophylls is uncertain because of the uncertainty concerning the structure and position of P-680, hence the question marks. Other question marks emphasize areas of doubt.

kinds of reaction center was not generally realized. In the early 1980s, however, it became clear from spectroscopic studies that the two kinds of reaction center had much in common [10]. It was recognized that the purple bacterial reaction center could serve as a useful model for understanding both structural and functional aspects of PSII. Many of the components of the PSII reaction center are homologues of those in the purple bacterial reaction center. Thus when

X-ray crystallography provided a model for the atomic structure of the bacterial reaction center in the mid-1980s [8], it was used directly as a structural model for PSII [5, 11–13]. This model was developed in more detail by amino acid sequence comparisons of the respective reaction center proteins [5, 13, 14]. The isolation of a minimal reaction center in PSII confirmed the identification of the reaction center proteins of PSII [15] and site-directed mutagenesis experiments verified aspects of the folding model of PSII that were derived from the amino acid sequence comparisons [16–18]. Several groups have applied computer modeling techniques to extend the comparison [17, 19]. Thus relatively detailed models of PSII exist in the literature; however, there are several aspects of PSII structure that remain poorly defined or at best ambiguous.

The marked similarities and specific differences present in the two types of reaction center have a bearing on the evolutionary origin of these features. In general, it is assumed that the characteristics that are in common were probably present in a common ancestor. It is thus useful to compare the structure and function of the two kinds of reaction center.

6.2.2 Photosynthetic Electron Transfer in Purple Bacteria: An Overview

The reaction center in purple bacteria is made up of three or four proteins depending on the species of the bacterium. In all species, there is a central core made up of two subunits known as L and M, which have molecular masses of around 32 kDa. All of the cofactors required for photochemical charge separation and the first electron transfer steps are located in these two subunits [8, 9]. The L- and M-subunits are structurally almost symmetric to each other [Fig. 6.2], both being made up of five membrane-spanning helices, and it seems clear that they have evolved from a common ancestral protein that gave rise to a reaction center core that was homodimeric [8, 20].

Purple bacterial reaction centers also have a protein subunit of 28.5 kDa that does not contain any chromophores, the H-subunit. This protein caps the electron acceptor side of the reaction center, insulating the quinone acceptors, particularly Q_A, from the aqueous phase. It has one membrane-spanning helix running down the side of the interface between the L- and M-subunits [8, 9]. No H-subunit counterpart has been reported in *Chloroflexus*. This, however, is not really surprising because this bacterium is capped by a large pigment–protein complex known as the chlorosome [21].

The fourth subunit, which is present in the majority of purple bacterial species and also in *Chloroflexus*, is the tetraheme cytochrome. This subunit protrudes into the aqueous phase at an angle that seems to vary from species to species [22].

In purple bacteria, the primary electron donor, that is, the only component that undergoes true photooxidation, is a special pair of bacteriochlorophyll molecules, $(BChl)_2$. There is some debate about whether the first electron acceptor is the nearby bacteriochlorophyll molecule or the bacteriopheophytin (see Fig. 6.2) [23]. Recent electron transfer theory [24] and experimental data [25] favor the involvement of the monomeric bacteriochlorophyll as a true redox component. In any case, in about 3 ps, a charge pair involving the oxidized $(BChl)_2$ and the reduced BPh is formed. The BPh^- donates the electron to a nearby quinone, Q_A, with a half-time of around 200 ps [23].

The kinetics and much of the structural information on the electron transfer components were established by spectroscopic methods. The crystallographic studies, as well as providing much greater structural detail, showed that the protein was heterodimeric and that both halves of the heterodimer contained symmetrically related cofactors. Electron transfer appears to occur only along one side of the heterodimer. Exactly how the protein achieves this directionality is the subject of much research and debate [23].

The semiquinone, Q_A^-, formed by electron transfer from the BPh^-, can donate to the second quinone, Q_B, in 10 to 400 μs (depending on the species), forming the Q_B semiquinone, Q_B^-. This state is stable until a second photochemical charge separation occurs, at which time it picks up a second electron from Q_A^-, and two protons from the aqueous phase via a series of proton-carrying amino acid residues and then, as the hydroquinone, is able to exchange with ubiquinone from a pool in the membrane [26, 27].

The positively charged form of the primary donor, $(BChl)_2^+$, is reduced by electrons coming from a cytochrome c heme with a half-time that can be as short as approximately 200 ns in species that have a tetraheme cytochrome bound to the reaction center [22]. The oxidized cytochrome is subsequently reduced by a soluble monoheme cytochrome c_2. In species lacking a bound tetraheme cytochrome, electron donation to $(BChl)_2^+$ occurs directly from a soluble monoheme cytochrome c_2 [28].

Photosynthetic electron transfer in purple bacteria is a cyclic system, with the electrons from the charge separation leaving the reaction center in the form of hydroquinone. The hydroquinone donates electrons to the cytochrome $b-c-$Rieske complex, which in turn reduces the cytochrome c_2, the electron donor to the purple bacterial reaction center (Fig. 6.1) [29]. In *Chloroflexus*, the electron donor to the reaction center-bound tetraheme cytochrome is a membrane-bound Cu protein known as auracyanin. Auracyanin is presumably reduced by electrons from a cytochrome $b-c-$Rieske complex [21].

Electrons can enter this cycle at various points coming from reducing substrates such as sulfur etc., depending on the species [30]. Electrons can also leave the cycle; for example, the soluble cytochrome c_2 can act as an electron donor to cytochrome oxidase in those species that can respire [31].

6.2.3 Photosynthetic Redox Reactions in the PSII Reaction Center: An Overview

In PSII, the charge separation reactions and electron acceptor side reactions can be considered to be more or less the same as those described in the purple bacterial reaction center, at least to a first approximation. The nature of the components involved in charge separation and accepting electrons, their kinetics, and their structural characteristics are for the most part similar in both kinds of reaction center (see Fig. 6.2). On a more detailed level, however, there are several features of PSII that could reflect important differences. The pigments are chlorins rather than bacteriochlorins; thus many of their physical properties are somewhat different, notably their colors and their redox properties. The special features of the primary reactions are discussed in Section 6.4.

The components and reactions of the electron *donor* side of PSII are quite different from those in the purple bacterial reaction center. These differences relate to the major unique function of PSII, namely its enzyme activity as the water-splitting/oxygen-evolving enzyme [17, 32, 33].

The first electron donor to the photooxidized chlorophyll, $P\text{-}680^+$, is Tyr-160 of the D1 polypeptide and is known as TyrZ. The oxidized $\text{TyrZ}^{\cdot}$ is reduced by electrons from a complex of probably four Mn ions. A symmetrically positioned tyrosine, TyrD, is present on the D2 polypeptide. It too can undergo oxidation, probably through donation to $P\text{-}680^+$, and is involved in redox reactions through a series of slow electron transfer reactions. The role of TyrD is unclear; however, it can oxidize the Mn cluster when in the lowest redox state of the charge accumulation cycle and thus the role of TyrD may be in redox poising the cluster or in the photooxidation/ligation events involved in the assembly of the cluster [17, 32, 33].

The locations of the redox active tyrosines are predicted with some confidence from the folding model of the reaction center protein. This confidence comes from the fact that the tyrosines are probably located on the membrane-spanning helices which form a common structural motif in both kinds of reaction centers. The location of the Mn is still open to debate. Early models suggested the Mn cluster would be symmetrical relative to the D1/D2 heterodimer. The justification for this was largely aesthetic. Several more recent models place the Mn cluster on the D1 protein relatively close to TyrZ. This is largely based on spectroscopic indications that the Mn is closer to TyrZ than to TyrD [17, 32–35] and on the fact that site-directed mutagenesis of amino acid residues, hunting for effects on O_2 evolution, have found a greater number of candidates on D1 than on D2 [16–19]. Very recently, indications have been obtained that the Mn may be less asymmetric than proposed in many models [36].

The Mn cluster is thought to play a dual role as the charge accumulator and the active site for water oxidation. The oxidation of water requires the extraction of four electrons from two molecules of water releasing molecular oxygen and four protons: $2\,H_2O \to O_2 + 4\,H^+ + 4\,e^-$. Since the reaction center photochemistry involves the generation of one charge separation, and hence one positive charge, per photon, a system for accumulating the positive charges must exist. It is commonly, but not unanimously, thought that the four-electron oxidation of water occurs after four positive charges have been accumulated. The enzyme is proposed to exist in five states (S_0 to S_4) in which the subscript represents the number of charges accumulated. Some or all of the transitions to higher "S states" represent the oxidation of the Mn cluster [17, 32, 33]. The S_2 to S_3 transition in the inhibited enzyme involves oxidation of an organic species, probably an amino acid, the UV/visible spectrum of which is similar to that of an oxidized histidine. It has been suggested that oxidation of this component may occur in the functional system. It could be the true S_3 state, or, more likely, a state in equilibrium with Mn oxidation in S_3, in which case it may represent an intermediate electron transfer component functioning as an electron carrier in other steps of the charge accumulation cycle [37].

Under normal conditions, only approximately two of the protons from water oxidation are released into the aqueous phase at the same step as the O_2. The rest of the protons from the oxidation of two water molecules probably serve to protonate amino acid groups that undergo deprotonation, releasing the protons on other steps in the charge accumulation cycle. Under some special conditions one proton is released on each step of the charge accumulation cycle but under normal circumstances the complex deprotonation pattern is observed. The deprotonation–protonation reactions are thought to be important for balancing out the electrostatic constraints on electron transfer associated with accumulating four positive charges [38]. In addition, deprotonation of water prior to its oxidation is thought to be thermodynamically important [39]. It is thought that calcium and chloride are involved in charge compensation and deprotonation reactions [37]. Surprisingly, the presence of a specialized light-harvesting complex, the CP26, has been linked with the proton release pattern but its mode of action is unclear [38].

Having provided general overviews of PSII and purple bacterial reaction centers, we shall now focus on some specific features of PSII. These will be compared to their counterparts, should they exist, in the purple bacterial reaction center. Where appropriate, we will discuss the evolutionary significance of the differences and similarities observed.

6.3　The Proteins of PSII

6.3.1　Introduction

An obvious difference between PSII and the purple bacterial reaction center is the large number of polypeptides in PSII that are required to maintain the in vivo activity. While the bacterial reaction center requires three basic subunits for its cytochrome *c*-quinone photooxidoreductase activity, PSII requires approximately seven protein subunits to perform its water–quinone photooxidoreductase activity and there are around 20 subunits in all that are associated with the PSII complex. This difference is not surprising because PSII contains not only all the main photochemical features of the bacterial reaction center (albeit with some important modifications) but also the complex hardware associated with water oxidation [17, 40–42].

6.3.2 The Heterodimeric Reaction Center

The central core of the reaction center is made up of the D1 and D2 subunits, which are the counterparts of the bacterial L- and M-subunits. Most of the cofactors involved in electron transfer are bound to these two subunits. Although the sequence homology between the bacterial reaction center proteins and their PSII counterparts is not very high, important amino acid residues are strictly conserved and the overall structures of the subunits are very similar. The basic structural motif of five membrane-spanning helices is present in each subunit. However, there are several marked differences in the protein structure that presumably represent differences in function [5, 17, 40, 41].

On the electron acceptor side of the D1 subunit there is an extra loop between the fourth and fifth transmembrane helices that has been implicated as a cleavage site associated with the uniquely rapid turnover of this protein. The rapid turnover of the D1 polypeptide is thought to reflect some kind of "burnout" reaction, intrinsic to the chemistry performed by PSII [17,40]. On the donor side, there are longer loops in D1 and D2 between helices 1 and 2 and in addition there are C-terminal extensions. In D1 at least, these features are thought to be involved in liganding the Mn cluster of the O_2-evolving enzyme [17–19].

It has been suggested recently, from detailed parsimony analysis, that the L- and M-subunits of purple bacteria developed from a homodimeric ancestor different from the ancestor of the D1/D2 heterodimer of PSII (Fig. 6.3) [3]. This would indicate that the features that differentiate Q_A from Q_B, and which are similar in both kinds of reaction centers, in fact are the result of parallel evolution (see however, Ref [91] for an alternative explanation involving so-called "concerted evolution").

The suggestion that D1/D2 and LM did not evolve from a common heterodimeric ancestor is somewhat ironic because it was comparisons of quinone acceptors that provided the first indications that the two centers might be structurally related. The marked functional and structural similarities of the quinones [12] led to a strong expectation that the reaction centers shared a common heterodimeric ancestor. It seems, however, that the properties of Q_A and Q_B in both kinds of reaction center need to be reassessed in light of the analysis of the amino acid sequences.

If we consider that the homodimeric ancestors of PSII and purple bacterial reaction centers carried out electron transfer to the quinone pool, then each of the two quinones present in the ancestral homodimer probably had properties that are predicted to be a hybrid of those found in Q_A and Q_B. Its Q_B-like properties included the ability to store up one electron until a second electron arrived from its neighboring quinone and, when fully reduced and protonated, the ability to exchange with an oxidized quinone from the pool (see, however, Ref. [72] for an alternative view of electron transfer via quinone in primitive reaction centers). Its Q_A-like properties included acting as an electron acceptor

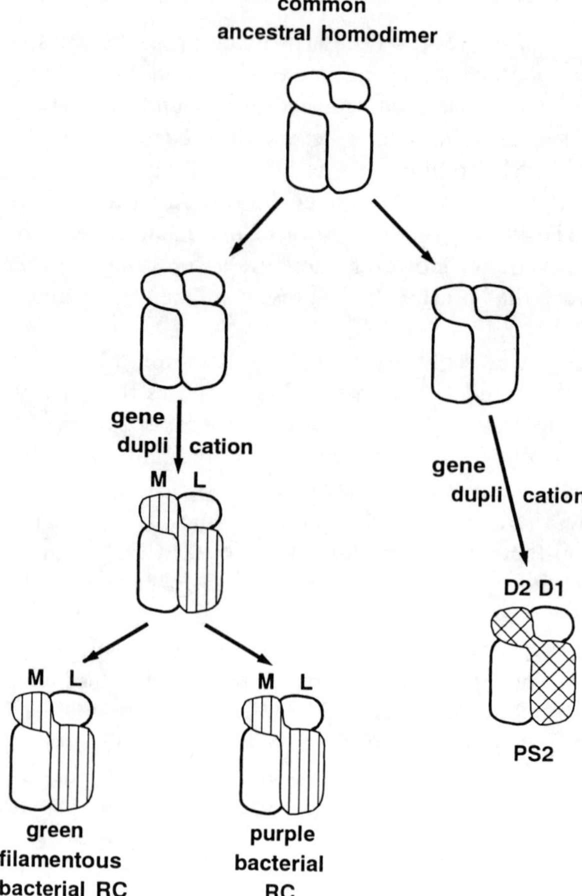

Figure 6.3   The proposed evolutionary pathway leading to LM and D1/D2 reaction centers, based on the properties of the reaction center proteins. This figure is based on a figure in Ref. [3]. The arrows indicate the occurrence of diversity. The common ancestor is assumed to have the basic structural motif for the chromophores as shown in Figure 6.2; however, it is assumed (1) that no differentiation into Q_A and Q_B is present and (2) that electron transfer occurs up both sides of the reaction center (see text).

from Ph^- and as an electron donor to its neighboring quinone. Thus specialization into Q_A and Q_B could have required rather minor changes. It is thus not too difficult to imagine a situation in which gene duplication took place separately in the respective ancestors of PSII and the purple bacterial reaction center and that a small number of similar key mutations occurred separately, leading to comparable Q_A and Q_B properties in both reaction centers.

The homodimeric ancestors of both types of reaction center would probably have been inefficient in charge separation compared to current heterodimeric systems. Assuming that both sites were occupied by oxidized quinones, the first charge separation would occur with a high quantum yield. The second charge separation would, however, have an even chance of occurring in the half of the dimer in which semiquinone was already present. Whenever this occurred, it is likely that rapid recombination of the primary radical pair would occur, just as occurs in existing reaction centers when Q_A^- is present. Although some reexcitation of P and thence further stable charge separation might occur, it seems likely that excitation energy would be lost. In other words, there would be a significant miss factor on the second excitation.

It is possible, however, that an electrostatic effect of the semiquinone could inhibit charge separation on its side of the dimer, thus favoring charge separation up the other side of the reaction center. In this way, a relatively efficient "alternating" homodimer could have existed. Such a reaction center might well have lost efficiency from the P^+Q^- back reaction. Thus homodimeric quinone-reducing reaction centers would have benefited greatly from a rapid electron donor system to P^+.

In the homodimeric centers in which two charge separations occurred, it seems likely that the two semiquinones would dismutate by an electron transfer from one to the other followed by protonation and exchange by an oxidized quinone from the pool. It is even possible that the first reduced quinone could have been stabilized by some kind of protonation event, thus favoring electron transfer to it from the subsequently formed semiquinone. The exchange of hydroquinone for quinone in a homodimer would mean that the site would be empty part of the time, resulting in another chance for photochemical misses to occur.

From this speculation on the properties of a homodimeric quinone-reducing reaction center (see Ref. [72] for earlier speculations), a number of features can be predicted that would make electron transfer relatively inefficient. It is clear then that gene duplication would provide an advantage in terms of photochemical efficiency by allowing (1) specialization of the quinones into Q_A and Q_B and (2) the directionality of charge separation to be defined. It seems likely that these two changes were linked. Is it possible that the mutations that differentiated Q_A from Q_B at the same time led to charge separation becoming unidirectional? It is perhaps easier to imagine that two sequential events took place. However, it is interesting to look at chimeric mutants in which two Q_A or two Q_B sites are present to see if the directionality of charge separation is affected.

A recently reported double-Q_A mutant showed no activity. However, a pseudo-revertant was isolated that showed an amino acid change away from the quinone binding site. The change in the revertant was close to the inactive (M-side) bacteriopheophytin and is thought to be related to the channel by which quinone exchange occurs [43]. It may be possible then that changes in this area of the protein in the ancestral (double-gene) homodimer affected quinone exchangeability and the pheophytin properties in such a way that directionality and quinone differentiation occurred at the same time.

6.3.3 The 43- and 47-kDa Polypeptides

The 43- and 47-kDa polypeptides (CP43 and CP47) each bear many chlorophylls that play a light-harvesting role. There is some debate over the actual number of chlorophylls and estimates range from around 12 to 25 chlorophylls per subunit. Both are thought to be made up of six transmembrane helices. Intriguingly, they both have massive loops between the fifth and sixth helices that protrude into the internal chloroplast space and could play important roles in water splitting. For the CP43, some residues on this loop appear to be required for O_2-evolving activity while for the CP47, part of the loop appears to be required for binding of the extrinsic 33-kDa polypeptide [40, 41, 44].

From their structural similarities, it seems likely that the CP43 and CP47 polypeptides evolved from a common ancestor. If so, then we might expect these polypeptides to be structurally symmetric to each other. An ancestral PSII could have been homodimeric in terms of its CP43/CP47 precursor proteins. It is also possible that these subunits are evolutionarily related to the antenna-containing peripheral part of the RCI-type reaction center [4]. We note that most folding models of the RCI reaction center subunits predict 11 membrane-spanning helices (see Chapter 7). Our detailed mathematical analyses predict that the addition of a six-helix core antenna (e.g., CP47 or CP43) to a five-helix reaction center (e.g., D1 or D2) would give rise to an 11 membrane-spanning helix, RCI-like protein. If so, some fundamental structural relationships between the central photochemical part and the peripheral antenna part of the PSI reaction center may be present between D1/D2 and CP43/CP47 in PSII.

Below we postulate the existence of a PSII ancestor that was capable of generating a very high oxidizing potential but that had a homodimeric core. Such a protein could also have been flanked by two identical symmetric antenna proteins that later became the CP43 and CP47 polypeptides. Given the indications that these polypeptides have some kind of functional role in O_2 evolution, it is possible that they played a similar role in a homodimeric O_2-evolving ancester, should it have existed.

6.3.4 The Extrinsic Subunits

Three major extrinsic proteins are associated with the internal surface of the PSII reaction center. A 33-kDa protein is present in the enzyme from all species studied. It does not seem to contain any cofactors and its role is not clear, although a structural role insulating the Mn complex from reductive attack had been inferred from studies of the enzyme in the absence of this polypeptide. A more sophisticated role for this protein may become clear as our understanding improves. Because activity in the enzyme occurs only in the presence of a high salt concentration in vitro, a role associated with chloride function might be inferred but as yet the situation is vague [17, 40, 42, 45].

Extrinsic subunits with molecular masses of 23 and 17 kDa are present in PSII from plants and algae. These polypeptides are related to calcium and chloride requirements and are likely to play a role in regulation of enzyme function as well as passively insulating the Mn from reductive attack. The 23- and 17-kDa polypeptides are not present in cyanobacteria; instead two extrinsic membrane proteins, a 12-kDa polypeptide and a cytochrome c-550, seems to be associated with the O_2-evolving enzyme [19]. The differences at this level between cyanobacteria and plants may be taken as a further indication of secondary, probably regulatory, roles for these extrinsic polypeptides [17,45]. Indeed a specific regulatory role has been suggested to occur in plants under some circumstances: Ca^{2+} release, presumably modulated by the 23-kDa polypeptide, may be involved in the down-regulation of PSII electron transfer that is required for optimization of electron flow as light intensities increase [46].

6.3.5 Other Subunits

There are indications in the literature for other proteins associated with PSII that may yet be shown to play important functional roles [40–42]. An intriguing case is an intrinsic membrane protein of around 30 kDa that shows weak amino acid oxidase activity and has been suggested to have a role in Mn binding [47]. Another case that has received much more attention is cytochrome b-559. This is probably a heterodimeric monoheme protein and is closely associated with the reaction center. There are no obvious counterparts to this cytochrome in other reaction centers, so it is difficult to speculate on its evolution. There is an on-going debate about whether one or two cytochromes per reaction center exists [17]. Despite precedence for pairwise symmetry, at the time of writing, we would say those in the one-per-reaction-center corner are slightly ahead on points, but there may still be several rounds to go.

The role of cytochrome b-559 is not clear. Consequently, suggestions for the function of cytochrome b-559 abound [16, 17, 48]. The following are examples that come to mind: (1) a range of protective effects, donating electrons and/or accepting electrons in times of need; (2) special roles associated with activation of the O_2-evolving enzyme, which may also be protective under these special conditions; (3) structural roles; and (4) some involvement in protonation events. The protective roles suggest that the association of cytochrome b-559 with PSII came as a secondary event to improve the efficiency of the system once the oxidizing capacity required for O_2 evolution had been attained.

There is no obvious H-subunit counterpart in PSII. There have been a number of suggestions of candidates [17] but because the H-subunit does not contain any cofactors it will not be easy to prove. As mentioned earlier, in bacterial reaction centers, the H-subunit is a bulky protein that caps the acceptor side of the reaction center and its main function is presumably to insulate Q_A from the outside medium. In PSII, extra loops of the D1 and D2 are present but no large globular capping protein has yet been identified. The need to regulate electron flow at the level of the iron may account for the absence of a H-subunit or at least may be reflected in different properties of a putative H-subunit in PSII (i.e., it might be less bulky and more weakly associated to the reaction center).

Some low molecular weight proteins have been found to be associated with the reaction center; however, no function has yet been determined for these and they are likely to be regulatory.

6.4 Charge Separation, Primary Reactions, and the Electron Acceptor Side of PSII

6.4.1 Kinetics of Charge Separation in PSII

There is disagreement about the kinetics of charge separation in PSII [7]. The earliest reports indicated that charge separation occurred in a few (3) ps as observed in purple bacterial reaction centers. Recently a 20-ps phase has been assigned to charge separation. This discrepancy has not yet been resolved but it has been pointed out that in bacterial reaction centers in which Tyr-210 of the M-subunit is specifically mutated to Leu, the charge separation kinetics become close to 20 ps. It was also noted that the equivalent amino acid in the PSII reaction is naturally a Leu [49]. Similarly, *Chloroflexus* also has Leu in the corresponding position in its reaction center and it too shows a relatively slow charge separation time [49b].

Whether the primary charge separation is 3 or 20 ps (see Ref. [7]), a recent electron transfer theory indicates that the distance between the electron transfer components must be small [24]. If this reflects an electron transfer step from chlorophylls positioned like those of the bacterial special pair to pheophytin, then an intermediate electron transfer component equivalent to the monomeric BChl is likely to be present. An alternative model suggests that P680 may involve a Chl that is structurally equivalent to the BChl monomer [6, 50] and is thus already close to Ph.

Electron transfer from Ph^- to Q_A has been measured to occur with a half-time of a few hundred picoseconds, a value very similar to that obtained for the equivalent step in purple bacterial reaction centers [51]. A tryptophan residue (M252) is present between the BPh and the quinone ring of Q_A. An equivalent tryptophan is conserved in all reported sequences of PSII and in purple bacterial reaction centers [5, 20] and thus would appear to be important in this electron transfer step.

6.4.2 The Shallow Trap in PSII

The yield of charge separation in PSII is low when Q_A is reduced whereas in purple bacteria it remains high. As a result of this, the fluorescence in PSII, and thus the loss of absorbed light energy, is very high when Q_A is reduced. This seems to be due mainly to the fact that the antenna in PSII have almost the same spectrum and hence energy as the primary electron donor chlorophyll. Therefore, the equilibrium, $Chl^*_{antenna}P\text{-}680 \rightleftharpoons Chl_{antenna}P\text{-}680^*$, results in a lowering of the effective energy level of the P-680*, making charge separation less energetically favorable than in the other reaction centers. This has a most marked effect when Q_A^- is reduced, conditions in which it is proposed that an electrostatic effect on the P-680$^+$Ph$^-$ radical pair makes charge separation energetically less favorable [7, 52]. Although there is no compelling reason to invoke any other explanations, other factors may contribute to this effect. (1) The energy between P* and the P$^+$Ph$^-$ radical pair may be smaller in PSII than in bacterial reaction centers in the equivalent state. (2) The first charge separation may be slower in PSII. (3) The electrostatic effect of Q_A^- on the P-680$^+$Ph$^-$ radical pair may be greater than between the corresponding states in purple bacteria.

The shallow trap of PSII may be an inherent and unavoidable consequence of a primary donor at 680 nm; however, the lower yield of charge separation can be seen as a beneficial protective mechanism preventing charge separation and associated damage effects when Q_A is reduced [53, 88].

6.4.3 The Pheophytin and Monomeric Chlorophylls

The spectroscopic characteristics of the pheophytin, its redox potential, and its reactivity in PSII are similar to those of its bacteriopheophytin counterpart in purple bacteria [10, 54]. The presence of a second pheophytin as a counterpart to the M-side inactive branch BPh in purple bacteria is indicated from spectroscopic measurements, pigment quantitation, and from the bacterial analogy model [15]. The protein environment around the pheophytin is similar to that around the BPh in the bacterial reaction center [5].

The presence of monomeric chlorophylls in PSII that are counterparts to those situated between $(BChl)_2$ and BPh in purple bacteria is inferred from the fact that the minimal reaction center seems to contain four chlorophyll molecules [15, 55, 56]. Electron paramagnetic resonance (EPR) has demonstrated the presence in PSII of a chlorophyll that has a ring tilt that is the same as that of the monomeric bacteriochlorophylls of the bacterial reaction center [50]. However, the His residues that ligate the monomeric bacteriochlorophylls and that are located on the short helices that run parallel to the membrane between helices three and four are not conserved in PSII [5, 20]. Nevertheless, in model studies, it has usually been assumed that the pigments in PSII are positioned as in the bacterial reaction center (see Fig. 6.2). It has been suggested, however, that a monomeric chlorophyll in PSII may play the role of the photooxidizable pigment rather than the first electron acceptor as in the bacterial reaction center [6, 50]. This suggested change from being the primary acceptor in purple bacteria to being (or at least being part of) the primary electron donor in PSII may be of major fundamental significance (see Section 6.5.1).

6.4.4 The Quinone–Iron Complex

The existence of the two quinones in association with a nonheme iron atom in both PSII and purple bacterial reaction centers was one of the most convincing lines of evidence that the two reaction centers had marked structural similarities. This was particularly the case from the EPR studies used to demonstrate the existence of the semiquinone–iron complexes in both systems [12, 57, 58]. The EPR signals arose from complex magnetic interactions between the semiquinone and the iron which are determined by distance and geometry of the interacting components. The presence of remarkably similar EPR signals in PSII indicated that the structure and geometry of these components and their protein environments were very similar in both systems.

Despite the close similarities, there are several properties of the quinone–iron complex in PSII that were not found in the bacterial reaction center. (1) The iron can be oxidized and has a midpoint potential at around 400 mV. This E_m is modulated by a range of components that probably bind directly to the iron [57, 59]. (2) Electron transfer reactions in the quinone–iron complex are modulated by the binding of anions to the iron and its close environment. The literature has focussed on the role of bicarbonate as the natural ligand and formate as its inhibitory replacement [60].

Both of these effects are thought to reflect a difference in the ligand environment around the iron in PSII compared to the bacterial system. The crystal structure of the bacterial reaction center showed that the iron was liganded by four histidine residues and a bidentate glutamate [5]. Amino acid sequence comparisons showed that the four equivalent histidines were present in PSII but owing to an extra stretch of loop inserted into the relevant region of the D1 polypeptide, it was less clear if the glutamate residue was conserved. It was thus suggested that formate /bicarbonate may bind to the iron in place of the glutamate [5, 12].

What is the significance of the differences in the ligand environment of the iron in PSII? It is known that bicarbonate binding (probably to the iron but also perhaps nearby) is required for normal electron transfer through the Q_B site and it is thought that this is related to quinone protonation [60]. It is known that other carboxylic acids can displace the bicarbonate from the site, thereby slowing electron transfer. Recently, carboxylic acids with potential physiological relevance have been shown to have marked effects [59]. It is thus possible to imagine that complex feedback mechanisms, involving CO_2 (the final electron acceptor), carboxylic acids (with different inherent charges) produced in the Calvin cycle and protons, may regulate electron flow out of PSII. The regulation may be achieved through the influence of the charge on the carboxylic acid affecting the pK of an amino acid group close to the quinone site. Electron transfer inhibition could result from several factors: perturbation of the protonation reactions associated with reduction of Q_B, changes in the relative redox potentials of the two quinones, or changes in the redox properties of the iron. No such regulation seems to exist in purple bacteria [61].

In PSII, the midpoint potential of the Q_A/Q_A couple seems to be modulatable. Under some conditions, notably under conditions of low pH, electron transfer through PSII is down-regulated owing to light-induced Ca^{2+} release. At the same time, the E_m of Q_A shifts to more positive potentials. This shift also occurs when the Mn cluster is damaged or removed [46, 62]. Illumination of this system is predicted to give rise to $P680^+Q_A^-$ formation and recombination.

It is expected that recombination via a true back reaction can give rise to a high yield of triplet formation [52] which, under aerobic conditions, will lead to the formation of singlet oxygen and hence to protein damage (see Ref. [40]). It is possible that the positive shift in midpoint potential that occurs on pH-induced down-regulation results in a state that favors recombination by a tunnelling reaction rather than by a true (triplet-generating) back reaction [62]. In support of this hypothesis, it is known that in purple bacteria, the potential of the quinone determines the nature of the $P^+Q_A^-$ back reaction [27].

In bacterial reaction centers, this regulation may not be necessary for several reasons. (1) Cyclic electron transfer has inherently different regulatory problems from those of the linear electron transfer system of which PSII is part. (2) The oxidizing power in PSII is much more likely to cause problems if unregulated than the weak oxidants in the bacterial reaction center. (3) Most species of bacteria have carotenoids close the reaction center bacteriochlorophylls and these effectively quench bacteriochlorophyll triplets. No good evidence for carotenoid triplet quenching in the PSII reaction center has yet been reported. It has been suggested that carotenoid quenching mechanisms are prohibited in PSII because the highly oxidizing $P-680^+$ generated will oxidize the carotenoid in competition with other donor side reactions [63]. Indeed, it has been shown that under certain conditions, carotenoid oxidation can take place in PSII [10, 63]. (4) The linear electron transfer chain of oxygenic photosynthesis involves the very reducing reaction center, PSI, that can easily reduce oxygen, forming superoxide. Regulation of electron flow out of PSII could be an important in vivo method of avoiding PSI-mediated oxygen reduction. Purple bacteria do not have a PSI to keep under control.

6.5 The Electron Donor Side

6.5.1 The Photooxidized Chlorophyll: P-680

One of the key functional differences in PSII compared to all other reaction centers is that the primary donor is much more oxidizing. The bacterial special pair has a midpoint around 450 to 500 mV depending on the species, whereas P-680 has an E_m that is estimated to be around 1100 mV [51, 55, 64]. It is possible that this greater oxidizing power is obtained as an inherent feature of chlorophyll in that (1) there is more energy available in the 680 nm photon than in the longer wavelength absorptions of bacteriochlorophyll and thus there is more scope for having a highly oxidizing photooxidized state and (2) a chlorophyll special pair is intrinsically more monomeric and monomers are intrinsically more difficult to oxidize than dimers. These arguments are fine until we think about the primary donor of PSI, P-700, which is also thought to be a chlorophyll special pair but has an E_m of around 450 mV. It seems likely then that the extra oxidizing power of P-680 is due to much more than it simply being made up of chlorophyll rather than bacteriochlorophyll.

Recent work on the bacterial reaction center has shown that the special pair can be made to be much more oxidizing by the addition of histidines that hydrogen bond to the bacteriochlorophylls [65]. It is possible that this strategy could have made P-680$^+$ more oxidizing. However, the presence of a sufficiently increased number of potential hydrogen bonding histidines is not evident in the PSII reaction center proteins. Alternatively the oxidizing power of the chlorophyll could be boosted by the proximity of other positive charges on amino acids, on other oxidized chromophores, or on nearby metal cations.

In recent work, the purple bacterial reaction center was engineered so that an 11 amino acid stretch around the special pair was changed to be identical to that found in PSII. Although the special pair was found to have spectroscopic properties indicating a much more monomeric structure, its midpoint potential did not become oxidizing [66]. This may indicate that it is not simply the amino acid sequence around the special pair that is responsible for the high potential.

The nature of P-680 as determined by spectroscopy is not clear (see Refs. [7], [47], [51], [55], [64], [67–69]). It is convenient to consider that a chlorophyll special pair is involved that is structurally similar to that in the bacterial reaction center. This viewpoint is favored by the presence of conserved histidines in the PSII reaction center (D1 His-198, D2 His-198: numbering as in, e.g., refs. [17] and [40]) that corresponds to those liganding the Mg^{2+} ions of the bacteriochlorophyll special pair [5,20]. Models of this kind are found throughout the literature and although they could be correct, they seem to reflect a desire for neatness rather than a consideration of the existing experimental data which indicate that significant differences exist between P-680 and (BChl)$_2$.

First, the exciton coupling between the pigments is estimated to be much weaker in P-680 than between the bacteriochlorophylls of (BChl)$_2$ [7, 53, 55, 64, 67–69]. Indeed, the couplings in P-680 have been estimated to be on the order of those occurring between the so-called monomeric BChl and the (BChl)$_2$. Thus, in bacterial reaction center terms, the chlorophylls of P-680 are monomers. This weaker coupling does not necessarily reflect a major structural difference between the two reaction centers for the following reasons: (1) A chlorophyll special pair with a structure identical to the (BChl)$_2$ would be predicted to have a weaker excitonic coupling due to the intrinsic properties of Chl compared to BChl: that is, less aromatic overlap (see Chapter 7) and a weaker oscillator strength. (2) Minor structural changes (a moving apart or rotation of the pigments) could result in much weaker excitonic coupling).

Second, the chlorophyll triplet state, which is formed by recombination of the radical pair, P-680$^+$Ph$^-$, is located on a pigment in PSII that is oriented at 30° to the membrane plane rather than being perpendicular to the membrane plane as found for the equivalent triplet in the bacterial reaction center. This suggested that P-680 is made up of a chlorophyll that is tilted at 30°, a situation quite different from that in the bacterial reaction center [50].

Two models were proposed to explain the 30° orientation of the P-680 [50]. First, it was suggested that P-680 could be a Chl that was structurally equivalent to the monomeric BChl of the bacterial reaction center, interacting weakly with the neighboring pigments [6, 50]. An alternative model was that one of the counterparts to the special pair is tilted by 30°, thereby diminishing the coupling and providing an essentially monomeric P-680 ([50]; see also Ref. [71]). However, significant rearrangements of the protein structure are probably required to tilt a big tetrapyrrole molecule that is slotted between membrane-spanning helices. Thus the model of P-680 in which it is made up of a counterpart of the monomeric bacteriochlorophyll was favored [50].

Because it had not been unambiguously determined if the triplet was actually on P-680 itself rather than having migrated to a different chlorophyll and because the triplet orientation data indicated a rotation difference between the chlorophyll monomer and the BChl monomers of the bacterial reaction center, these models were put forward with some caution [6, 50]. However, recent spectral hole-burning experiments on the P-680 triplet led to the conclusion that the triplet was localized on P-680 itself (See Ref. [69]) and it was found that a rotation of the triplet axes would be expected in chlorophyll relative to bacteriochlorophyll (Ref. [70], pointed out in Ref. [69]). These results then seem to remove some of the ambiguity in the "30°-tilted-Chl" model for P-680. The weak coupling to nearby pigments has been suggested to involve another chlorophyll molecule, suggested to be a counterpart to one of the special pair [50] and a pheophytin [68, 69].

If P-680 is mainly made up of a counterpart of a monomeric bacteriochlorophyll, what would be the evolutionary consequences? First we should ask which kind of primary donor system came first: the special pair type or the monomeric chlorophyll type? Because the dimeric reaction center may have evolved from a monomeric precursor [3, 72], an early homodimeric reaction center may be imagined that contained two monomeric donors and that was a common ancestor of the PSII and special pair type reaction centers. We do not favor this scenario for several reasons. These include the fact that the special pair motif is present in all types of reaction center other than PSII and that PSII itself contains features that may reflect the residual special pair (i.e., conserved histidines). Instead, it seems more likely that the PSII-type monomeric electron donor evolved from an ancestor that contained a special chlorophyll pair.

The key event in the evolution of PSII could have been the mutation that changed the role of the monomeric chlorophyll from an electron acceptor to an electron donor. This could have provided a sudden quantum jump in the oxidizing potential which made water oxidation a possibility. It is not clear what this mutation was, although it seems likely that both the special pair and the monomeric chlorophyll(s) were affected. The fact that the histidine residues that ligand the monomeric BChls in the bacterial RC are absent in both D1 and D2 is consistent with a change occurring at this level prior to gene duplication. Thus a double-P-680 homodimer should be imagined (an object prone to severe photodamage, some might argue). Of course, it is possible that some kind of modification of the special pair of chlorophylls could have occurred in a heterodimeric ancestor that led to the monomer chlorophyll becoming the donor; even so, we are attracted by the sheer bizarreness of the double-P-680 homodimeric reaction center.

If we assume that recognizable structural counterparts of the special pair exist in PSII, what then could be their role if the primary electron donor involves a structural counterpart to a monomeric chlorophyll? One possibility is that one or other of the counterparts of the special pair is coupled to the monomeric chlorophyll [50, 68, 69] and may contribute to the excited state, cation, and triplet states of P-680. Another possible role for the special pair counterparts is as secondary electron donors. If they are involved in forward electron transfer, we must assume that they have undergone modifications to make them more oxidizing. Alternatively, they may be used as a side-pathway of lower potential donors as a way of supplying electrons to the overoxidized system when the electron donation from water is nonfunctional or kinetically limited.

Direct electron donation from TyrZ to a monomeric P-680$^+$ is inferred in this model [53]. Based on the structure of the bacterial reaction center, the oxidizable tyrosines (see below) are predicted to be almost as close to the monomeric chlorophylls as they are to the counterparts of the special pair.

6.5.2 Tyrosines

The existence of the two redox-active tyrosines in symmetrically related positions on D1 and D2 indicates that this feature was inherited from a homodimeric ancestor. The role of the tyrosines as donors presumably arose only when the primary donor had attained its very high oxidizing power and hence the system may already have been capable of oxidizing water. The existing heterodimer has clearly undergone several modifications that left the two symmetric tyrosines with quite different properties. One (TyrZ) is rapidly oxidized and reduced during forward electron transfer whereas the other (TyrD) is oxidized relatively slowly and remains stable in the oxidized form for hours [32]. If the monomeric chlorophyll, which is closer to TyrZ, is the primary donor, then distance factors could contribute to the kinetic differences between TyrZ and TyrD.

The positions of the tyrosines are not between a putative chlorophyll special pair and the lumen; therefore they are not obvious choices as electron carriers from the lumen to the putative special chlorophyll pair. On the other hand, electron donation from the lumen to a monomeric P-680$^+$, a counterpart to the monomeric BChl, would seem more likely to occur via the Tyr.

The role of TyrZ seems to be as a rapid electron donor associated with rapid deprotonation [32, 38]. Both tyrosines could have played similar roles in the homodimeric ancestor, at least at first. The tyrosines are far apart; hence it seems likely that their electron donors were independent. Mn^{2+} ions could have donated electrons to both tyrosines; thus we might expect to find evidence of metal binding sites close to each tyrosine. A possible metal binding site involving Asp-170 of D1 (for numbering see, for example, Ref. [17]) has been pointed out. Indeed this amino acid has been implicated in O_2 evolution in site-directed mutagenesis studies [17, 18]. However, spectroscopic studies indicate that the *paramagnetic* Mn cluster is further away than this site [32, 36]. It is possible that this site is a relic of an electron-donating metal-binding site but now has some other role, possibly as the Ca^{2+} binding site. It is even possible that this site still allows donation of electrons from Mn^{2+} to TyrZ˙ under some conditions, perhaps even under functionally relevant circumstances, for example, during photoactivation, the process by which the Mn cluster is assembled in the PSII reaction center protein [73]. Given the poor knowledge of the Mn cluster, we have to admit grudgingly the possibility that two separate dimers could be present; thus the site closer to TyrZ could be occupied by a nonparamagnetic dimer under the conditions studied. In short, a lot is still possible given the lack of hard information.

6.5.3 Multiple Donation Paths in PSII

The great oxidizing power of $P\text{-}680^+$, as well as providing the advantage of being able to oxidize water, brings with it the disadvantage of being able to oxidize a range of other components that could be detrimental: other chlorophylls, carotenoids, amino acids, etc. To avoid such oxidation problems, it was suggested that $P\text{-}680^+$ would have to be rather distant from the nearest chlorophyll [64]. This would lead to a decrease in the efficiency of excitation transfer from the nearest antenna chlorophyll. It was suggested that this could be overcome by Ph, which is not as easily oxidizable, being the immediate antenna for P-680 [64]. However, modifications of the chlorophylls other than distancing them from P-680 could also limit their ability to donate electrons. Indeed, it is not unreasonable to suggest that chlorophylls close to P-680 could also have shifted potentials as a result of the same influences that are responsible for the very high potential in P-680 itself.

It is thought that multiple donor pathways for P-680 may be used as protective mechanisms, injecting electrons into the system to reduce the lifetimes of potentially damaging oxidative species under conditions in which electron donation from water is kinetically insufficient. A donation pathway involving chlorophyll, cytochrome *b*-559, and Q_B has been suggested to work as just such a protective mechanism [74]. Also, donation from TyrD to the more oxidizing components of PSII could represent another protective donation pathway. In the latter case at least, the protective donation pathway could have been adapted from the ancestral donation route that at one time represented a more central role in electron transfer.

6.5.4 The Mn Cluster: How Did the Current O_2 Evolving Enzyme Evolve?

The structure and the location of the Mn cluster remain unsure [17, 32, 33]. Indications exist that a number of amino acids in the lumenal exposed sections of D1 and at least one amino acid in D2 (Glu-69) could be involved. Among the best candidates as Mn ligands are Ala-344, which forms the C-terminal of D1 and the close by residues, His-332, His-337, and Asp-342 [75]. Being on the terminal portion of a relatively long peptide chain protruding from the membrane-spanning area, this does not provide much useful information on the whereabouts of the cluster. A Mn binding role for other polypeptides that are obligatory for oxygen evolution has not been ruled out. These include cytochrome b-559 and the CP47 [17] and other "dark-horse" proteins such as that showing residual amino acid oxidase activity [47].

There is some consensus that the cluster is likely to be made up of four Mn. This is based on X-ray absorption and EPR spectroscopy [17, 33, 34, 76]. Although this is the best interpretation of the current data, it would be a mistake to accept this as a hard fact. Ambiguities remain, and it would not be too surprising if the Mn cluster turned out to be, for example, two separate dimers or a separate trimer and monomer system.

Without good knowledge of the structure or location of the Mn cluster, evolutionary insights are difficult. In what follows, we shall follow a line of thought based on an evolutionary theme. This is one of many such lines of thought and is given as an example that should not be taken too seriously, unless of course it turns out to be useful or even correct.

It has been suggested that gradual shifts in the potential of P^+ occurred as easily oxidizable electron sources were used up. Eventually the oxidizing power was enough so that water could be oxidized. Specific scenarios have been suggested involving the gradual use of more and more oxidizing electron sources, including nitrogen compounds or Fe^{2+} [77, 78]. The difficulty, as pointed out by those authors, is that there are no existing examples of photosynthetic species that have midpoint potentials for the primary donor between those of the bacterial reaction center and of P-680. The missing links are still missing.

An alternative scenario is that chlorophyll-containing reaction centers were able to oxidize Mn^{2+} very early in photosynthetic evolution. This would require the reaction center to be capable of generating a strong oxidant. We consider a sudden jump in the redox potential of the primary donor could have occurred, by, for example, the use of the monomeric chlorophyll as P-680 instead of a special pair of chlorophylls (see Section 6.5.1).

It is reasonable to assume that Mn^{2+} acted as an electron donor to the reaction center before the sophisticated multinuclear Mn cluster of the present-day enzyme was developed. The oxidized Mn, as Mn^{3+}, was probably reduced by reductants other than water. But accumulation of Mn^{3+} could have led to complex Mn chemistry involving water and leading to multimer formation.

The tendency for Mn^{3+} to form complexes in water has been demonstrated in the chemistry literature [79, 80]. In general, reduction of such systems to Mn^{2+} leads to disintegration of the complexes.

Above, we proposed the existence of a homodimeric reaction center that was capable of generating oxidizing power high enough to oxidize Tyr and Mn. Should we consider two isolated Mn donation sites or could there have been a centrally positioned Mn cluster? If there were two symmetric, isolated sites, then one of them could have given rise to an asymmetric positioned cluster in the current enzyme while the other has been lost. If, on the other hand, a central Mn cluster was originally present then clearly the donation pathway via TyrD has been lost. Because of the ambiguities existing concerning the position of the Mn cluster in the modern enzyme, it is not yet possible to resolve these and related questions. However, the first of these possibilities is perhaps the most attractive from an evolutionary standpoint.

A fairly wild yet intriguing idea cropped up in our debates on the cluster structure. Electron micrographs indicate that PSII is made of face-to-face dimers of the reaction center core [81]. If an ancestral homodimer was capable of oxidizing Mn (as Mn dimers?) at separate sites on both parts of the reaction center homodimer, and association of two reaction centers could have allowed the formation of two interreaction center bridged Mn tetramers. Although such a mixture does not fit very well with current stuctural information, notably the fact that monomers of PSII appear to be functional in O_2 evolution [81], we just throw the idea in here to amuse those readers with a taste for the outlandish. The significance of dimerization of the reaction center has yet to be understood. From a structual viewpoint, a face-to-face dimer could influence the interpretation of some spectroscopic data, especially the EPR relaxation enhancement of TyrD˙: if the Mn cluster is very assymetrically positioned, being much closer to TyrZ than to TyrD, is it possible that in a dimeric reaction center, TyrD is closer to the Mn cluster on the other half of the dimer? (To understand the context of the last sentence see Refs. [32] and [36].)

Probably the simplest structure potentially capable of oxidizing water is a Mn dimer. Many models have been proposed in which a Mn dimer is the only charge accumulating device [82]. A major problem for such a dimer might be that the four-electron oxidation of water would lead to Mn^{2+} formation which would be unstable and liable to fall apart. The addition of a second high valence Mn dimer would greatly improve the efficiency of such a process.

In primitive, Mn dimer-containing reaction centers, the partial oxidation of water to peroxide, a diffusable toxic species, could also have been a major limitation. Here again, the presence of more than two high-valence Mn cations could have provided a nearby site for peroxide oxidation, while maintaining the valence of the Mn at a higher, more stable level.

The idea of the Mn cluster being made up of two dimers, each sequentially performing a two-electron oxidation of water, has been raised earlier in different forms [e.g., Refs. [17], [79], [84]. Their arrival, like that of Noah's animals, two by two [85], during the evolution of PSII is an idea that provides some scope for further speculation.

It has been suggested that one (or both) of the dimers may have been derived from an existing catalase-type enzyme [83, 84]. The existence of catalase-type enzymes prior to the appearance of the O_2-evolving enzyme, as we know it, is possible. Catalase could have been necessary to deal with peroxide already present in the environment from both photolytic and primitive photosynthetic water oxidation. Thus, the structure of the binuclear catalase may be incorporated into the water oxidizing enzyme. Indeed, the μ-oxo dicarboxylato bridged dimer, which is thought to be present in catalase, can be incorporated into current popular models of the tetranuclear Mn cluster of the O_2-evolving enzyme without conflicting with current experimental evidence [17, 76]. It is also possible that a recognizable catalase motif may exist in lower S-states but that at higher S-states the ligation of the cluster changes. Another interesting thought, related to the "two-by-two" hypothesis given above, is that the two dimers may be found on different protein subunits, one of which may be a relic of a previously separate enzyme (see Ref. [47]).

The photoactivation process [73] by which the Mn cluster is assembled into the reaction center may reflect in some way the events that took place during evolution. Here again, there are signs that the Mn atoms are put in two by two: first the binding and oxidation of two Mn^{2+} ions, perhaps forming a $Mn^{3+}-Mn^{3+}$ dimer, followed by the association of two more Mn atoms, perhaps associated with some major refolding of the protein.

The existing O_2-evolving enzyme has associated with it a good deal of complexity which has been rationalized in terms of mechanisms protecting against premature water oxidation leading to the formation of partially oxidized intermediates (peroxide) and/or over-reduction of the cluster. The extrinsic polypeptides and the Ca^{2+} and Cl^- ions have been proposed to contribute to such a protective mechanism [17, 35]. Indeed, when the enzyme is damaged by heat, or removal of its extrinsic polypeptides or Cl^-, peroxide thought to have originated from water oxidation has been detected [17]. Thus, in evolutionary terms, these protective devices probably arrived later. However, it is also possible that Ca^{2+} and Cl^- play more direct roles in the water chemistry, for example, controlling the redox potential of the Mn cluster, etc. [17, 86].

6.6 Conclusions

In this section, we summarize some of the points discussed in the preceding sections. The summary contains less of the provisos and qualifications that we generally serve up with such stark speculations. For these, the reader is referred to the previous sections and to Ref. [87].

The differences between PSII and the purple bacterial reaction center and be divided into several categories: (1) donor side differences directly associated with the highly oxidizing electron donor (P-680) and the hardware of the O_2-evolving enzyme (TyrZ, Mn cluster, etc.); (2) differences associated with the assembly of the Mn cluster; and (3) differences associated with regulation of electron transfer and protective mechanisms that are assumed to be necessary owing to the existence of the highly oxidizing chlorophyll, the water oxidizing capacity, and the assembly of the Mn cluster. In general, the features in the first two categories are thought to predate those in the third. However, since the O_2-evolving enzyme appears to be rather complex in structure, it is likely that it evolved ito its present form via a number of intermediate, less sophisticated forms. Thus certain protective or regulatory devices may have been incorporated into the hardware of later forms (the two-by-two model of the evolution of the Mn cluster is an example of this; see section 6.5.4).

A fourth category of differences are those associated with the position of PSII at the start of an electron transfer chain that contains a second photochemical reaction center. These may include features that could have evolved before and after the appearance of O_2 evolution (see Fig. 6.4).

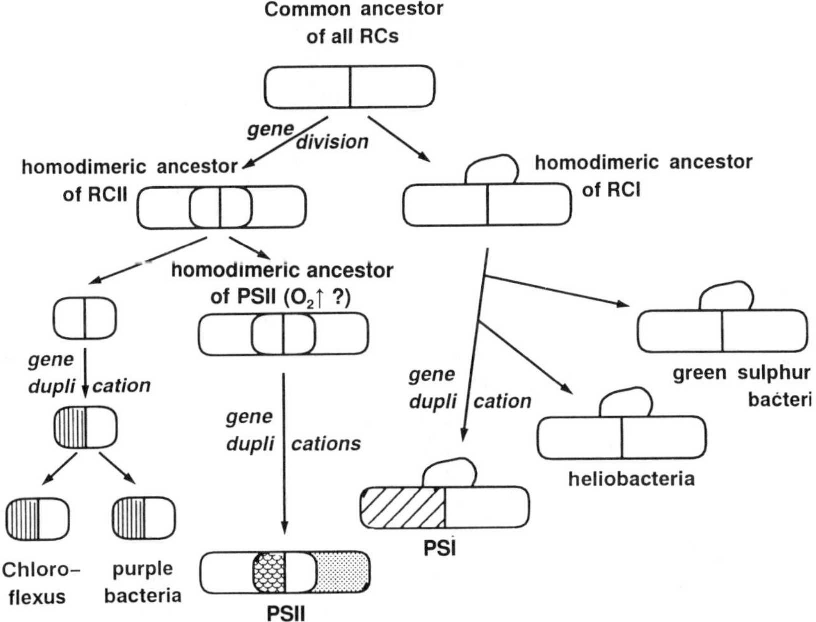

Figure 6.4 A possible evolutionary pathway for photosynthetic reaction centers based on the following hypotheses: (1) the small, RCII-type reaction centers evolved from the bigger, RCI-type, and (2) the CP43 and CP47 are evolutionarily related to the peripheral reaction center antenna of RCIs. It is not specified whether the common ancestor contains a nonheme iron or a [4Fe–4S] protein (F_X); however, the latter seems more likely. The gene division step may be linked to the change from F_X to Fe. The extra subunit on the RCI-type reaction centers represents the protein containing the F_A and F_B iron–sulfur clusters. Should it be proven that the tetraheme cytochromes exist as reaction center bound subunits in heliobacteria or green sulfur bacteria, this subunit might then be expected to be present in the common ancestor (see Ref. [4]). The properties of the homodimeric ancestor of PSII are speculated upon in the text. The scheme incorporates conclusions that are reached in Chapter 7 on the evolution of RCI-type reaction centers. This tree is based on the properties of the proteins themselves; see Chapter 7 for species-derived evolution trees.

The most recent common ancestor of purple bacterial reaction centers and PSII is thought to have been a homodimeric reaction center containing a special pair of Chl or BChl (see Chapter 7), monomeric BChls and BPhs, quinones as electron acceptors, and an iron atom associated with the quinones. Electron transfer to the quinone pool probably occurred by quinone exchange. Electrons were donated to the photooxidized special pair probably from a cytochrome heme.

When the separate gene duplications occurred on the way to form the heterodimeric ancestors of the existing heterodimeric RCII-type reaction centers, major differences had already evolved between the homodimers (see Fig. 6.5). In particular, the proto-PSII homodimer is thought not only to have switched to chlorophyll (if it was not using it already) but also to have attained a very oxidizing electron donor owing to a switch in function of the monomeric chlorophyll(s) from being the primary acceptor(s) to being the primary donor(s). As a consequence, a change in function of the special pair of chlorophylls must also have occurred. (See Section 6.5.1 for an alternative, less favored, model in which P-680 was formed by a mutation in which one of the chlorophyl special pair is tilted to 30°). The increased oxidizing power generated allowed Tyr radicals to be photooxidized on each side of the homodimer.

It is suggested that the oxidized tyrosines could have been reduced by a wide range of substrates, among which was Mn^{2+}. Mn^{3+} and Mn^{4+} could have been stabilized as dimers in primitive Mn binding sites on the surface of the reaction center. Such dimers, which we assume could accumulate charges up to the $Mn^{4+}-Mn^{4+}$ state, could have oxidized water to peroxide and to O_2 but with low efficiency owing, in part, to the over-reduction of the dimers and to the damage caused by released peroxide. We savor the idea of the possible existence of a rather exotic protein: an oxygen-evolving homodimeric proto-PSII reaction center containing two P-680s and two separate Mn dimers, although less outlandish options are of course possible and will certainly be more palatable to the squeamish.

The addition of a second Mn dimer (possibly by the capture of a preexisting manganoprotein, or by the association of the two reaction center-associated dimers perhaps as a result of a seismic change in the luminal-side proteins, or simply by the binding of two more Mn ions) could have improved the efficiency of the enzyme by allowing four-electron reduction of the cluster while maintaining high-valence Mn throughout the cycle, thus minimizing Mn loss. This scenario results in a dimer-of-dimer-type tetramer, which is the currently favored structure.

Gene duplication led to the appearance of a heterodimeric reaction center in which Q_A and Q_B functions were differentiated along with loss of electron transfer up the Q_B side of the reaction center. This also led to a change in role for one of the oxidizable tyrosines, from being TyrZ-like to TyrD-like. If separate, primitive Mn clusters were bound to each side of the ancestral homodimer, then the appearance of the heterodimeric reaction center, along

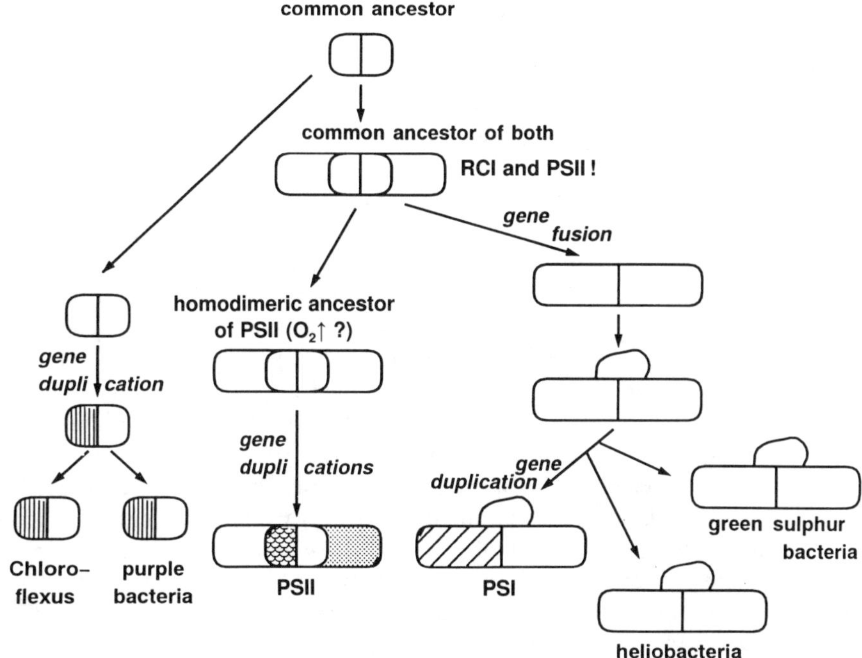

Figure 6.5 A possible evolutionary pathway for reaction centers based on the following hypotheses: (1) the big RCI-type reaction centers evolved from the smaller, RCII-type reaction centers, and (2) the CP47 and CP43 are evolutionarily related to the peripheral reaction center antenna of RCIs. Another version of this tree could be drawn in which the common ancestor of all reaction centers contained the homodimeric proto CP43/47 core antenna proteins. The pros and cons of this alternative version will not be debated here. The version we have chosen to show illustrates one simple model in which the first common ancestor is as simple as possible, that is, without associated core antenna. The scheme incorporates conclusions that are reached in Chapter 7 on the evolution of RCI-type reaction centers. For other details see the legend to Figure 6.4. This scheme is less favored than that in Figure 6.4.

with the associated electron transfer directionality, would probably have consigned the Mn site, on what became the TyrD-side of the reaction center, to Nature's dustbin.

The greatly increased oxidizing capacity of P-680$^+$ brought with it the problem of oxidizing damage to nearby components. This led to adaptations resulting in a series of fail-safe mechanisms in the form of sidepath donation systems and a mechanism for the fast replacement of the D1 protein when the inevitable photodamage occurred. The generation of chlorophyll triplet in PSII from charge recombination and the associated damage resulting from its reaction with O$_2$ may have led to adaptations that limit this problem: redox modulation of Q$_A$, a rapid triplet quenching mechanism [88], and the effect of

Q_A^- on charge separation. The association of several proten subunits with PSII probably reflects the accumulation of devices, some just passive insulation, others active in aspects of function or regulation, that improved the efficiency of the enzymic activity. A recurring question in PSII research (and in this chapter) is to distinguish those features, both proteins and cofactors, that are required for catalysis from those required for regulation, protection, etc. Answers to this question will be very helpful in understanding the function of the enzyme and in future efforts to trace its evolutionary origins.

The linear electron transfer pathway from an extrinsic substrate to PSI is a special situation that requires regulatory mechanisms, some of which could have been acquired before efficient O_2 evolution appeared. This is also the case for the regulation of the use of light energy between the two photosystems. A wide range of features of PSII may be assigned to such regulatory devices, including electron transfer inhibition at the level of the quinones via the bicarbonate and calcium effects, phosphorylation of certain subunits, the reversible association of certain subunits, and so forth.

We know of no examples of species that use only PSII as a reaction center. It seems possible that the development of an oxygen evolving reaction center was advantageous only as part of a linear, PSI-containing, electron transfer chain. An endless supply of electrons (from water), at the expense of all the associated paraphernalia for protection and repair that characterize PSII, might be worthwhile only when the electrons can be used directly in a second, low-potential, photoreaction to reduce useful substrates. From this viewpoint, a RCI-like reaction center is seen as having existed prior to the evolution of an xygen evolving PSII and PSII is seen as having evolved in a species in which the former was already present.

This brings us to the question of the evolutionary development of all the known reaction centers. We have previously avoided the question of whether RC1-type reaction centers developed from RCII-type or vice versa [4] and a discussion relevant to this question is provided in Chapter 7. There are, however, new elements given in the present chapter that have repercussions for this debate.

The CP43 and CP47kDa polypeptides may have been present as a homodimeric precursor in the ancestral homodimer reaction center of the PSII. It is suggested that these two proteins may be evolutionarily related to the core antenna which exists as an integrated part of the large subunits that make up the PSI-type reaction centers. This suggestion implies a closer relationship between PSII and PSI than is generally assumed and also implies that these two photosystems evolved within the same membrane and did not evolve in isolation, only to have themselves thrown together by some kind of gene transfer event (see Refs. [3], [88] and Chapter 7).

If the smaller RCIIs evolved from bigger RCIs (Fig. 6.4), we may assume that they did so by a gene division that led to a separation of the reaction center core from the peripheral reaction center antenna. These subunits have been maintained as a skirt of protein around the PSII reaction center and have been adapted in the luminal loop region to play some kind of important role in oxygen evolution. The purple bacteria must have disposed of this protein as excess baggage, preferring direct interactions with small-antenna proteins. This would also mean that the most recent common ancestor for RCIIs contained symmetric, and probably homodimeric, precursors of the CP47 and CP43.

If we now look at the alternate evolutionary pathway (Fig. 6.5), namely that the RCIs evolved from small RCIIs, then a gene fusion of the small homodimeric reaction center core with the peripheral antenna can be imagined to obtain the fat RCI-type structure. This implies that the most recent common ancestor of PSII and purple bacteria is also a common ancestor of RCIs and that a small, RCII-type reaction center with separate peripheral antenna, as in PSII, is the most recent common ancestor of PSII and the RC1s but that this ancestor is *not* shared by purple bacterial reaction centers. In other words, in this evolutionary pathway, PSII is more closely related to RCIs than to purple bacteria! If this is too difficult to swallow and if we accept the suggestion that the CP47 and CP43 are related to the peripheral reaction center antenna of RCIs, then we are led to favor the first evolutionary scheme described previously in which the common ancestor of all reaction centers was like RCI in terms of its reaction center protein size.

In Figures 6.4 and 6.5 and the associated discussion given above, we have anticipated the information and arguments given in Chapter 7. Comparisons of RCI and RCII are given in detail and their evolutionary origins are discussed against a background of the current thinking on phylogenetic relationships between species. Chapter 7 should also be consulted for a discussion of the difficulties in constructing a species-based evolutionary tree compared to those based on the properties of the reaction centers themselves (e.g., Figs. 6.4 and 6.5). A wider variety of evolutionary scenarios than previously proposed can in principle explain the extant photosynthetic organisms (see Chapter 7).

Since this chapter was written, the suggestion that the CP47 and CP43 proteins of PSII are evolutionarily related to the peripheral antenna part of the RCI-type reaction center has received independent support. Vermaas [90] performed computer-assisted comparisons of amino acid sequences and found significant regions of homology between the CP47 and the reaction center of heliobacteria. The evolutionary significance was discussed as were the structural ramifications. The conclusions drawn by Vermaas [90] were similar to those given here and in Chapter 7.

Acknowledgments

We would like to thank A. Boussac, K. Brettel, W. J. Coleman, J.-J. Girerd, A. Ivancich, G. Johnson, A. Krieger, W. Leibl, P. Mathis, T. Mattioli, A. Seidler, P. van Vlict, F. J. E. van Mieghem, S. Un, and E. Weis for stimulating discussions and H. J. Van Gorkom and G. Renger for useful comments on the manuscript.

References

1. Schopf, J. W., Walker, M. R. In *The Biology of the Cyanobacteria*, Carr, N. G., Whitton, B. A. (eds.), Blackwell, Boston, 1992; pp. 543–583.

2. Awramik, S. M. *Photosynth. Res.* 1992; *33*, 75–89.

3. Blakenship, R. E. *Photosynth. Res.* 1992; *33*, 91–111.

4. Nitschke, W., Rutherford, A. W. *Trends Biochem. Sci.* 1991; *16*, 241–245.

5. Michel, H., Diesenhofer, J. *Biochemistry* 1988; *27*, 1–7.

6. Rutherford, A. W. In *Light-Energy Transduction in Photosynthesis: Higher plants and Bacterial models*, Stevens, S. E., Bryant, D. A. (eds.), American Society of Plant Physiology, Rockville, MD, 1988; pp. 163–177.

7. van Grondelle, R., Dekker, J. P., Gilbro, T., Sunstrom, V. *Biochim. Biophys. Acta* 1994; *1187*, 1–65.

8. Deisenhofer, J., Michel, H. *Science* 1989; *245*, 1463–1473.

9. Feher, G., Allen, J. P., Okamura, M. Y., Rees, D. C. *Nature (Lond.)* 1989; *339*, 111–116.

10. Mathis, P., Rutherford, A. W. In *New Comprehensive Biochemistry: Photosynthesis*, vol. 15, Amesz, J. (ed.), Elsevier, Amsterdam, 1987; pp. 63–96.

11. Rutherford, A. W. *Biochem. Soc. Trans.* 1986; *14*, 15–17.

12. Rutherford, A. W. In *Progress in Photosynthesis Research, Vol. 1*, Biggins, J. (ed.), Martinus-Nijhoff, Dordrecht, 1987; pp. 277–283.

13. Trebst, A. *Z. Naturforsch.* 1986; *41c*, 240–245.

14. Barber, J., Marder, J. B. *Biotechnol. Genet. Eng. Rev.* 1986; *4*, 355–405.

15. Satoh, K. In *The Photosynthetic Reaction Centre* 1, Deisenhofer, J., Norris, J. R. (eds.), Academic Press, San Diego, 1993; pp. 289–318.

16. Pakrasi, H. B., Vermaas, W. F. J. In *The Photosystems: Structure Function and Molecular Biology*, Barber, J. (ed.), Elsevier, Amsterdam, 1992; pp. 231–257.

17. Debus, R. *Biochim. Biophys. Acta*, 1992; *1102*, 269–352.

18. Diner, B. A., Nixon, P., Farchaus, J. W. *Curr. Opin. Struct. Biol.* 1991; *1*, 546–554.

19. Vermaas, W. F. J., Styring, S., Schroder, W. P., Andersson, B. *Photosynth. Res.* 1993; *38*, 249–264.

20. Komiya, H., Yeates, T. O., Rees, D. C., Allen, J. P., Feher, G. *Proc. Natl. Acad. Sci. USA*, 1988; *85*, 9012–9016.

21. Blankenship, R. E., Fuller, R. C. *Encycl. Plant Physiol.* 1986; *19*, 390–399.

22. Nitschke, W., Dracheva, S. M. (1994) In *Anoxygenic Photosynthetic Bacteria*, Blankenship, R. (ed.), Kluwer, Dordrecht, 1994; pp. 775–805.

23. Deisenhofer, J., Norris, J. R. (eds.) *The Photosynthetic Reaction Centre*, Vol. II, Academic Press, San Diego, 1993.

24. Moser, C. C., Keske, J. M., Warncke, K., Farid, R. S., Dutton, P. L. *Nature* 1992; *355*, 796–802.

25. Arlt, T., Schmidt, S., Kaiser, W., Lauterwasser, C., Meyer, M., Scheer, H., Zinth, W. *Proc. Natl. Acad. Sci. USA*, 1993; *90*, 11757–11761.

26. Okamura, M. Y., Feher, G. *Annu. Rev. Biochem.* 1992; *61*, 861–896.

27. Shinkarev, V. P., Wraight, C. A. In *The Photosynthetic Reaction Centre 1*, Deisenhofer, J., Norris, J. R. (eds.), Academic Press, San Diego, 1993; pp. 193–255.

28. Tiede, D. M., Dutton, P. L. In *The Photosynthetic Reaction Centre I*, Deisenhofer, J., Norris, J. R. (eds.), Academic Press, San Diego, 1993; pp. 257–288.

29. Crofts, A. R., Wraight, C. A. *Biochim. Biophys. Acta* 1983; *726*, 149–185.

30. Brune, D. *Biochim. Biophys. Acta* 1989; *975*, 189–221.

31. Daldal, F., Zannoni, D. *Microbiol. Rev.* 1994 (in press).

32. Babcock, G. T., Barry, B. A., Debus, R. J., Hoganson, C. W., Atamian, M., McIntosh, L., Sithole, I., Yocum, C. F. *Biochemistry* 1989; *28*, 9558–9565.

33. Rutherford, A. W., Zimmermann, J. L., Boussac, A. In *The Photosystems: Structure, Function and Molecular Biology*, Barber, J. (ed.), Elsevier, Amsterdam, 1992; pp. 179–229.

34. Brudvig, G. W., Beck, W. F., de Paula, J. C. *Annu. Rev. Biophys. Biophys. Chem.* 1989; *18*, 25–46.

35. Rutherford, A. W. *Trends Biochem. Sci.* 1989; *14*, 227–232.

36. Un, S., Brunel, L.-C., Brill, T., Zimmermann, J.-L., Rutherford, A. W. *Proc. Natl. Acad. Sci. USA* 1994; *91*, 5262–5266.

37. Boussac, A., Rutherford, A. W. *Biochem. Soc. Trans.* 1994; *22*, 353–358.

38. Lavergne, J., Junge, W. *Photosynth. Res.* 1993; *38*, 279–296.

39. Krishtalik, L. I. *Bioelectrochem. Bionerg. 1990; 23*, 249–263.

40. Andersson, B., Styring, S. *Current Topics in Bioenergetics*, 16, Lee, C. P. (ed.), Academic Press, San Diego, 1991; pp. 1–81.

41. Vermaas, W. F. J., Ikeuchi, M. In *The Photosynthetic Apparatus*, Bogorad, L., Vasil, I. K. (eds.), Academic Press, San Diego, 1991; pp. 26–111.

42. Ghanotakis, D. F., Yocum, C. F. *Annu. Rev. Plant. Physiol.* 1990; *41*, 255–276.

43. Coleman, W. J., Youvan, D. *Nature* 1993; *366*, 517–518.

44. Bricker, T. *Photosynth. Res.* 1990; *24*, 1–13.

45. Murata, N., Miyao, M. *Trends Biochem. Sci.* 1985; *10*, 122–124.

46. Krieger, A., Weiss, E. *Photosynth. Res.* 1993; *37*, 117–130.

47. Pistorius, E. K. In *Electron and Proton Transfer in Chemistry and Biology, Studies in Physical and Theoretical Chemistry*, Vol. 78, Muller, A. (ed.), Elsevier, Amsterdam, 1993; pp. 237–251.

48. Cramer, W. A., Furbacher, P. N., Szczepaniak, A., Tae, G. S. In *Current Research in Photosynthesis*, Vol. III, Baltscheffsky, M. (ed.), Kluwer, Dordrecht, 1990; pp. 221–230.

49. Hastings, G., Durrant, J. R., Barber, J., Porter, G., Klug, D. R. *Research in Photosynthesis*, Vol. II, Murata, N. (ed.), Kluwer, Dordrecht, 1992; pp. 247–250.

49. b. Becker, M., Nagarajan, V., Middendorf, D., Parson, W. W., Martin, J. E., Blankenship, R. E. *Biochim. Biophys. Acta* 1991; *1057*, 299–312.

50. van Mieghem, F. J. E., Satoh, K., Rutherford, A. W. *Biochim. Biophys. Acta* 1991; *1058*, 379–385.

51. Renger, G. In *The Photosystems: Structure, Function and Molecular Biology*, Barber, J. (ed.), Elsevier, Amsterdam, 1992; pp. 45–99.

52. van Gorkom, H. J. *Photosynth. Res.* 1985; *6*, 97–112.

53. van Mieghem, F. J. E., Rutherford, A. W. *Biochem. Soc. Trans.* 1993; *21*, 986–991.

54. Klimov, V. V., Krasnovsky, A. A. *Photosynthetica* 1981; *15*, 592–609.

55. Seibert, M. In *The Photosynthetic Reaction Centre 1*, Deisenhofer, J., Norris, J. R. (eds.), Academic Press, New York, 1993; pp. 319–356.

56. Barber, J. In *Light-Energy Transduction in Photosynthesis: Higher Plants and Bacterial Models*, Stevens, S. E., Bryant, D. A. (eds.), American Society of Plant Physiology, Rockville, MD, 1988; pp. 178–196.

57. Diner, B. A., Petrouleas, V. *Biochim. Biophys. Acta* 1988; *895*, 107–125.

58. Evans, M. C. W., Nugent, J. H. A. In *The Photosynthetic Reaction Centre 1*, Deisenhofer, J., Norris, J. R. (eds.), Academic Press, San Diego 1993; pp. 391–415.

59. Petrouleas, V., Sanakis, Y., Deligiannakis, Y., Diner, B. A. In *Research in Photosynthesis*, Vol. II, Murata, N. (ed.), 1992; pp. 119–122.

60. Blubaugh, D., Govindjee, *Photosynth. Res.* 1988; *19*, 85–128.

61. Govindjee, Van Rensen, J. J. S. In *The Photosynthetic Reaction Centre 1*, Deisenhofer, J., Norris, J. R. (eds.), Academic Press, San Diego, 1993; pp. 357–389.

62. Johnson, G. N., Rutherford, A. W., Krieger, A. *Biochim. Biophys. Acta* 1995; *229*, 202–207.

63. Telfer, A., De Las Rivas, J., Barber, J. *Biochim. Biophys. Acta* 1991; *1060*, 106–114.

64. Van Gorkom, H. J., Schelvis, J. P. M. *Photosynth. Res.* 1993; *38*, 297–301.

65. Williams, J. C., Alden, R. G., V. H. Coryell, V. H., Lin, X., Murchison, H. A., Peloquin, J. M., Woodbury, N. W., Allen, J. P. In *Research in Photosynthesis*, Vol. I, Murata, N. (ed.), Kluwer, Dordrecht, 1992; pp. 377–380.

66. Coleman, W. J., Mattioli, T., Youvan, D., Rutherford, A. W. *Biochemistry* 1995 (submitted).

67. Hillman, B., Brettel, K., van Mieghem, F. J. E., Kamlowski, A., Rutherford, A. W., Schlodder, E. *Biochemistry* 1995; *34*, 4814–4827.

68. Schelvis, J. P. M., van Noort, P. I., Aartsma, T. J., van Gorkom, H. J. *Biochim. Biophys. Acta* 1994; *1184*, 242–250.

69. Kwa, S. PH.D. Thesis, Free University of Amsterdam.

70. van der Voss, R., van Leeuwen, P. J., Braun, P., Hoff, A. J. *Biochim. Biophys. Acta* 1992; *1140*, 184–198.

71. Noguchi, T., Inoue, I., Satoh, K. *Biochemistry* 1993; *32*, 7186–7195.

72. van Gorkom, H. J. In *Photosynthesis*, Vol. 15, Amesz, J. (ed.), Elsevier, Amsterdam, 1986; pp. 343–350.

73. Tamura, N., Cheniae, G. M. In *Light-Energy Transduction in Photosynthesis: Higher Plant and Bacterial Models*, Stevens, S. E., Bryant, D. A. (eds.), American Society of Plant Physiologists, Rockville, MD, 1988; pp. 227–242.

74. Buser, C. A., Diner, B. A., Brudvig, G. W. *Biochemistry* 1988; *31*, 11449–11459.

75. Nixon, P. J., Trost, J. T., Diner, B. A. *Biochemistry* 1992; *31*, 10859–10871.

76. Klein, M., Sauer, K., Yachandra, V. K. *Photosynth. Res.* 1993; *38*, 265–277.

77. Olson, J. M. *Science* 1970; *168*, 438–446.

78. Olson, J. M., Pierson, B. K. *Int. Rev. Cytol.* 1987; *108*, 209–248.

79. Pecoraro, V. L. (ed.) *Manganese Redox Enzymes*, VCH, New York, 1992; pp. 197–231.

80. Thorp, H. H., Brudvig, G. W. *New. J. Chem.* 1991; *15*, 479–490.

81. Boekema, E. J., Boonstra, A. F., Dekker, J. P., Rogner, M. *J. Bionerg. Biomembr.* 1994; *26*, 17–29.

82. Renger, G. *Photosynth. Res.* 1993; *38*, 229–247.

83. Vincent, J. B., Christou, G. In *Advances in Inorganic Chemistry*, 33, Academic Press, San Diego, 1989; pp. 197–255.

84. Gerasimenko, V. V., Gol'fel'd, M. G., Khanulov, S. V., Andreyeva, N. Y., Barynin, V. V., Grebyenko, A. I. *Biofizika* 1989; *34*, 618–622.

85. Genesis, 6–7, *The Bible.*

86. Dismukes, G. C. In *Bioinorganic Catalysis*, Reeijk, J. (ed.), Marcel Dekker, New York, 1992.

87. Roget, P. M. In *Roget's Thesaurus of English Words and Phrases*, Lloyd, S. M. (ed.), Penguin Books, New York, 1986; pp. 190–193.

88. van Mieghem, F. J. E., Brettel, K., Hillman, B., Kamlowski, A., Rutherford, A. W., Schlodder, E. *Biochemistry* 1995; *34*, 4798–4813.

89. Pierson, B. K., Olson, J. M. In *Microbial Mats*, Cohen, Y., Rosenburg, E. (eds.), American Society for Microbiology, Washington DC, 1989; 402pp.

90. Vermaas, W. F. J. *Photosynth. Res.* 1994; *41*, 285–294.

91. Dover, G. A., Linares, A. R., Bowen, T., Hancock, J. M. *Methods Enzymol.* 1993; *224*, 525–541.

The FeS-Type Photosystems and the Evolution of Photosynthetic Reaction Centers

*W. Nitschke, T. Mattioli,
and A. W. Rutherford*

7.1 Introduction

For more than a hundred years, the taxonomy of microorganisms was based mainly on morphology and ecology. The discovery of new species entailed only localized changes in the topology of phylogenetic trees. However, in the early 1980s, molecular biology and biomathematics invaded the field and led to a dramatic change in our understanding of phylogenetic relationships, in particular with respect to microorganisms. By the end of the 1980s, the general outlines of the new phylogenetic map of the microbial world had become generally accepted. The world of all living organisms is now considered to consist of three distinct "kingdoms," that is, those of the Eukarya, the Archaea, and the Bacteria. Whereas the eukaryotes were previously recognized as a distinct grouping, the former "prokaryotes" are now distinguished into two different kingdoms, the evolutionary distance between which is comparable to that between either of the new kingdoms and the Eukarya [1] (Fig. 7.1). No photosynthetic organisms have so far been found in the archaebacterial kingdom (the light-driven ion pumps of halobacteria are not related to the photosynthetic reaction centers performing light-induced charge separation as discussed in this chapter).

Among eukaryotes, by contrast, photosynthesis is widespread from unicellular algae to the vegetation of the Amazonean rain forest. In all of these organisms, however, the photosynthetic processes occur within organelles that have been recognized as those of "enslaved" descendents of cyanobacteria, that is, of species that belong to the kingdom of the Bacteria. The ability to use light

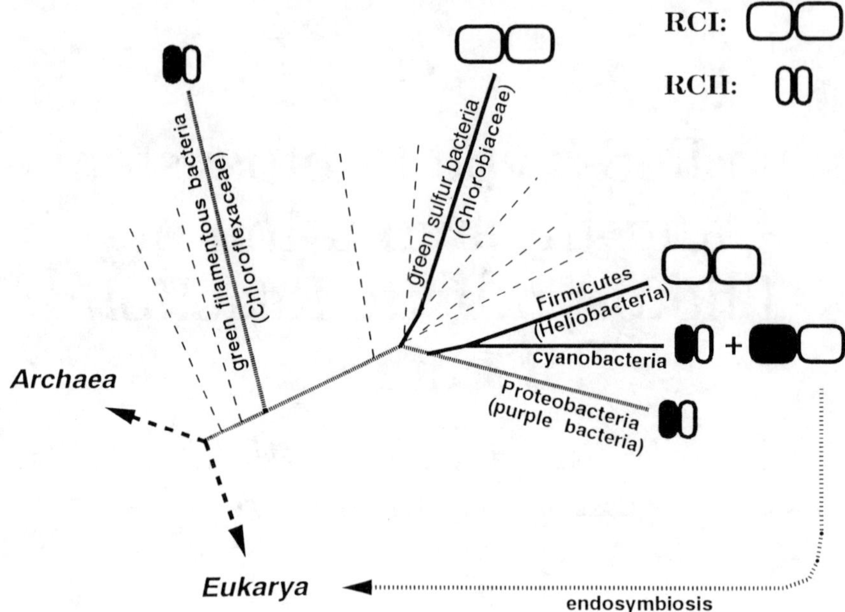

Figure 7.1   Phylogenetic relationship of photosynthetic species within the Bacteria (the former Eubacteria) based on 16S rRNA comparisons [1]. The tree represents the topology of these relationships and does not precisely reflect absolute evolutionary distances between species. The type of RC found in the respective phyla is indicated by the fat (RCI) or slim (RCII) dimers. Differently shaded halves of the dimers denote heterodimeric RCs. Branches marked by dashed lines represent phyla where no photosynthetic representatives have been discovered so far (these are in clockwise order starting from the root of the Bacteria: aquificales [98], thermotogales, deinococci and relatives, spirochetes, bacteroides–flavobacteria, planctomyces and relatives, chlamydiae). Note, that recently a tree with a somewhat different topology has been published [1a].

energy for metabolic needs has therefore been "imported" into the Eukarya and does not represent a genuine evolutionary achievement of this kingdom (see Fig. 7.1).

The actual origin of photosynthesis seems indeed to be located within the Bacteria. Five out of 12 branches (the so-called phyla) of the bacterial phylogenetic tree contain photosynthetic representatives (Fig. 7.1). Moreover, phototrophic species (i.e., *Chloroflexus*) are found close to the root of this tree, which has prompted several authors to propose a photosynthetic organism as the common ancestor of all Bacteria (see Ref. [1]).

Within the kingdom of the Bacteria, the detailed architecture of photosynthetic electron transport chains varies considerably between species. The photosynthetic reaction centers themselves, however, show much less variability. All reaction centers studied so far can be attributed to one of two distinct groups, that is, they can be classed as either a RCI- or a RCII-type reaction center (adopting the nomenclature introduced by Pierson and Olson [2]). RCI- (or FeS-) type photosystems are typically made up of a pair of core proteins in the 70- to 80-kDa range, a close antenna situated directly on these core subunits, and FeS centers as terminal electron acceptors. Apart from photosystem I (PSI) found in O_2-evolving organisms, RCI-type photosystems are present in species belonging to the Chlorobiaceae and to the Heliobacteria (belonging to the Gram-positive phylum, recently renamed the phylum of the "Firmicutes" [3]; see Fig. 7.1). RCII- (or quinone-) type RCs are characterized by a core consisting of a dimer of subunits each in the 30-kDA range and each possessing a two-electron-gating quinone as the terminal electron acceptor (see Chapter 6).

Most bacterial species found so far use only one of these two types of reaction centers for driving their electron transport chains. Only in the cyanobacteria, a relatively "typical" RCI-type photosystem (i.e., PSI) functions in series with the water splitting PSII, which is a very atypical version of an RCII-type reaction center (owing to specialization or old age; see Chapter 6).

7.2 The Properties of an RCI-Type Photosystem

7.2.1 Peptide Composition

Photosystem I from chloroplasts and cyanobacteria seems to be composed of more than 12 different protein subunits with apparent molecular weights ranging from 4 kDa to 80 kDa [4, 5]. However, only the two largest subunits, PsaA and PsaB, and the small (8 kDa) PsaC protein are considered to carry redox centers involved in primary charge separation [4] and therefore form the "functional core" of PSI. Some of the additional subunits probably serve to optimize the docking of soluble redox carriers (PsaD, PsaE, and PsaF [6–8]) or to mediate cyclic electron flow around PSI (PsaE [9]), whereas the function of others is unknown.

In contrast to this rather complicated picture of PSI, the purified RCIs from green sulfur and heliobacteria appear much less complex [10–21]. The isolated heliobacterial RC even migrates on gels as a single protein [14, 15]. Nevertheless, many of these isolated RCIs differ considerably from the in vivo state with respect to essential functional properties. This may be related to depletion of important additional subunits that are possibly homologous to subunits of PSI. Improved purification protocols for the green sulfur and heliobacterial RCs might reveal such additional proteins, yielding information on which subunits are essential for all RCIs and which ones serve to optimize the specific functions of the RC within the respective organisms (e.g., see Ref. [20]).

Whereas the presence of terminal FeS acceptors related to centers F_A and F_B (bound by the PsaC protein; see below) in PSI has been shown spectroscopically both in green sulfur and in heliobacteria (reviewed in Ref. [21]), the protein harboring these FeS centers has for some time escaped detection. Recently, the gene coding for the putative subunit homologous to PsaC in green sulfur bacteria has been found adjacent to the gene of the RC core protein [22] and the presence of the respective gene product in the isolated RC has been demonstrated [23].

In Heliobacteria, this kind of direct evidence for a PsaC-related subunit is still missing. Circumstantial evidence provided by depletion studies has been reported [24].

Green sulfur bacteria and heliobacteria might indeed contain significantly fewer subunits than does PSI; however, there is one protein present both in green sulfur and in heliobacteria that is clearly absent in PSI. Whereas re-reduction of the photooxidized primary donor P-700 in PSI is mediated by a soluble electron carrier (plastocyanin or cytochrome c_6), green sulfur and heliobacteria use a membrane- (and probably even RC-) associated cytochrome for this purpose. This detailed nature of this cytochrome is still a matter of debate (see below).

7.2.1.1 *Primary Sequences*

Analogies between the reaction centers of green sulfur bacteria and PSI were inferred already in the 1960s from their common ability to directly reduce $NAD(P)^+$, which indicated the presence of terminal electron acceptors substantially more reducing than the E_m of the $NAD(P)^+/NAD(P)H$ couple ($-320\,mV$) [25].

Evolutionary relatedness of the reaction centers from green sulfur bacteria, heliobacteria, and PSI was suggested from the extensive and detailed resemblances of the electron acceptors with respect to their chemical nature, electron transfer characteristics, and spectral properties (see below). Conclusive evidence for this relationship was provided by the elucidation of the primary sequences of their core proteins [22, 26, 27]. These primary sequences showed that the core of the reaction centers from green sulfur bacteria and heliobacteria were made up of proteins with molecular weights of 70 to 80 kDa, which is more than twice as large as those from RCII-type photosystems but rather similar in size to the PsaA/PsaB core proteins of PSI (for green sulfur bacteria, a large, PSI-like core was predicted from biochemical evidence already in 1984 [10]). Secondary structure prediction algorithms applied to the various RCI-type sequences (i.e., those of PSI, the green sulfur, and the heliobacterial RC) yield between 8 and 11 transmembrane helices, with significant interspecies differences close to the N- and C-termini but with a consistent common arrangement of transmembrane helices and hydrophilic stretches around the conserved sequence patches [22, 26–28]. Whereas the overall sequence homology between the core proteins from green sulfur, heliobacteria, and PSI is only around 20%, a number of highly conserved sequence domains stand out. The most obvious of these, a stretch of 12 amino acids long, was considered a likely candidate for the region ligating the FeS cluster F_X in PSI some time ago [4] and has been demonstrated to contain at least one of the four cysteine ligands to the F_X cluster [29]. The sequence region initially proposed to bind the special pair P-700 in PSI [30] also turned out to be conserved between the different RCI-type core proteins, adding weight to the proposal that the respective amino acids are structurally and/or functionally important. The function of these residues as ligands to P-700 may have to be reassessed based on the hypothesis in Chapter 6, in which RCI is seen as a combination of a five-helix RCII with a six-helix antenna equivalent to CP43/CP47. If this were correct, then the special pair ligands would be expected in helix 10 and, indeed, fully conserved histidines (His-792) are possible candidates (see sequence comparison in Ref. [27]). Recent studies using site-directed mutagenesis in fact exclude the originally proposed residues as ligands for P-700 in PSI [31]. Furthermore, the "antenna + RCII" model has recently found support (1) by sequence comparisons between the heliobacterial RC and the core antenna proteins in PSII [32] and (2) by the demonstration that a histidine in helix X of PSI is a likely ligand for P-700 [33].

Other sequence stretches found to be conserved between the three different RCI-type reaction centers were not anticipated and are therefore promising targets for site-directed mutagenesis in an attempt to elucidate their function (for a detailed description of the respective parts of the sequence, see Refs. [22], [26], [27]).

One quite remarkable outcome of the sequencing studies was the finding that only one protein (i.e., the *pscA* and *pshA* gene product, where "c" and "h" stand for *Chlorobium* and *Heliobacterium*, respectively) seems to be present in the core of the reaction centers from both green sulfur and heliobacteria. Since the F_X-binding stretches in these RCs, however, provide only two of the four ligands required for binding of an FeS center (just as in the PsaA and PsaB subunits of PSI), the core must be made up of two identical subunits. The photosystems of green sulfur and heliobacteria therefore seem to be homodimeric reaction centers, structures previously proposed as precursors of extant reaction centers but considered to have been lost during the evolution of photosynthesis.

7.2.1.2 The F_A/F_B-Binding Subunit

In all RCI photosystems, the terminal electron acceptors seem to be a pair of [4Fe–4S] centers, situated together on an extrinsic protein that is attached to the actual "RC core" on the n- (negative) side of the membrane (for a review, see Ref. [21] and also the more recent work in Refs. [12], [13], [23], and [18]). In PSI, this subunit (called F_A/F_B protein or *psaC* gene product) has a molecular weight of about 8 kDa and probably evolved from a soluble 2[4Fe–4S] ferredoxin [34]. In *Chlorobium*, a gene (named *pscB*) coding for a 2[4Fe–4S] protein has been found downstream of the RC core gene [26]. Since the respective gene product was shown to be present in the isolated RC from *Chlorobium* [23], it most probably represents the homolog of the PsaC subunit in *Chlorobium*. Surprisingly, in the *Chlorobium* F_A/F_B subunit, a separate N-terminal domain has apparently been added to the 2[4Fe–4S] domain, yielding a calculated molecular mass of about 20 kDa for this subunit. No protein or gene of a respective F_A/F_B subunit has been found in Heliobacteria yet [14]. Nevertheless, the light-induced reduction of FeS centers of the F_A/F_B type in *H. chlorum* [35] and their dissociability from the RC core [24] strongly indicate the presence of such a subunit.

7.2.2 The Chain of Electron Acceptors

7.2.2.1 Energetic Constraints in RCI and RCII

The main task of the chain of redox centers within photosynthetic reaction centers is to move rapidly and efficiently the negative charge to the other side of the membrane, away from its positive partner, P^+, thereby lowering the probability of charge recombination and of wasting the energy. In RCII-type photosystems, the transmembrane electron transfer ultimately takes the electron to an acceptor at an E_m of about -100 mV (Q_A) (see Chapter 6 and Fig. 7.2). This electrochemical potential is far too high to allow the negative charge on such an acceptor to be directly used as a reducing equivalent in the dark

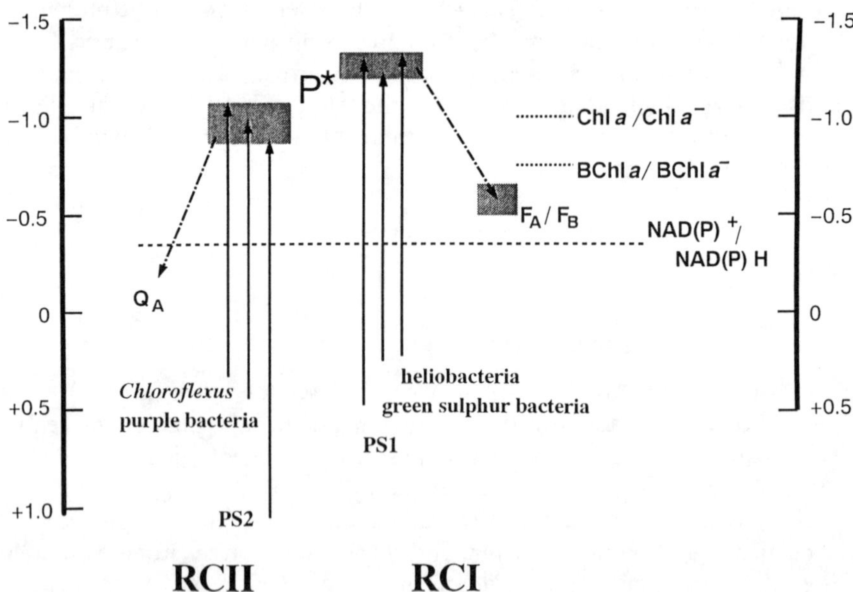

Figure 7.2 Comparison of the electrochemical properties of RCI and RCII photosystems. The redox midpoint potentials for reduction of Chl *a* and Bchl *a* shown are those obtained for the isolated pigments in acetonitrile [48].

reactions which are mainly mediated by the $NAD(P)^+/NAD(P)H$ couple with a redox midpoint potential of $-320\,mV$ (see Fig. 7.2).

By contrast, in RCI-type photosystems, electrons finally end up at FeS centers with E_m values between -500 and $-650\,mV$ (for reviews, see Refs. [4] and [36]). This leaves a sufficiently large drop of free energy to drive the intermediate electron transport events (involving soluble components such as Fd and FNR) that allow the eventual reduction of $NAD(P)^+$ to $NAD(P)H$ (Fig. 7.2).

Correspondingly, the starting point of the electron transfer cascade, that is, the first excited state of the primary donor, P^*, must be at a level of free energy allowing ejection of an electron toward significantly more reducing acceptors in RCIs than in RCIIs. The redox potential of this excited state cannot be determined directly. However, it can be estimated from the midpoint potential for the reaction in the ground state (P/P^+) and the energy of the photon inducing charge separation. E_m values for the heliobacterial RC, green sulfur bacterial RC, and PSI are $+220\,mV$, $+250\,mV$, and $+470\,mV$, respectively (see Refs. [4], [16], [17]). The absorption wavelengths of 798 nm, 840 nm, and 700 nm, respectively, for these three systems translate into $1.55\,eV$, $1.48\,eV$, and $1.77\,eV$. Therefore, the redox potential of P^*/P^+ turns out to be relatively constant around $-1.3 \pm 0.05\,eV$ for all RCI photosystems studied so far, whereas the redox midpoint potential of the ground state of P varies by up to $300\,mV$ for the respective systems (Fig. 7.2).

With regard to similarities of the chain of electron acceptors functioning between P and the terminal FeS clusters in the three representatives of RCI-type photosystems, the degree of consensus in the literature varies with the specific redox centers under consideration. Some of these centers are undoubtedly homologous (and in fact astonishingly similar) in all these species, whereas, for example, the existence and possible role of the quinone acceptor A_1 are still under debate (see below).

7.2.2.2 The Terminal Electron Acceptors F_A and F_B

It is now well established that both in green sulfur and in heliobacteria there are at least two different FeS centers operating as terminal electron acceptors in very much the same way as centers F_A and F_B in PSI [3, 6, 11, 12, 24, 35, 37, 38]. The detailed attribution of the spectroscopic features of these centers with respect to their counterparts in PSI is still somewhat controversial (for a discussion of this topic, see Ref. [21]).

The estimated redox midpoint potentials of F_A and F_B in the different RCI-type photosystems range from $-500\,mV$ to $-650\,mV$ [36].

As described previously, F_A and F_B are located on an extrinsic subunit attached to the hydrophobic RC core on the n-side of the membrane.

All other redox centers involved in light-induced electron transfer within the RC complex seem to be located on the core proteins.

7.2.2.3 Center F_X

Spectroscopic evidence for an F_X-type FeS cluster has been presented for green sulfur bacteria [37] and for heliobacteria [24]. Whereas the electron paramagnetic resonance (EPR) spectral features and orientations reported for green sulfur bacteria correspond to what is known for center F_X in PS1, the evidence for the heliobacterial center F_X relies so far essentially on kinetic optical data [24].

7.2.2.4 A_1, the Quinone Acceptor

In PSI, the role of a quinone as electron acceptor (A_1) prior to or in parallel with F_X [39] was recognized some time ago (reviewed in Ref. [4]). In both green sulfur and heliobacteria this acceptor seems to be rather elusive. It has even been reported for the heliobacterial RC that no quinone acceptor participates in electron transfer within the RC [40]. However, because the back reaction kinetics of the control sample in this work resembles that seen in an F_A/F_B-depleted preparation [24], the functional intactness of the control sample is doubtful. Moreover, spectral evidence for a quinone-type acceptor has been presented both for green sulfur [37, 41, 42] and heliobacteria [43] and the presence of quinones in a purified RC core of heliobacteria has been demonstrated [14].

7.2.2.5 The "Primary" Acceptor A_0, a Chlorophyll a-Related Molecule in All RCIs

By contrast to the situation for the acceptor A_1, the primary acceptor A_0 has been identified in all types of RCIs (reviewed in Ref. [36]; see furthermore Ref [44]) and has surprisingly been found to be a chlorophyll a-related molecule in all of these systems. This was particularly unexpected for green sulfur and heliobacteria, in which the reaction center was initially considered to contain only BChl a and BChl g, respectively. Therefore, a special chlorophyll molecule is apparently used as primary acceptor in these two bacterial RCIs and this chlorin molecule happens to be chemically very similar to Chl a, that is, the "bulk" chlorophyll of PSI. Evolutionary reasons for this fact have previously been discussed based on the so-called Granick hypothesis [45, 46]. In the framework of the Granick hypothesis it is assumed that the complexity of biosynthetic pathways mirrors phylogeny, that is, that increasing complexity of such pathways is found in more recently evolved organisms. Following this idea, it was speculated that these special chlorophyll a-type molecules might be a relic from putative ancestral Chl a-only containing photosynthetic organisms [36]. Recent comparative sequence studies on enzymes involved in the biosynthesis of Chl and BChl, however, strongly favor BChl as being used by the ancestral photosynthetic reaction centers [47], in contradiction to the Granick hypothesis.

What then could be the reason for this stubborn preference of the RCIs for a chlorophyll molecule in the A_0 site? One possible explanation is that functional requirements forced green sulfur and heliobacteria to use a specialized molecule (i.e., Chl a) instead of BChl a and BChl g, respectively, in the site of the primary electron acceptor. Indeed, the free energy of reduction to the anion is roughly 300 mV more negative for chlorophyll than for bacteriochlorophyll (-1040 mV versus -750 mV, both in acetonitrile [48]; see Fig. 7.2). Although it is admittedly well established that proteins can significantly modify E_m values of redox centers, a shift of almost 500 mV necessary to make BChl a suitable for accepting electrons from P* and delivering them to subsequent acceptors might nevertheless be too difficult to achieve for these RCs without resulting in deleterious side effects.

7.2.2.6 *The Primary Donor P*

In contrast to the situation for the primary acceptor A_0, the optically determined P/P^+ difference spectra of the three groups of RCI photosystems show that the primary donor is formed by molecules identical or chemically equivalent to the bulk chlorophylls of the respective systems, that is, BChl a in green sulfur bacteria, Chl a in PSI, and BChl g in heliobacteria. The specification "chemically equivalent" is necessary because epimeric forms (i.e., the 13^2-epimer) of the respective BChls have been proposed to constitute P both in PSI [49] and in heliobacteria [50]. Because chlorin reductase, the enzyme catalyzing the final step in synthesis of the chlorin moiety [51], was recently shown to be unable to produce the epimer form, a biological activity resulting in 13^2-Chl a has yet to be discovered. In any case, a primary donor constituted from the epimeric form of the BChls does not seem to be a RCI-specific feature because no epimeric BChl has been found in green sulfur bacteria [52].

In purple bacterial (RCII-type) RCs it was demonstrated already some time ago that the primary donor is actually a dimer of BChls, whereas a monomeric nature of P-700 seemed possible in part on the basis of measurements of the spin-polarized triplet ^{3}P-700 in PSI [53, 54]. More recent studies of the triplet state of P-700, however, were interpreted to favor a dimeric structure also for P-700 [55, 56]. The monomeric appearance of ^{3}P-700 has been shown to be observed only at cryogenic temperatures, whereas at higher temperatures the triplet spectrum is indicative of delocalization onto or charge transfer between the two halves of a chlorophyll dimer [56]. Data obtained by Raman spectroscopy have furthermore led to the proposal that P-700 is structurally very similar to the special pair in purple bacteria [57]. ENDOR and Raman data obtained on the green sulfur bacterial cation radical P-840$^+$ have shown a perfectly symmetric structure of P-840 [58, 59].

That RCI-type photosystems contain a special pair as primary donor has already been proposed some years ago on the basis of the fact that the zero-field splitting (ZFS) parameters of the triplet state of the green sulfur bacterial door (^{3}P-840) were indicative of a dimeric structure of P-840 [37, 60].

Data with respect to the spin-polarized triplet ^3P have also been reported for heliobacteria [35, 61, 62].

The ZFS parameters and the orientation of the triplet axes reflect some of the structural characteristics of the pigments carrying the triplet. Essential relationships between the triplet, optical, and molecular axes are visualized in Figure 7.3, using the example of the purple bacterial special pair. The triplet z-axis is perpendicular to the plane(s) of the parent BChl molecule(s) [64]. Since the planes of the two BChl a molecules constituting the special pair are both roughly perpendicular to the membrane plane, the triplet z-axis lies parallel to the membrane. The x- and y-axes of a triplet localized on an isolated BChl in solution have been shown to correspond to the optical Q_y and Q_x-axes, respectively [64]. In turn, the relationship of the optical transitions to molecular axes is shown in Figure 7.3; Q_y is roughly collinear with a line through the pyrrol nitrogens of ring I and ring III both in BChl and in Chl (note that the triplet x-axis is rotated by $55°$ in Chl a with respect to BChl a; see below). The orientation of the x- and y-axes in the dimer, however, depends on the degree of triplet delocalization over and charge transfer character between the two halves of the dimer. In *R. viridis*, where charge transfer is significant, the triplet x- and y-axes are determined by the molecular axes of the individual bacteriochlorin molecules, that is, they are about collinear with the Q_y- and Q_x-axes of the BChls (Fig. 7.3, top scheme, shown to the right of the dimer). In *R. sphaeroides*, where delocalization dominates charge transfer character, the triplet axes correspond to the vector sums of the optical transitions, that is, they are perpendicular and parallel to the symmetry axis of the dimer (C_2), which coincides with the membrane normal (this situation is shown to the left of the dimer) [63–65].

The only common feature of the triplets in the three different RCIs, that is, ^3P-840 (in green sulfur bacterial), ^3P-700 (in PSI) and ^3P-798 (in heliobacteria) is the orientation of the triplet z-axis. This axis was found to be parallel to the plane of the membrane in all three RCIs, indicating that the plane of the BChls carrying the triplet is perpendicular to the membrane just as in the case of the special pair in purple bacteria.

However, with respect to the remaining two axes, ^3P-700 and ^3P-800 turned out to be rather different from each other in that x and y are parallel and perpendicular, respectively, to the membrane in ^3P-840, whereas they lie at oblique angles in ^3P-700 (see Fig. 7.3). In the heliobacterial ^3P-798, these two axes could not be determined so far. Furthermore, the degree of narrowing of the triplet spectrum is also quite heterogeneous between the three systems (see values of $r = D[\text{RC}]/D[\text{BChl}]$ in Fig. 7.3).

Assuming that the analogies to the purple bacterial special pair extend to the mutual arrangement of the two halves of the dimer, these differences can be rationalized as follows. The triplet on ^3P-840 in green sulfur bacteria would be symmetrically delocalized onto both halves of the dimer, resulting in x- and y-axes that represent the average over the molecular axes of the individual molecules as in the case of *R. sphaeroides*. In fact, it seems difficult to imagine

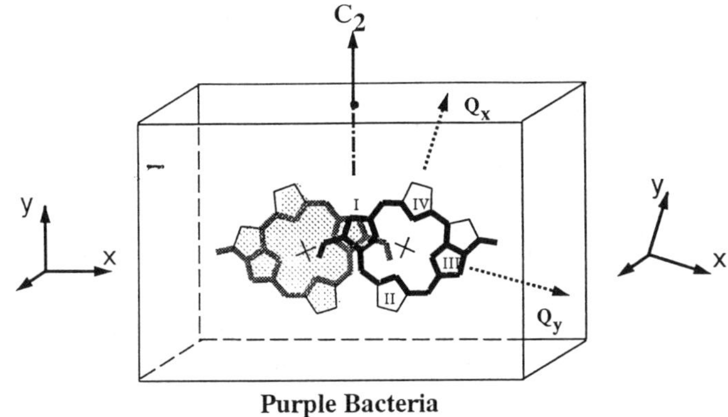

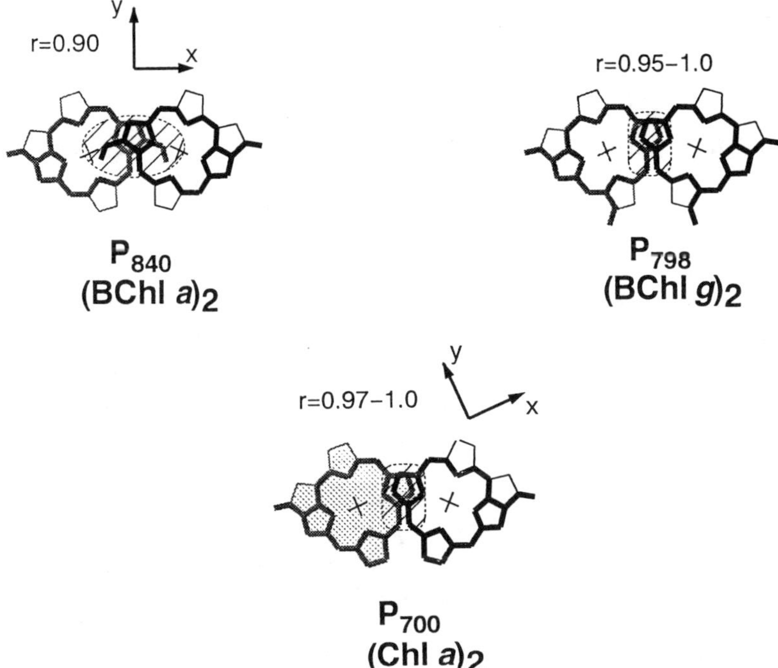

Figure 7.3 Proposed structures for the primary donors of green sulfur bacteria (P-840), heliobacteria (P-798), and PSI (P-700) based on the X-ray structure of the purple bacterial primary donors. The relationship between optical, triplet, and molecular axes is detailed for the purple bacterial special pair in the top scheme. The orientations of the triplet x- and y-axes shown to the right and to the left of the special pair correspond to the results as obtained for *R. viridis* and *R. sphaeroides*, respectively. In P-700 and the purple bacterial special pair, one half of the dimer is shaded to indicate that asymmetry between the two BChl molecules is induced by the nonidentical proteins of the heterodimeric core, PsaA/PsaB, and L/M, respectively.

how anything other than a completely symmetric situation could arise in the homodimeric RC of green sulfur bacteria. By contrast, the obliquely oriented x- and y-axes of ^{3}P-700 suggest a significant localization of the triplet state on one half of the dimer in the heterodimeric PSI. As mentioned previously, in Chl a the x- and y-axes make an angle of 55° and 43°, respectively, with the optical Q_y transition [65]. The experimentally obtained orientations of the triplet x- and y-axes would therefore yield an inclination of the Q_y direction with respect to the membrane of about 25°, that is, rather close to the value of 15° found in the X-ray structure of the purple bacterial RC.

Thus, in both green sulfur bacteria and PSI, the orientations of the spin-polarized triplet axes are in line with a geometry of the primary donor, similar to that of the purple bacterial special pair. Molecular orbital calculations furthermore show that the optical spectroscopic parameters of P-798 are in line with a dimeric structure [66, 67]. If this is so, what could be the reason for the pronounced difference between PSI, the helio-, and the green sulfur bacterial RC with respect to the degree of narrowing of the triplet spectrum? A clue for this problem could come from the study of the heliobacterial primary donor. BChl g is chemically intermediate between BChl a and Chl a. It does not possess the acetyl substituent at position 3 of ring I, characteristic for BChl a, but has a vinyl group instead (typical for Chl a) in this position. Furthermore, ring II is only partially included in the chlorophyll macrocyclic π-system in BChl g, whereas it is fully included in that of Chl a and fully excluded in that of BChl a. The reduction factor of the triplet ZFS parameters of the heliobacterial ^{3}P-798 closely resembles that of ^{3}P-700, indicating that the presence or absence of delocalization or charge transfer character might be related to the chemical differences on ring I modulating the extent of overlap of the π-systems of the two halves of the dimer. In Figure 7.3, circled areas represent the extent of overlap of the π-systems of the constituents of the special pair. The expected greater overlap for P-840 is based on the fact that BChl a possesses a conjugated acetyl carbonyl group at the C3 position of ring I that extends the conjugated system of the macrocycle, at this position, to a greater degree than do the vinyl groups of Chl a and BChl g. This effect is most easily seen in the absorption spectra of BChl a and Chl a derivatives modified at the C3 position [68, 69]. In both cases, the presence of an acetyl group at the C3 position of BChl a or [3-acetyl]-Chl a exhibits a Q_y absorption band that is significantly red-shifted compared to its [3-vinyl]-BChl a or Chl a counterparts, respectively.

7.2.3 Location of the Chromophores

The purple bacterial RC, which is also characterized by a dimeric core, contains two symmetric chains of chromophores, only one of which is predominantly used for forward electron transfer. The high selectivity for only one of the two branches is brought about by the protein environment, which is rather asymmetric between the two core subunits as there are two different proteins (L/M or D1/D2) forming the heterodimeric core.

Such detailed structural information is not yet available for the RCI-type reaction centers. Obviously, either the pigments form two symmetric branches situated on the two respective sides of the dimer or they are lined up in the contact region between the two halves of the dimer. Two lines of evidence argue strongly against the latter hypothesis. (1) Two quinones, considered to play the role of the electron acceptor A_1, have been found per P-700 [70], reminiscent of the quinones Q_A and Q_B symmetrically located on the two halves of the dimer in RCII photosystems. (2) No electron density attributable to chlorophylls or quinones can be seen in the 6 Å electron density map of PSI from *Synechococcus* along the symmetry axis of the protein, whereas two chlorophyll molecules are proposed to be symmetrically located with respect to this axis and in a position equivalent to that of the so-called accessory bacteriochlorophylls in the purple bacterial reaction center [71]. Only the special pair and the FeS center F_X are seen to be located on the symmetry axis [71], in striking resemblance to the positions of the special pair and the nonheme iron in the purple bacterial reaction center (for a discussion of this resemblance, see Refs. [21] and [72].

Therefore, it seems most likely that the electron acceptors are located on the two halves of the dimer rather than between them and that at least some of the electron acceptors exist in two copies.

Provided that no posttranslational modification of one of the two halves of the dimeric RC in green sulfur and heliobacteria occurs, the presence of a homodimer strongly indicates that *all* pigments not located on the axis of symmetry must be present twice and positioned symmetrically on the two halves of the dimer. In PSI, some asymmetry could in principle be brought about by the inequality of two individual proteins forming the heterodimer. The initial interpretation of the 6 Å electron density map proposing only single copies of the electron acceptors A_0 and A_1 and locating them on different halves of the dimer [71] seemed to substantiate this possibility. However, this proposed structure is in conflict with the following arguments:

(1) The structure proposed the presence of only one quinone whereas a large body of experimental evidence shows that there are two distinguishable quinones present in PSI [70, 73–77].

(2) EPR data have been interpreted to indicate the presence of two symmetric A_0 acceptors in PSI [77].

(3) Estimations of distances between P, A_0, and A_1 based on photovoltage measurements are more consistent with a purple bacterial RC-type motif [5, 77].

(4) The homology between the two halves of the dimer in purple bacterial (i.e., RCII-type) RCs is about 25%. This rather low sequence homology results in unilateral electron transfer. The location of pigments and the overall structure of the complex, however, nevertheless are extraordinarily symmetric. By contrast, the homology between *psaA* and *psaB* in PSI being as high as 50%

makes it almost tempting to call PSI a quasi-homodimer. (Note: The sequence identity of homologous proteins in different organisms, such as, for example, the L-subunits of different purple bacteria, is around 45%.) It therefore seems unlikely that such a high symmetry with respect to protein could result in the very asymmetric structure proposed by Krauss et al. [71].

On the basis of the results obtained for green sulfur bacteria and heliobacteria, we therefore favor a model featuring two symmetric electron transfer chains (Fig. 7.4), similar to that proposed previously [72]. Recently, an

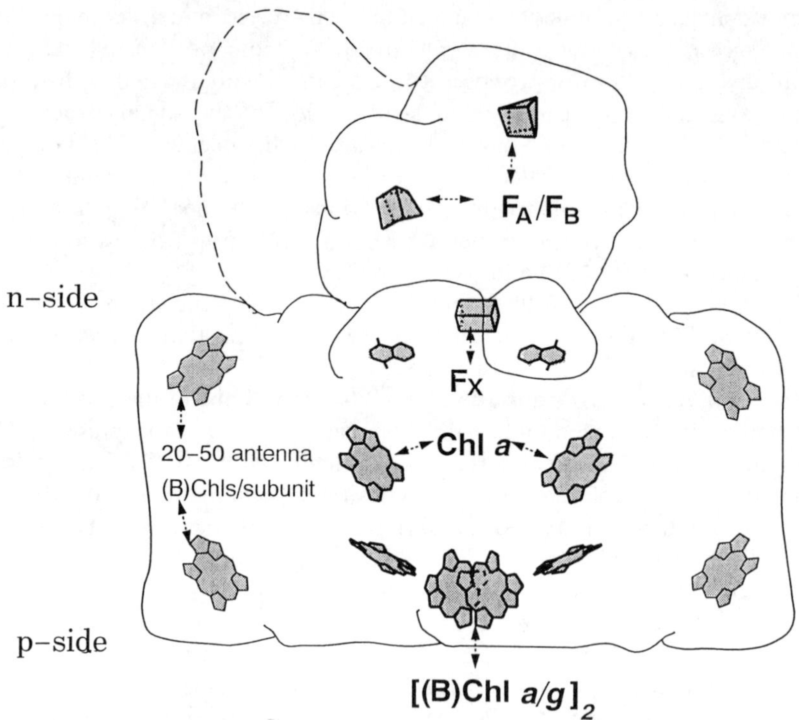

Figure 7.4   A hypothetic picture of the arrangement of cofactors and proteins involved in binding of these cofactors within the RCI photosystem. The position and orientation of the special pair and the four monomeric BChls are those seen in the purple bacterial RC for the four BChls and two BPhes. The menaquinones are in the position of Q_A and Q_B, however, rotated to account for the results obtained by ESP [99]. The FeS clusters F_X and F_A/F_B were positioned with respect to P as reported in the 6 Å-crystal structure of PSI [71]. The depicted orientation of cluster F_X follows from the synthesis of data reported in Refs. [37], [102], [103]. The dashed extension of the F_A/F_B-binding subunit indicates that this protein might be roughly twice as large in green sulfur bacteria than in PSI [22]. The mutual orientations of the two clusters F_A/F_B are taken from the structure of the 2[4Fe–4S] protein of *P. aerogenes* [101]. The four "outer" BChls are intended to represent the core antenna without any emphasis on position or orientation. The n- and p-side denote the negative and positive sides of the energized membranes, respectively.

improved structural model of PSI at 4.5 Å was presented [78]. This structure shows a much more symmetric arrangement of cofactors than the previous one and begins to resemble the model proposed in Ref. [72].

The additional chlorophylls at a rather small angle with respect to the membrane adjacent on both sides of the special pair were originally inferred [72] from the analogy with the purple bacterial type RCs (see below) and in fact showed up already in the 6 Å electron density map of PSI [71].

7.3 Evolutionary Implications

The homologies between the complete amino acid sequences of the three types of RCI are all roughly around 20%. The two subunits of the PSI heterodimer, by contrast, show significantly higher homology with each other (about 45% identity and 55% conservation) in all oxygenic organisms studied. Therefore, the gene duplication that led to the heterodimeric PSI reaction center most probably took place only after the divergence of the green sulfur bacterial, heliobacterial, and PSI phylogenetic lines. The exact topology of the divergence region from the common RCI ancestor to the three different RCI lines cannot be deduced unambiguously from these sequence comparisons owing to the small differences in mutual homology, as discussed in Ref. [79].

However, a large fraction of the sequences from the three species show almost no mutual homology. In the respective regions of the sequence, the proposed alignments are therefore mainly based on the superposition of predicted membrane-spanning stretches. Such an alignment globally makes perfect sense; however, it is not detailed enough to be useful for comparisons aiming at determining phylogenetic relationships. A misalignment by only one amino acid residue will wipe out all residual information with respect to evolutionary relatedness in this specific sequence stretch. If, as in the case of the RCI photosystems, such misalignments are liable to occur in the major part of the sequence, the information content actually present will be artificially diminished down to the level of noise. In fact, restricting the sequence comparisons to regions sufficiently conserved to allow for a reliable alignment (e.g., 641 to 660, 680 to 720, and 791 to 803 in the numbering of the sequence alignment presented in Ref [27]) yields a rather clear picture of the evolutionary interrelationship of the RCI photosystems. The heliobacterial sequence shows 40% and 48% identity toward PsaA and PsaB, respectively, whereas these values are 32% and 39% between the *Chlorobium* and the PsaA/PsaB sequences. *H. mobilis* and *C. limicola* are 42% identical, whereas an identity of 58% is found between PsaA and PsaB in this very sequence region. These values yield a phylogenetic tree as depicted in Figure 7.5 with green sulfur bacteria branching off first and a closer relationship between Heliobacteria and PSI. Most interestingly, however, both the green sulfur and the heliobacterial RC are more closely related to PsaB than they are to PsaA. The most straightforward interpretation of this seems to assume that *psaB* corresponds

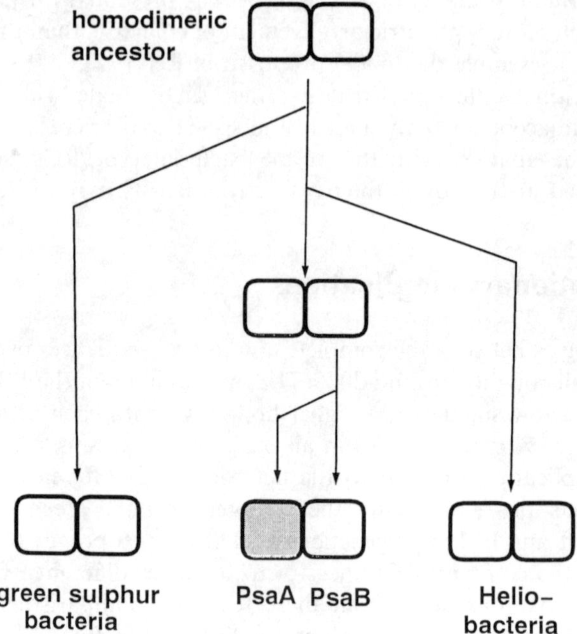

Figure 7.5   Evolutionary relationship of the three types of RCI photosystems discovered so far.

to the ancestral gene present in the homodimeric ancestor of PSI. If PSI should turn out to perform unilateral electron transfer (see below), then it seems rather likely that PsaB is the half of the heterodimer that harbors the cofactors of the active branch.

7.4 Functional Implications

The almost coincident discoveries of homodimeric reaction centers in two bacterial species left the PSI community surprised. Whereas the electron transfer on both sides of the PSI RC had been discussed partly based on the quasi-homodimeric nature of the reaction center and from functional considerations [72], true homodimeric RCs were considered only as necessary evolutionary intermediates on the way to extant RCs but apart from that had only an "archaeopterix-like" status in the minds of scientists interested in evolution of reaction centers. Frequently, the "inherent instability" of homodimeric RCs was pointed out and the ancestral species having to rely on homodimers during the early days of photosynthesis were pitied.

Suddenly it became clear that many photosynthetic bacteria using homodimeric RCs are alive and well and living next to us. One is therefore led to reconsider whether the previous "all-good-RCs-are-heterodimeric-RCs" dogma was not based on a bias from research on purple bacterial RCs. Furthermore, unilaterality of electron transfer using only one branch of the redox centers, which is well documented in RCII-type photosystems, was quite frequently assumed also for PSI. The provocative existence of homodimeric RCs together with the relatively high internal homology of the heterodimeric PSI as mentioned previously, however, make it necessary to consider the possibility of bilateral electron transfer within this class of photosystems, that is, even in PSI itself.

One might argue that the asymmetrically attached F_A/F_B protein (see Ref. [71]) could induce lateral asymmetry. However, (1) we do not know whether the same arrangement is present in green sulfur and heliobacteria and (2) electron transfer in PSI might proceed in parallel down to the level of center F_X and only then result in preferential reduction of the closer of the terminal FeS centers. Clearly, in RCI there is no such obvious functional constraint requiring unilaterality equivalent to the double reduction of Q_B in the RCII-type photosystems.

7.5 RCI Versus RCII

As pointed out previously and in Chapter 6, all photosynthetic reaction centers known so far are either of the RCI or of the RCII type, a classification based on both structural and functional criteria. The obvious question arising from the presence of two quite different types of RC is whether they are evolutionarily related. The fundamentally different reactions occurring at the terminal electron acceptors (Fe–S clusters capable of direct reduction of $NAD(P)^+$ in the aqueous phase versus a two-electron gating quinone acceptor complex delivering reducing equivalents into the membrane phase) seem to argue against an evolutionary relatedness. However, such considerations based on functional characteristics can be misleading as evidenced by the example of PSII, which, although it clearly belongs to the RCII group, mediates redox reactions on its oxidizing side that are radically different from those of the purple bacterial RC. Nature therefore managed to evolve the structure of the RCII ancestor common to both purple bacterial RCs and PSII into structures fulfilling very different tasks. Thus, a more detailed comparison is required when trying to find common characteristics of RCI and RCII photosystems.

Such common characteristics have been discussed previously [2, 72, 79, 80, 85, 86] and therefore only a short summary and updating of the respective data will be presented here.

A striking analogy between RCI and RCII was inferred from the nature of the chromophores as well as the sequence and kinetics of reduction of the redox centers during charge separation. During the last few years, these similarities became increasingly obvious with respect to the primary donor, the primary acceptor, and the electron transfer between these two. Further electron transfer toward the secondary electron acceptors also appears to proceed with rather similar kinetics in RCII (about 250 ps) and RCIs (about 600 ps in green sulfur and heliobacteria [81,82] and about 20 to 50 ps in PSI [83,84]). As detailed previously, there is a large body of evidence favoring a quinone to fulfill the role of this secondary acceptor in RCI-type photosystems, just as in purple bacterial RCs and in PSII (however, see Ref. [40]).

The only redox components different in RCI- and RCII-type photosystems are the FeS center F_X and the so-called nonheme iron, respectively. Based on the idea of a common structural principle for RCI and RCII we had suggested previously that F_X might occupy the same position in the RCIs as does the nonheme iron in RCII-type RCs ([72], see also Ref. [85]), thereby "only" requiring locally different folding of the polypeptide chain and the correct amino acids for the two different redox centers. In fact, the low-resolution crystal structure of PSI shows center F_X in a position only a few angstroms displaced toward the aqueous phase with respect to that of the nonheme iron in purple bacterial RCs [71]. F_X in RCIs and the nonheme Fe^{2+} in RCIIs, both bridging the two core subunits of their respective reaction centers, have symmetrically positioned ligands. This indicates that F_X and the nonheme iron were present in their respective homodimeric ancestors and that the change from F_X to Fe^{2+} (or, less likely, vice versa) occurred in homodimeric reaction centers.

The most obvious "RCI-only" redox centers, the FeS clusters F_A and F_B, are not associated with the RC core but are located on an extra protein subunit (see Fig. 7.4) and therefore do not affect considerations concerning the similarities of RCI and RCII cores.

The presence of RC-bound cytochrome donors in green sulfur and heliobacteria, originally considered to be related to the purple bacterial tetraheme cytochromes [72], was taken as further evidence for a common evolutionary origin of these types of RC. Subsequently, a monoheme cytochrome has been proposed to be the bound donor to the green sulfur bacterial RC [87]. Recently, the existence of a tetraheme cytochrome in green sulfur bacteria has been demonstrated [88]. At present, this question seems not yet unambiguously settled (for a more detailed discussion, see Ref. [89]) and further work is required.

Relics of a common origin of RCI and RCII have been looked for on the level of the primary sequences of the core proteins [30, 90, 91], yielding conflicting models. Whereas Robert and Moënne-Loccoz [30] and Margulies [90] assume that the sequence of the archetypical RCII-type RC is embedded within the sequence of the RCIs with RCI-specific flanking sequence stretches at the N- and C-termini of the RCI core protein, Otsuka et al. [91] deduce from their results that the theme of the RCII is repeated twice to form the (nearly twice as long) RCI sequence. However, the model arrived at by the latter authors is in striking conflict both with functional data on PSI and with the presently available crystal structure. An alternative model (see Chapter 6) envisages the RCII motif as making up the five C-terminal helices of the RCI reaction centers.

At present, the support for a common origin of RCI and RCII photosystems brought about by sequence comparisons is rather limited. Moreover, it is not obvious that the large evolutionary distance between the two types of RC has not wiped out close to all direct sequence homology. It seems more promising to us to look for analogous structural motifs and similarities of their sequential order within the respective proteins. A higher resolution of the PSI crystal structures will yield crucial information with respect to the presence or absence of such common structural motifs.

7.6 Before RCI and RCII?

Phylogenetic trees based on ribosomal RNAs are now considered to be the best experimentally achievable approximation to the evolution of species [1]. A rather extensive phylogenetic tree of the photosynthetic species based on these rRNA molecules has been constructed (Fig. 7.1). Prior to the use of rRNA, the study of the evolution of enzymes relied on comparisons of their amino acid sequences and it was tacitly assumed that the evolution of the enzymes was correlated to the evolution of species.

Data sets for the relationships between the photosynthetic reaction centers are now available and can be used to construct protein-based trees. A comparison between these two kinds of evolutionary trees, however, shows major discrepancies.

The evolution of the RCII proteins is incompatible with the evolution of photosynthetic species. Cyanobacteria (containing PSII) are more closely related to the proteobacteria [100] (where one finds the purple bacterial RCII) than to the green filamentous bacteria, which are considered to be the earliest photosynthetic organisms in the phylogenetic tree, yet in the protein-based tree are closely related to purple bacteria.

Since the L/M subunits of the purple bacterial core and the D1/D2 subunits of the PSII core show a higher sequence homology with each other than L/M with D1/D2, it seems likely that the divergence between the two lines (i.e., purple bacteria and PSII) occurred from a homodimeric ancestor and that subsequent independent gene duplications in the two lines gave rise to the heterodimeric RCIIs (see Chapter 6). From their position in the phylogenetic tree, the green filamentous bacterial RC would therefore be expected to have evolved from a homodimeric RC, independent of PSII and the purple bacterial RC, that is, showing again higher internal homology than homology with subunits of the other RCs. This is clearly not the case. Instead, the L-subunit of *Chloroflexus* is more closely related to the L-subunit of purple bacteria than to its "own" M-subunit and the equivalent is true for the M-subunits (for a detailed discussion, see Ref. [79]). Lateral gene transfer from or to *Chloroflexus* or the phenomenon of "parallel evolution" have been invoked to rationalize this seemingly paradoxical situation ([79,92]; see also Ref. [93]). Most recently, even the validity of the evolutionary tree for the L/M/D1 and D2 proteins has been challenged [94].

The situation is even worse when one tries to reconcile the evolutionary relationship between RCI- and RCII-type photosystems with the phylogenetic tree. The earliest photosynthetic organisms branching off in the eubacterial tree are the green filamentous bacteria containing an RCII as mentioned previously. The next photosynthetic branch seen in the tree is formed by the green sulfur bacteria, characterized by the possession of an RCI, followed by the proteobacteria (again RCII), the heliobacteria (RCI), and finally the cyanobacteria containing both types of photosystems. Two distinct hypotheses for the evolution of RCs have been put forward trying to reconcile this problem.

First, Pierson and Olson [2] proposed that the divergence into RCI- and RCII-type photosystems happened before the branching-off of the Chloroflexaceae and that the ancestor common to all extant photosynthetic species already contained both types of RC. Owing to adaptation to specific environments, the subsequent photosynthetic phyla of green filamentous bacteria, green sulfur bacteria, proteobacteria, and heliobacteria would have lost either of the two RCs and only the cyanobacteria kept them both. This scenario would certainly reconcile the phylogenetic tree and the pattern of distribution of RCI/RCII within the tree.

Alternatively, a decoupling of the evolution of reaction centers from that of species by allowing for lateral gene transfer would also be able to rationalize the data. In such a model, however, lateral gene transfer must be a very frequent event because not only should it have transferred RCII to or from green filamentous bacteria as discussed previously but it must also explain the interpositioning of the purple bacterial RCII between the green sulfur bacterial and the heliobacterial RCI as well as the existence of RCI and RCII in the same species (the cyanobacteria).

It is obvious then that both hypotheses in their rigorously exclusive version do not seem fully convincing. However, several mixed models are possible assuming lateral gene transfer events but also the early existence of species containing both types of RC. The phylogenetic tree in Figure 7.1 can be reconciled with the protein-based tree of Figure 6.4 in Chapter 6 in the following models in which a minimal number of gene transfer events are assumed. (1) Transfer of the photosynthetic RC genes occurred from the green filamentous bacteria to proteobacteria. This model would require the presence of both types of RC in species predating the branching off point of *Chloroflexus* in Figure 7.1. (2) Gene transfer from proteobacteria to *Chloroflexus*. This model requires the presence of both types of RC in the same species only prior to the branching off of the proteobacteria. Prior to this point, RCIs were the only RCs present.

The picture apparently becomes less complicated if one disregards the phylogenetic relationship between species, focussing only on the evolution of reaction centers. An intriguingly simple evolutionary tree for photosynthetic reaction centers was proposed by Blankenship [79] (represented in Fig. 7.6a). In this evolutionary scenario, the ultimate ancestor was a monomeric RC that evolved to form different monomeric ancestors to RCI and RCII. The RCI and RCII then evolved separately and their presence in the same species is a result of gene transfer. Several aspects of this model are unappealing to us. First, the existence of monomeric reaction centers, although considered by several authors, may not be an obligatory precursor of a dimeric reaction center. Based on the structure of existing reaction centers, it is difficult to imagine how a monomeric ancestor could function. The three-dimensional model of the purple bacterial RC shows that the two halves of the dimer are structurally interdependent with respect to many fundamental aspects of protein–protein and of protein–pigment interactions. For example, in purple bacterial RCs the quinone acceptor in one half of the dimer is close to the pheophytin in the opposite subunit. In an isolated subunit, electron transfer from pheophytin to the quinone would have to span too great a distance to compete with back reactions. Furthermore, the "special" role in charge separation of the "special pair" of BChls sandwiched between the two halves of the dimer also indicates the importance of the dimeric structure of the RC. Therefore, it seems possible that photosynthesis is intrinsically correlated to the dimeric enzymes, that is, that photosynthetic reactions came into being through the dimerization of monomeric proteins possibly involved previously in other metabolic processes (Fig. 7.6b).

A second criticism of the model in Figure 7.6a is the virtually complete separation of the evolution of RCI and RCII. The many similarities between the two types of reaction centers argue against this and in favor of the idea that RCI and RCII evolved in the same species from an ancestral homodimeric reaction center. These evolutionary processes might be more complex than those discussed so far.

The transition from a homo- to a heterodimer necessarily involves gene duplication. Shortly after this duplication had happened (i.e., when the two genes do not yet differ significantly), a heterogeneous population of RC must have existed, more specifically, two distinct homo- and one heterodimeric population of RC within the same organism (see Fig. 7.6b). Furthermore, there is no reason to assume that such a gene-duplication event occurred only once within the same organism. Therefore, the complexity of possible scenarios (i.e., owing to coexistence of various homo- and heterodimeric RCs in one species) can be significant. For example (as shown in Fig. 7.6b), a population of homodimeric RCI photosystems might evolve (via gene duplication) into the above described heterogeneous (still RCI-like) state. If one of the two genes was subsequently split (at some appropriate position to still allow dimerization), then the ensemble of states could become as complex as that shown on the fourth row in Figure 7.6b. Three of the four states shown resemble

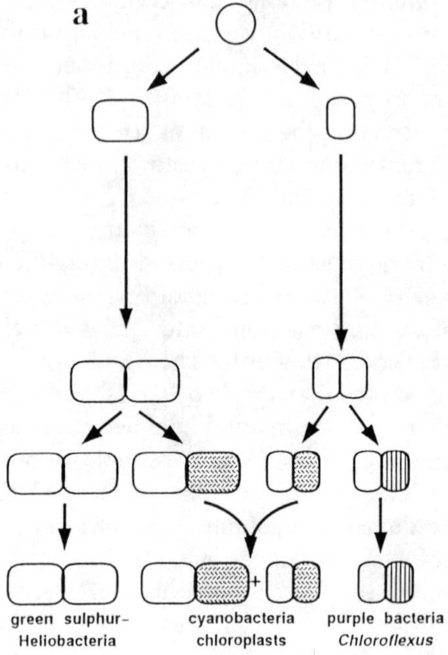

a
green sulphur–
Heliobacteria
cyanobacteria
chloroplasts
purple bacteria
Chloroflexus

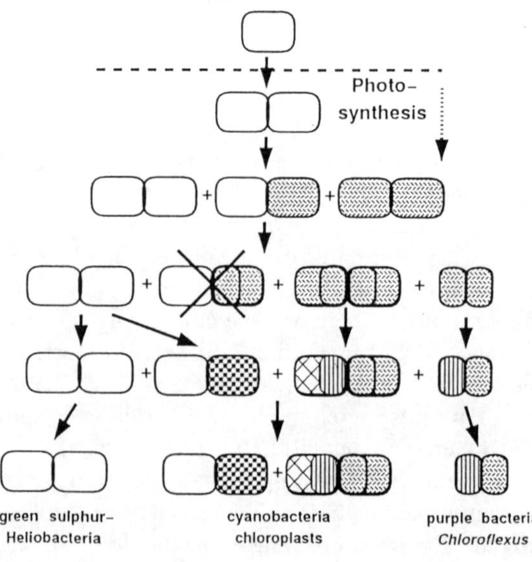

b
Photo–
synthesis
green sulphur–
Heliobacteria
cyanobacteria
chloroplasts
purple bacteria
Chloroflexus

ancestors of RCI- and both RCII-type reaction centers. Adaptation to environmental constraints could select out species that simply lost one or two of these three reaction centers, resulting in species reminiscent of what we find today.

This specific example has been elaborated in detail only to illustrate the underlying ideas. When the reader sits down with a pencil and a sheet of blank paper (an eraser might be helpful) then he or she will rapidly find out that a large number of different combinations are able to result in the present world of photosynthetic reaction centers.

However, the message taken from the above described considerations should rather be that the equations describing this problem are at present underdetermined. An extra constraint on the construction of these trees is the hypothesis in the previous chapter that the PSII core antenna proteins (CP47/CP43) are evolutionarily related to peripheral antenna parts of the RCI protein. Specific protein-based evolutionary trees using this constraint are given in Chapter 6. So far, the photosynthetic RCs are the only complexes for which these kinds of evolutionary problems can be formulated. Comparable data on further components of the photosynthetic apparatus (such as antenna proteins, RC-associated cytochromes, or cytochrome $b–c$ complexes) will decrease the number of free variables and thereby also the number of possible scenarios.

Relevant data on these complexes are already becoming available (e.g., antenna proteins [93] or cytochrome $b–c–$FeS complexes [95]), so that improvements in the understanding of the evolution of photosynthetic RCs are to be expected in the near future [101 103].

Acknowledgments

We would like to thank Drs. R. E. Blankenship, G. Drews, F. A. W. Kleinherenbrink, A. Kolpasky, W. Leibl, B. Robert, and H. van Gorkom for stimulating discussions as well as Drs. U. Feiler, J. H. Golbeck, G. Hauska, U. Liebl, and U. Mühlenhoff for their help with the manuscript. Important informations and ideas concerning the chemical properties of chlorophylls were contributed by Drs. S. Creuzet and J. Reichert.

Thanks are also due to Dr. W. Haehnel for providing access to the SG workstations on which several of the figures were produced. Figures 7.3 and

Figure 7.6 Conceivable pathways for the evolution of photosynthetic reaction centers. Scheme (a) is a modified version of the scheme presented in Ref. [79] which is based on the idea that the divergence of RCI and RCII was on the level of monomeric RCs and that the cyanobacterial system developed from a fusion of RCI and RCII containing genetic information. Scheme (b) considers that photosynthesis might have come into being due to dimerization and that the divergence into RCI and RCII might have occurred already in this organism.

7.4 are based on the coordinates of the *R. viridis* RC and the 2[4Fe–4S] protein from *Peptococcus aerogenes* obtained from the Brookhaven Data Bank ([96,97], entry 1PRC and 1FDX).

References

1. Woese, C. R. *Microbiol. Rev.* 1987; *51*, 221–271.

1. a. Olsen, G. J., Woese, C. R., Overbeek, R. *J. Bacteriol.* 1994; *176*, 1–6.

2. Pierson, B. K., Olson, J. M. In *Microbiol Mats*, Cohen, Y., Rosenberg, E. (eds.); American Society for Microbiol., Washington, DC, 1989; pp. 402–427.

3. Wayne, L. G., Brenner, D. J., Colwell, R. R., Grimost, P. A. D., Kandler, O., Krichevsky, M. I., Moore, L. H., Moore, W. E. C., Murray, R. G. E., Stackebrandt, E., Starr, M. P., Trüper, H. G. *Int. J. Syst. Bacteriol.* 1987; *37*, 463–464.

4. Golbeck, J. H., Bryant, D. *Curr. Top. Bioenerg.* 1991; *16*, 83–177.

5. Mühlenhoff, U., Haehnel, W., Witt, H., Herrmann, R. G. *Gene* 1993; *127*, 71–78.

6. Zanetti, G., Merati, G. *Eur. J. Biochem.* 1987; *169*, 143–146.

7. Hippler, M., Ratajczak, R., Haehnel, W. *FEBS Lett.* 1989; *250*, 280–284.

8. Rousseau, R., Sétif, P., Lagoutte, B. *EMBO J.* 1993; *12*, 1755–1765.

9. Yu, L., Golbeck, J. H., Zhao, J., Schluchter, W. M., Mühlenhoff, U., Bryant, D. A. In *Research in Photosynthesis*, Vol. II, Murata, N. (ed.), Kluwer, Dordrecht, 1992; p. 685.

10. Hurt, E., Hauska, G. *FEBS Lett.* 1984; *168*, 149–154.

11. Feiler, U., Nitschke, W., Michel, H. *Biochemistry* 1992; *31*, 2608–2614.

12. Oh-Oka, H., Kakutani, S., Matsubara, H., Malkin, R., Itoh, S. *Plant Cell Physiol.* 1993; *34*, 93–100.

13. Kjaer, B., Jung, Y. S., Yu, L., Golbeck, J. H., Scheller, H. V. *Photosynth. Res.* 1994; *41*, 105–114.

14. Trost, J. T., Blankenship, R. E. *Biochemistry* 1989; *28*, 9898–9904.

15. van de Meent, E. J., Kleinherenbrink, F. A. M., Amesz, J. *Biochim. Biophys. Acta* 1990; *1015*, 223–230.

16. Amesz, J. In *Anoxygenic Photosynthetic Bacteria*, Blankenship, R. E., Madigan, M. T., Bauer, C. E. (eds.), Kluwer, Dordrecht, 1995; pp. 687–697.

17. Feiler, U., Hauska, G. In *Anoxygenic Photosynthetic Bacteria*, Blankenship, R. E., Madigan, M. T., Bauer, C. E. (eds.), Kluwer, Dordrecht, 1995; pp. 665–685.

18. Kusomoto, N., Inoue, K., Nasu, H., Sakurai, H. *Plant Cell Physiol.* 1994; *35*, 17–25.

19. Oh-Oka, H., Kamei, S., Matsubara, H., Iwaki, M., Itoh, S. *FEBBS Lett.* 1995; *365*, 30–34.

20. Hager-Braun, C., Xie, C., Jarosch, U., Herold, E., Büttner, M., Zimmermann, R., Deutzmann, R., Hauska, G., Nelson, N. *Biochemistry* 1995; *34*, 9617–9624.

21. Golbeck, J. H. *Proc. Natl. Acad. Sci. USA* 1993; *90*, 1642–1646.

22. Büttner, M., Xie, D.-L., Nelson, H., Pinther, W., Hauska, G., Nelson, N. *Proc. Natl. Acad. Sci. USA* 1992; *89*, 8135–8139.

23. Illinger, N., Xie, D.-L., Hauska, G., Nelson, N. *Photosynth. Res.* 1993; *38*, 111–114.

24. Kleinherenbrink, F. A. M., Chiou, H.-C., LoBrutto, R., Blankenship, R. E., *Photosynth. Res.* 1994; *41*, 115–123.

25. Buchanan, B. B., Evans, M. C. W. *Biochim. Biophys. Acta* 1969; *180*, 123–129.

26. Büttner, M., Xie, D.-L., Nelson, H., Pinther, W., Hauska, G., Nelson, N. *Biochim. Biophys. Acta* 1992; *1101*, 154–156.

27. Liebl, U., Mockensturm-Wilson, M., Trost, J. T., Brune, D., Blankenship, R. E., Vermaas, W. F. J. *Proc. Natl. Acad. Sci. USA* 1993; *90*, 7124–7128.

28. Fish, L. E., Kück, U., Bogorad, L. In *Molecular Biology of the Photosynthetic Apparatus*, Steinback, K. E., Bonitz, S., Arntzen, C. J., Bogorad, L. (eds.), Cold Spring Harbor, New York, 1985; pp. 111–120.

29. Warren, P., Smart, L. B., McIntosh, L., Golbeck, J. H. *Biochemistry* 1993; *32*, 4411–4419.

30. Robert, B., Moënne-Loccoz, P. In *Current Research in Photosynthesis*, Baltscheffsky, M. (ed.), Kluwer, Dordrecht, 1990; pp. 65–68.

31. Cui, L., Bingham, S. E., Kuhn, M., Käß, H., Lubitz, W., Webber, A. N. *Biochemistry* 1995; *34*, 1549–1558.

32. Vermaas, W. F. J. *Photosynth. Res.* 1994; *41*, 285–294.

33. Webber, A. N., Xu, H., Bingham, S. E., Käß, H., Kuhn, M., Lubitz, W. In *Xth International Photosynthesis Congress* 1995; Abstract P-5-042.

34. Matsubara, H., Saeki, K. *Adv. Inorg. Chem.* 1992; *38*, 223–280.

35. Nitschke, W., Sétif, P., Liebl, U., Feiler, U., Rutherford, A. W. *Biochemistry* 1990; *29*, 11079–11088.

36. Lockau, W., Nitschke, W. *Physiol. Plantarum* 1993; *88*, 372–381.

37. Nitschke, W., Feiler, U., Rutherford, A. W. *Biochemistry* 1990; *29*, 3834–3842.

38. Miller, M., Liu, X., Snyder, S. W., Thurnauer, M. C., Biggins, J. *Biochemistry* 1992; *31*, 4354–4363.

39. Sétif, P., Brettel, K. *Biochemistry* 1993; *32*, 7846–7854.

40. Kleinherenbrink, F. A. M., Ikegami, I., Hiraishi, A., Otte, S. C. M., Amesz, J. *Biochim. Biophys. Acta* 1993; *1142*, 69–73.

41. Nitschke, W., Feiler, U., Lockau, W., Hauska, G. *FEBS Lett.* 1989; *218*, 283–286.

42. Muhiuddin, L. P., Rigby, S. E. J., Evans, M. C. W., Heathcote, P. In *Xth International Photosynthesis Congress* 1995; Abstract P-5-039.

43. Brok, M., Vasmel, H., Horikx, J. T. G., Hoff, A. J. *FEBS Lett.* 1986; *194*, 322–326.

44. Feiler, U., Albouy, D., Pourcet, C., Mattioli, T., Lutz, M., Robert, B. *Biochemistry 1994; 33*, 7594–7599.

45. Granick, S. *Ann. NY Acad. Sci.* 1957; *69*, 292–308.

46. Mauzerall, D. *Photosynth. Res.* 1992; *32*, 163–171.

47. Burke, D. H., Hearst, J. E., Sidow, A. *Proc. Natl. Acad. Sci. USA* 1993; *90*, 7134–7138.

48. Watanabe, T., Koyayashi, M. In *Chlorophylls*, Scheer, H. (ed.), CRC Press, Boca Raton, FL, 1991; pp. 287–315.

49. Maeda, H., Watanabe, T., Kobayashi, M., Ikegami, I. *Biochim. Biophys. Acta* 1992; *1099*, 74–80.

50. Kobayashi, M., van de Meent, E., Erkelens, C., Amesz, J., Ikegami, I., Watanabe, T. *Biochim. Biophys. Acta* 1991; *1057*, 89–96.

51. Helfrich, M., Schoch, S., Lempert, U., Cmiel, E., Rüdiger, E. *Eur. J. Biochem.* 1994; *219*, 267–275.

52. Watanabe, T., Kobayashi, M. In *Current Research in Photosynthesis*, Baltscheffsky, M. (ed.), Kluwer, Dordrecht, 1990; Vol. II, pp. 109–112.

53. Frank, H. A., McLean, M. B., Sauer, K. *Proc. Natl. Acad. Sci. USA* 1979; *76*, 5124–5128.

54. Rutherford, A. W., Mullet, J. E. *Biochim. Biophys. Acta* 1981; *635*, 225–235.

55. Rutherford, A. W., Sétif, P. *Biochim. Biophys. Acta* 1990; *1019*, 128–132.

56. Sieckmann, I., Brettel, K., Bock, C., van der Est, A., Stehlik, D. *Biochemistry* 1993; *32*, 4842–4847.

57. Moënne-Loccoz, P., Robert, B., Ikegami, I., Lutz, M. *Biochemistry* 1990; *29*, 4740–4746.

58. Rigby, S. E. J., Thapar, R., Evans, M. C. W., Heathcote, P. *FEBS Lett.* 1994; *350*, 24–28.

59. Feiler, U., Albouy, D., Mattioli, T. A., Robert, B. *Biochemistry* 1995; *34*, 11099–11105.

60. Swarthoff, T., Gast, P., Hoff, A. J. *FEBS Lett.* 1981; *127*, 83–86.

61. Fischer, M. *Biochim. Biophys. Acta* 1990; *1015*, 471–481.

62. Vrieze, J., Van de Meent, E. J., Hoff, A. J. In *The Photosynthetic Bacterial Reaction Center: Structure, Spectroscopy and Dynamics*, Breton, J., Verméglio, A. (eds.), Plenum Press, New York, 1993; pp. 67–78.

63. Norris, J. R., Budil, D. E., Gast, P., Chang, C.-H., El-Kabbani, O., Schiffer, M. *Proc. Natl. Acad. Sci. USA* 1989; *86*, 4335–4339.

64. Thurnauer, M. C., Norris, J. R. *Chem. Phys. Lett.* 1977; *47*, 100–105.

65. van der Vos, R., van Leeuwen, P. J., Braun, P., Hoff, A. J. *Biochim. Biophys. Acta* 1992; *1140*, 184–198.

66. Hanson, L. K., Fajer, J. *J. Am. Chem. Soc.* 1987; *109*, 4728–4730.

67. Thompson, M. A., Fajer, J. *J. Phys. Chem.* 1992; *96*, 2933–2935.

68. Struck, A., Cmiel, E., Katheder, I., Schaefer, W., Scheer, H. *Biochim. Biophys. Acta* 1992; *1101*, 321–328.

69. Smith, J. R. L., Calvin, M. *J. Am. Chem. Soc.* 1966; *88*, 4500–4506.

70. Schoeder, H.-U., Lockau, W. *FEBS Lett.* 1986; *199*, 23–27.

71. Krauss, N., Hinrichs, W., Witt, I., Fromme, P., Pritzkow, W., Dauter, Z., Betzel, C., Wilson, K. S., Witt, H. T., Saenger, W. *Nature* 1993; *361*, 326– 331.

72. Nitschke, W., Rutherford, A. W. *Trends Biochem. Sci.* 1991; *16*, 241– 245.

73. Takahashi, Y., Hirota, K., Katoh, S. *Photosynth. Res.* 1985; *6*, 183– 192.

74. Ziegler, K., Lockau, W., Nitschke, W. *FEBS Lett.* 1987; *217*, 16– 20.

75. Malkin, R. *FEBS Lett.* 1986; *208*, 343– 346.

76. Biggins, J., Mathis, P. *Biochemistry* 1988; *27*, 1494– 1500.

77. Heathcote, P., Hanley, J. A., Evans, M. C. W. *Biochim. Biophys. Acta* 1993; *1144*, 54–61.

78. Schubert, W. D., Klukas, O., Krauß, N., Saenger, W., Fromme, P., Witt, H. T. In *Xth International Photosynthesis Congress* 1995; Abstract PL-13.

79. Blankenship, R. E. *Photosynth. Res.* 1992; *33*, 91–111.

80. van Gorkom, H. J. In *Photosynthesis*; Amesz, J., (ed.), Elsevier, Amsterdam, 1987; pp. 343–350.

81. Nuijs, A. M., Vasmel, H., Joppe, H. L. P., Duysens, L. N. M., Amesz, J. *Biochim. Biophys. Acta* 1985; *807*, 24–34.

82. Nuijs, A. M., van Dorssen, R. J., Duysens, L. N. M., Amesz, J. *Proc. Natl. Acad. Sci. USA* 1985; *82*, 6865–6868.

83. Hecks, B., Wulf, K., Breton, J., Leibl, W., Trissl, H. W. *Biochemistry* 1994; *33*, 8619–8625.

84. Hastings, G., Kleinherenbrink, F. A. M., Lin, S., McHugh, T. J., Blankenship, R. E. *Biochemistry* 1994; *33*, 3193–3200.

85. Golbeck, J. H. *Biochem. Biophys. Acta* 1987; *895*, 167–204.

86. Prince, R. C. In *The Bacteria*, Vol. XII, Krulwich, T. (ed.), Academic Press, San Diego, 1990; pp. 111–149.

87. Okkels, J. S., Kjaer, B., Hansson, Ö., Svendsen, I., Moeller, B. L., Scheller, H. V. *J. Biol. Chem.* 1992; *267*, 21139–21145.

88. Albouy, D., Feiler, U., Sturgis, J., Nitschke, W., Robert, B. In *Xth International Photosynthesis Congress* 1995; Abstract P-8-023.

89. Nitschke, W., Dracheva, S. M. In *Anoxygenic Photosynthetic Bacteria*, Blankenship, R. E., Madigan, M. T., Bauer, C. E. (eds.), Kluwer, Dordrecht, 1995; pp. 775–805.

90. Margulies, M. M. *Photosynth. Res.* 1991; *29*, 133–147.

91. Otsuka, J., Miyachi, H., Horimoto, K. *Biochim. Biophys. Acta* 1992; *1118*, 194–210.

92. Blankenship, R. E. *Antonie van Leeuwenhoek* 1994; *65*, 311–329.

93. Nagashima, K. V. P., Shimada, K., Matsuura, K. *Photosynth. Res.* 1993; *36*, 185–191

94. Lockhart, P. J., Larkum, A. W. D. In *Xth International Photosynthesis Congress* 1995; Abstract P-18-005.

95. Schütz, M., Zirngibl, S., leCoutre, J., Büttner, M., Xie, D.-L., Nelson, N., Deutzmann, R., Hauska, G. *Photosynth. Res.* 1994; *39*, 163–174.

96. Bernstein, F. C., Koetzle, T. F., Williams, G. J. B., Meyer, E. F. Jr., Brice, M. C., Rodgers, J. R, Kennard, O., Shimanouchi, T., Tasumi, M. *J. Mol. Biol.* 1977; *112*, 535–542.

97. Abola, E. E., Bernstein, F. C., Bryant, S. H., Koetzle, T. F., Weng, J. In *Crystallographic Databases*, Allen, F. J., Bergerhoff, G., Sievers, R. (eds.), Data Commission of the International Union of Crystallography, Bonn, 1987; pp. 107–132.

98. Burggraf, S., Olsen, G. J., Stetter, K. O., Woese, C. R. *System. Appl. Microbiol.* 1992; *15*, 352–356.

99. Füchsle, G., Bittl, R., van der Est, A., Lubitz, W., Stehlik, D. *Biochim. Biophys. Acta* 1993; *1142*, 23–35.

100. Stackebrandt, E., Murray, R. G. E., Trüper, H. G. *Int. J. Syst. Bacteriol.* 1988; *38*, 321–325.

101. Adman, E. T., Sieker, L. C., Jensen, L. H. *J. Biol. Chem.* 1973; *248*, 3987–3996.

102. Prince, R. C., Crowder, M. S., Bearden, A. J. *Biochim. Biophys. Acta* 1980; *392*, 323–337.

103. Gloux, J., Gloux, P., Hendrix, J., Rius, G. *J. Am. Chem. Soc.* 1987; *109*, 3220–3224.

8

Origin and Evolution of the Proton-Pumping NADH:Ubiquinone Oxidoreductase (Complex I)

Thorsten Friedrich and Hanns Weiss

8.1 Introduction

The proton-pumping NADH:ubiquinone oxidoreductase catalyzes the electron transfer from NADH to ubiquinone linked with a proton translocation according to the overall equation:

$$NADH + Q + 5H_n^+ \rightarrow NAD^+ + QH_2 + 4H_p^+$$

where Q refers to ubiquinone, and H_n^+ and H_p^+ to the protons taken up from the negative inner and delivered to the positive outer side of the membrane. Thus, a protonmotive force is generated that is utilized mainly for ATP synthesis. The enzyme in mitochondria is traditionally called complex I [1] whereas its counterpart in bacteria is often referred to as NADH dehydrogenase I [2]. For simplicity, we will use the term complex I throughout this chapter for both the mitochondrial and the bacterial enzyme. Characteristic features of complex I are its prosthetic groups, namely one FMN and six to eight iron–sulfur [Fe–S] clusters, its sensitivity to a number of naturally occurring compounds, for example, the insecticide rotenone or the antibiotic piericidin A, and its large number of subunits. The bacterial complex comprises 14 different subunits [3], whereas the mitochondrial complex contains also many additional polypeptides [4, 5]. In the genome of chloroplasts and cyanobacteria, most of the genes encoding the subunits of the bacterial complex I have been found [3]. Efforts, however, failed to demonstrate the existence of the genes encoding the NADH dehydrogenase part of the complex.

Chloroplasts and cyanobacteria may, therefore, contain a homologue to complex I equipped with a different electron input device. This complex may work as a ferredoxin : plastoquinone oxidoreductase participating in cyclic photosynthetic electron transfer [3, 6].

In this chapter, some ideas concerning the evolution and origin of complex I will be discussed. For a more comprehensive treatment of our knowledge on the assembly, the structure and the (mostly unknown) mechanism of complex I, the reader is referred to other recent reviews [1, 3, 5, 7–10].

8.2 Occurrence of Complex I

8.2.1 The *Minimal* Complex I of Purple Bacteria

Within the domain of Bacteria, complex I has been shown to occur in three of the four subdivisions of the purple bacteria, namely in *Paracoccus denitrificans* [8], *Rhodobacter sphaeroides* [11], *Rhodobacter capsulatus* [12], and *Zymomonas mobilis* [13] (α-subdivision), in *Alcaligenes eutrophus* (β-subdivision) [14], and in *Escherichia coli* (γ-subdivision) [3, 15, 16]. There is no evidence that any member of the δ-subdivision or bacteria other than purple bacteria contains a complex I. A NADH : menaquinone oxidoreductase has been isolated from the thermophilic (eu)bacterium *Thermus thermophilus* [17]. It is considered to be a proton pump and contains at least 10 different subunits with one FMN and several Fe–S clusters. Sequences are not yet known. In bacteria, the non-proton-pumping NADH : quinone oxidoreductase, called NADH dehydrogenase II, occurs much more frequently. It is a single polypeptide enzyme with FAD as the only redox group [2, 8, 18] (Fig. 8.1).

No evidence exists for the presence of a complex I in Archaea, the third domain of organisms [19]. The recently isolated $F_{420}H_2$: quinone oxidoreductase from *Archeoglobus fulgidus* (of the kingdom of Euryarchaeota) might turn out to be a proton-pumping enzyme related in part to complex I. The enzyme is made up of several polypeptides with one or two FAD molecules and several Fe–S clusters as prosthetic groups [20].

Most data concerning the bacterial complex I are obtained from *E. coli* [3, 15, 16]. It contains 14 complex I genes that are organized in the so-called *nuo* operon (from *N*ADH : *u*biquinone *o*xidorductase) localized at minute 49 of the chromosome [15].

Expression of the *nuo* operon is regulated by O_2, nitrate, and by C_4-dicarboxylates [21]. Growth on substrates supplying high amounts of NADH, for example, glucose or glycerol, does not stimulate the level of expression. The reason might be that maximal energy conversion is performed only under conditions of NADH limitation. Three quinones are present in *E. coli* at different growth conditions, namely ubiquinone, menaquinone, and demethylmenaquinone [22, 23], of which only ubiquinone, with restrictions also demethylmenaquinone, but not menaquinone can be used by complex I for

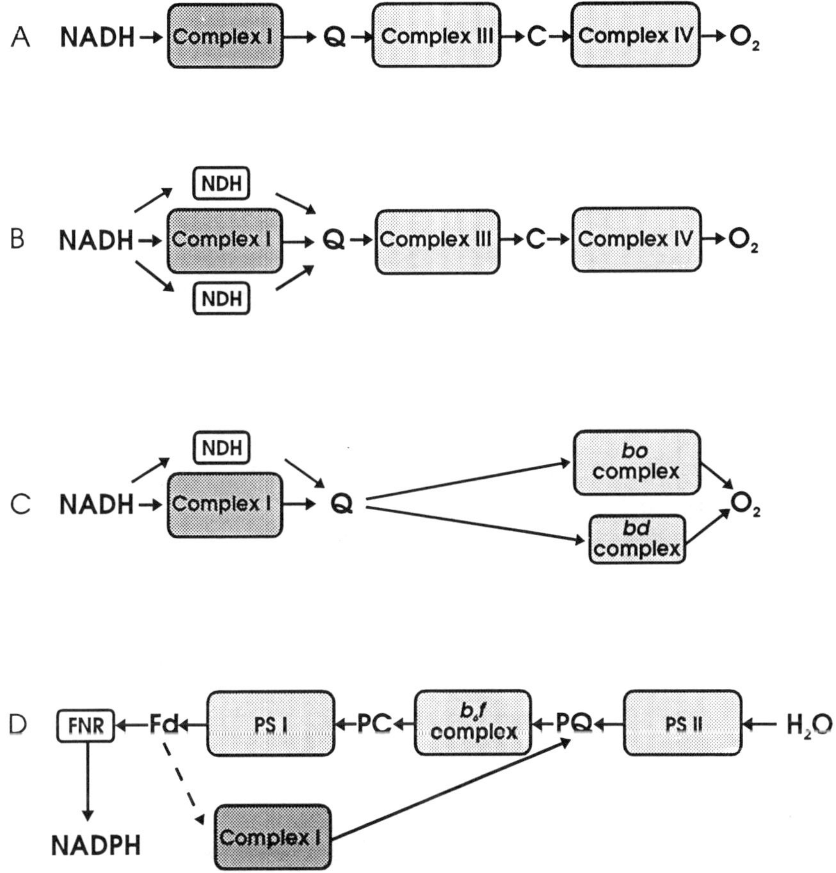

Figure 8.1 Scheme of electron transfer chains of (A) mammalian mitochondria, (B) plant and fungal mitochondria, (C) *E. coli* cytoplasmic membrane, and (D) plant chloroplast and cyanobacterial membrane. Hatched boxes represent energy transducing complexes, white boxes non-energy-conserving enzymes. Complex I is drawn as a dark hatched box. The following abbreviations are used: Complex I, NADH:ubiquinone oxidoreductase; complex III, ubiquinol: cytochrome *c* oxidoreductase; complex IV, cytochrome *c* oxidoreductase; NDH, non-proton-pumping NADH:ubiquinone oxidoreductase; *bo*, cytochrome *bo* ubiquinol oxidase; *bd*, cytochrome *bd* ubiquinol oxidase; PSI, photosystem I; PSII, photosystem II; b_6f, cytochrome b_6f plastoquinol:plastocyanine oxidoreductase; FNR, ferredoxin:NADPH reductase; Q, ubiquinone; C, cytochrome *c*; PQ, plastoquinone; PC, plastocyanine; Fd, ferredoxin.

thermodynamic reasons. Electron transfer from NADH to ubiquinone or demethylmenaquinone provides a free energy of $-84\,\mathrm{kJ/mol}$ or $-69\,\mathrm{kJ/mol}$ corresponding to a difference of midpoint redox potentials of $436\,\mathrm{mV}$ or $356\,\mathrm{mV}$, respectively. This is sufficient to transport two protons per electron across the membrane. The energy gap between NADH and menaquinone

Table 8.1 Nomenclature and Properties of Homologous Complex I Subunits of *E. coli (E.c.)* [15], *B. taurus (B.t.)* [5], and the Plastidial Complex I of *O. sativa (O.s.)* [45]

Designation of the Subunit			Molecular Mass			Membrane Helices			Predicted Function
E.c.	*B.t.*	*O.s.*	*E.c.*	*B.t.*	*O.s.*	*E.c.*	*B.t.*	*O.s.*	
NuoA	ND3	NDH-C	16.3	13.1	13.9	3	3	3	
NuoB	PSST	NDH-K	25.1	20.1	27.7	1	0	0	1 × [4Fe–4S]
NuoC	30(IP)	NDH-J	21.5	26.4	18.6	0	0	0	
NuoD	49(IP)	NDH-H	45.9	48.9	45.7	0	0	0	
NuoE	24(FP)	not found	18.6	23.7	—	0	0	—	1 × [2Fe–2S]
NuoF	51(FP)	not found	49.6	48.4	—	0	0	—	NADH-binding; FMN; 1 × [4Fe–4S]
NuoG	75(IP)	not found	91.2	77.1	—	0	0	—	1 × [4Fe–4S]; 1(2*) × [2Fe–2S]
NuoH	ND1	NDH-A	36.3	35.7	40.4	8	8	8	Ubiquinone-binding
NuoI	TYKY	NDH-I	20.4	20.2	21.1	1	0	0	2 × [4Fe–4S]
NuoJ	ND6	NDH-G	19.9	19.1	19.4	5	5	5	
NuoK	ND4L	NDH-E	11.2	10.8	11.3	3	3	3	
NuoL	ND5	NDH-F	66.3	68.3	82.6	13	11	13	
NuoM	ND4	NDH-D	56.5	52.1	56.5	12	12	12	
NuoN	ND2	NDH-B	45.9	39.7	56.8	10	8	9	

The additonal binuclear FeS cluster of the *E. coli* complex on subunit G for which so far no counterpart has been found in the mitochondrial complex is indicated by an asterisk.

($-47\,$kJ/mol) would not be sufficient for this purpose. Electron transfer from complex I should therefore be possible to the oxidases and nitrate reductases via dihydro-ubiquinone, but not to fumarate reductase via dihydro-menaquinone.

The *E. coli* complex was isolated in an alkylglucoside detergent at slightly acidic pH [16]. The preparation contains the 14 subunits encoded by the *nuo* genes. One noncovalently bound FMN, three binuclear, and three tetranuclear Fe–S clusters were found (Table 8.1). At slightly alkaline pH, the isolated *E. coli* complex I falls into three fragments. A NADH dehydrogenase fragment with three subunits (NuoE-G) and the FMN and 5 Fe–S clusters (Table 8.1; Fig. 8.2) is obtained in water-soluble form. An amphipathic fragment containing four subunits (NuoB-D and I) and one electron paramagnetic resonance (EPR)-detectable visible Fe–S cluster is presumed to connect the NADH dehydrogenase fragment with the third membrane intrinsic fragment, which is composed of seven subunits (NuoA, H, and J–N), these being the homologues of the mitochondrially encoded subunits of the eucaryotic complex I. The subunit composition of the three fragments is to some extent reflected by the order of the genes in the *nuo* locus (Fig. 8.2).

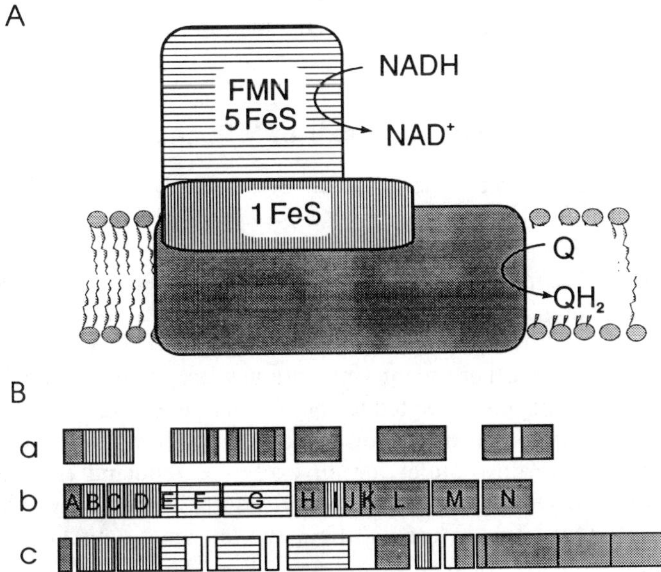

Figure 8.2 (A) Arrangement of the *E. coli* complex I fragments and (B) order of the corresponding genes in (a) *Oryza sativa*, (b) *E. coli*, and (c) *P. denitrificans*. The NADH dehydrogenase fragment and the genes encoding the subunits of this fragment are hatched horizontally, the connecting fragment and its genes are hatched vertically, and the membrane fragment and its genes are shown in dark gray. The homologues in (a) and (c) are depicted in the same manner. The white boxes denote introns in (a) and URFs in (c). The clustered genes in (a) depict the transcriptional units.

If *P. denitrificans* is grown aerobically, it expresses a complex I that in terms of EPR spectroscopic properties of the Fe–S clusters is very similar to the mitochondrial complex I [24, 25]. The genes of the *P. denitrificans* complex I are clustered [25] in the same order as the *nuo* genes in *E. coli* [15]. Besides the 14 genes encoding the counterparts of the *E. coli* complex I, six unidentified reading frames (URFs) were found in the cluster (Fig. 8.2). Whether these are accessory complex I subunits specific for *P. denitrificans* cannot be decided because the intact complex has not been isolated yet from this bacterium.

A complex I gene cluster has also been characterized for *R. capsulatus* [26]. The cluster contains the 14 complex I genes in the same order as in *E. coli* and *P. denitrificans* and six URFs that, however, are not related to the *P. denitrificans* URFs. Typical EPR signals characteristic for complex I have been detected in membranes of *A. eutrophus* [14] and *Rhodobacter sphaeroides* [11]. Evidence for complex I in *Z. mobilis* is given by its piericidin A-sensitive NADH oxidase activity [13].

In summary, purple bacteria seem to make a minimal complex I composed of 14 different subunits that together give a molecular mass of about 530 kDa. Seven subunits are peripheral (or predominantly peripheral) proteins, including all subunits with the binding motifs for NADH, FMN, and all Fe–S clusters. The remaining seven subunits are membrane intrinsic proteins. Apart from the proposed binding site for ubiquinone at one of these subunits [27] (see below) nothing is known about their function (Table 8.1).

8.2.2 The Advanced Complex I of Mitochondria

A mitochondrial complex I is found in most eucaryotic organisms. Known exceptions are the fermentative yeasts *Saccharomyces cerevisae, Schizosaccharomyces pombe,* and *Kluyveromyces lactis,* which do not contain a complex I [28]. Whereas complex I is the only NADH:ubiquinone oxidoreductase in animals, some fungi and plants contain in addition two non-proton-pumping, alternative NADH:ubiquinone oxidoreductases in their mitochondria, one facing the matrix, the other the intermembrane space of mitochondria (Fig. 8.1) [29]. The former enzyme is related to the NADH dehydrogenase II of bacteria [18, 30]. Its affinity for NADH is lower than that of complex I, and it possibly operates as an overflow outlet for an excess of reducing equivalents. The above-mentioned yeasts oxidize mitochondrial NADH only by means of this enzyme [29]. Much less is known about the alternative NADH:ubiquinone oxidoreductase facing the intermembrane space. The enzyme is sensitive to EGTA (ethylene-glycol-bis(β-amionoethylether)-tetraacetate) and specific for the hydrogenation in the β-position of NADH.

Preparations of the mitochondrial complex I exist for several animals, bovine, rat, pigeon [1, 5], for the fungi *Neursopora crassa* [4, 31], and *Aspergillus niger* [32], and for the higher plants *Vicia faba, Beta vulgaris,* and *Solanum tuberosum* [33–35]. Sodium dodecyl sulfate gel electrophoresis resolves the preparations into a large number of subunits, rendering nonspecialists suspicious of the purity of the preparations.

The sequences of (all?) 41 subunits of the bovine complex I [5, 9] and of so far 27 subunits of the *Neurospora crassa* complex I [36, 37] are known. Among them are those 14 subunits that make up the minimal bacterial complex I. Seven membrane intrinsic subunits, the homologues of the *E. coli* NuoA, H, and J–N, are mitochondrially endoded in all above-mentioned species [38]. All other subunits are nuclear-encoded in animals and fungi with the following exceptions: mitochondrially encoded are also the homologues of the *E. coli* NuoC and D in plants [39]; the homologues of NuoB, C, and D in the ciliate *Paramecium aurelia* [40]; and the homologue of NuoG in the slime mold *Dictyostelium discoideum* [41]. In the flagellate *Trypanosoma brucei*, the homologues of NuoB, C, and I subunits are encoded on the maxi-circle DNA [42].

The large group of additional subunits of which no counterpart exists in bacteria and which are all nuclear-encoded is a specific feature of the mitochondrial complex I. Most of them show no relationship to other proteins, and several even show no sequence similarity to complex I subunits from other species. This might mean that many of these additional subunits emerged late in evolution when animals and fungi had already diverged. However, it cannot be excluded that some of them are diverging so fast that a possible homology cannot be seen. The function of most of these subunits is obscure (see below).

8.2.3 The *Alien* Complex I of Cyanobacteria and Chloroplasts

The genes for 11 of the 14 subunits of the minimal bacterial complex I have been found also in the plastidial genomes of *Marchantia polymorpha* [43], *Nicotiana tabacum* [44], and *Oryza sativa* [45], and in the genome of the cyanobacterium *Synechocystis sp.* PCC6803 [46–48]. The genes are organized in four transcriptional units in the same order as in purple bacteria (Fig. 8.2.). All efforts failed to demonstrate the existence of the genes encoding the remaining three minimal subunits [3]. Remarkably, these three subunits make up the (highly conserved) NADH dehydrogenase part of complex I with the FMN and four Fe–S clusters (Table 8.1).

It is therefore proposed that complex I in cyanobacteria and plastides should be equipped with a different electron input device. Since it has been shown that the plastidial complex is located on the stroma thylakoids in close connection to photosystem I [49], ferredoxin might be the electron donor. In this case, the complex would work as a ferredoxin:plastoquinone oxidoreductase (Fig. 8.1). Alternatively, NADPH provided by the ferredoxin:NADPH oxidoreductase might be the electron donor, and the complex would operate as a NADPH:plastoquinone oxidoreductase. Both possibilities imply that the complex participates in a cyclic electron flow passing the electrons from photosystem I back to the plastoquinone pool [3, 6, 50]. In this respect, it is noteworthy that a photosystem I gene is included into one of the four transcriptional units encoding the plastidial complex [44] (Fig. 8.2).

8.3 Relationship of Complex I to Other Bacterial Enzymes

8.3.1 Relationship to Electron Transfer Enzymes

A soluble NAD$^+$-reducing hydrogenase found in the chemolithotrophic purple bacterium *Alcaligenes eutrophus* [51] and in the cyanobacteria *Anabaena variabilis* [52] and *Anacystis nidulans* [52] contains a structural unit of about 900 amino acids that is also part of complex I [4, 5, 15, 25, 53]. The bacterial hydrogenase is composed of four subunits: α, β, γ, and δ. The β/δ dimer is engaged in the splitting of hydrogen and the α/γ dimer exhibits NADH oxidoreductase (diaphorase) acitivity (Fig. 8.3). The diaphorase dimer bears a noncovalently bound FMN and several Fe–S clusters [55]. The N-terminal part of α is homologous to NuoE of complex I, while the C-terminal part of α corresponds to NuoF. The γ subunit is homologous to the first N-terminal 200 amino acids of NuoG (Fig. 8.3; Table 8.2). The hydrogenase genes are arranged in the so-called *hox* loci in the same order as the corresponding complex I genes in bacteria (Fig. 8.1).

Another relationship is that to the formate hydrogenlyase of *E. coli* [55, 56]. This enzyme, which is only expressed under strictly anaerobic conditions, couples the oxidation of formate by the formate dehydrogenase H with proton reduction by the hydrogenase 3 (Fig. 8.3). The formate hydrogenlyase genes are organized in the *hyc* operon which contains eight different genes designated

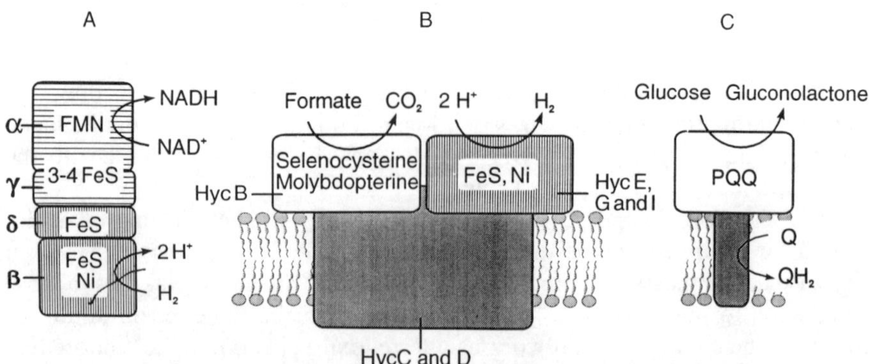

Figure 8.3 Schemes of electron transfer enzymes related to parts of complex I. (A) NAD$^+$-reducing hydrogenase of *A. eutrophus* (modified from Tran-Betcke et al. [51]), (B) formate hydrogenlyase of *E. coli* (modified from Sauter et al. [56]), and (C) glucose dehydrogenase of *A. calcoaceticus*. The parts of these enzymes related to complex I parts are hatched or dark gray as in Fig. 8.2. The nonrelated parts are white.

Table 8.2   Relationship of Complex I Subunits to Subunits in Other Bacterial Electron Transfer Enzymes and Transporters

	Name of the Enzyme/Transporter and Designation of the Related Subunit						
E. coli Complex I	*A. eutrophus* or *A. variabilis* NAD$^+$-Reducing Hydrogenase	*E. coli* Formate Hydrogenlyase	Bacterial Hydrogenases	Bacterial Glucose Dehydrogenase	*R. meliloti* K$^+$/H$^+$ Antiporter	*Bacillus* C-125 Na$^+$/H$^+$ Antiporter	Bacterial PTS
NuoE	α N-term.						
NuoF	α C-term.						
NuoG N-term	γ						
NuoB	δ	HycG	Small subunit				
NuoC	β	HycE N-term.					
NuoD	β	HycE C-term.	Large subunit				
NuoH		HycD		N-term.			
NuoI		HycF					
NuoL		HycC			PhaA	ORF1	Sugar
NuoM					PhaD	ORF4(?)	permease

213

as *hyc*A–H [55]. *Hyc*A is a regulatory gene, the gene product HycB is the small subunit of the formate dehydrogenase H, and HycH is a processing enzyme [56]. The remainder *hyc* gene products constitute the structural framework of the hydrogenase 3 which is related to a part of complex I (Fig. 8.3; Table 8.2). HycE, the large subunit of the hydrogenase 3, is related to NuoD [15]. It has also been proposed that the *hyc*E gene stems from fusion of the *nuo*G and *nuo*D genes [57]. The N-terminal 150 amino acids of HycE show sequence similarity to NuoC while the remaining part of HycE is related to NuoD. The conserved amino acids in HycE that ligate the Ni atom [58] are not found in NuoD. HycG, the small subunit of the hydrogenase 3, is related to NuoB [15]. HycF, a typical 2 [4Fe–4S] ferredoxin, is related to NuoI. This type of ferredoxin, which is found in many bacteria, contains two similar moieties that have most likely arisen by gene duplication. They may be the evolutionary precursors of the [4Fe–4S] [3Fe–4S], the [4Fe–4S], and the [3Fe–4S] ferredoxins [59]. HycC and HycD are related to NuoL and NuoH, respectively. They are membrane proteins presumed to be involved in proton translocation and ubiquinone binding (see below). In summary, complex I contains the homologues of the large subunit present in all known hydrogenases, and that of the small subunit of hydrogenases that is not present in the hydrogenase 2 of *E. coli* [60]. This implies that the β/δ dimer (the hydrogenase part) of the above mentioned NAD^+-reducing hydrogenase is also related to complex I.

Glucose:ubiquinone oxidoreductase (glucose dehydrogenase) is a widely spread bacterial enzyme oxidizing glucose to gluconolactone in the periplasm and delivering the electrons via pyrroloquinoline–quinone to ubiquinone [61]. The single-polypeptide enzyme is inhibited by a number of naturally occurring compounds in a competitive manner similar to complex I [27, 62]. The N-terminal region of the glucose dehydrogenase consists of five membrane spanning helices and contains the ubiquinone binding site [63]. This region shows sequence similarity to a consensus sequence derived from all known homologues of NuoH [27].

8.3.2 Relationship to Transporters

Rhizobium meliloti contains seven genes (called *pha*) that are important for nitrogen fixation. If one of these genes is disrupted, the bacterium becomes sensitive to high potassium concentrations and therefore cannot survive in plant cells (P. Putnoky, personal communication). The genes code for membrane intrinsic proteins that may constitute a new type of K^+/H^+ antiporter that is not related to the well-known Na^+/H^+ antiporter of *E. coli* [64, 65] (encoded by *nha*A and B). A gene system related to the *R. meliloti pha* system was found in *Bacillus* C-125, where it codes for an Na^+/H^+ antiporter [66]. While PhaA and ORF1 show 40% sequence identity to each other, both exhibit 30% sequence identity to NuoL. PhaD has 26% sequence identity to NuoM, the corresponding ORF in *Bacillus* C-125 has not been sequenced yet.

A gene related to *phaA*, ORF1, or *nuo*L was also found in *Bacillus stearother-mophilus* [67] and in *Bacillus subtilis* (EMBL Database D21199), probably encoding a component of an antiporter.

As the genes encoding the antiporter proteins show a greater degree of sequence identity to each other than one of them to the corresponding complex I subunits (P. Putnoky, personal communication) it is likely that they are early progenitors of the corresponding complex I genes. The gene order of the *pha* locus does not correspond to that of the *nuo* locus.

It has been proposed that NuoL and M and the sugar permeases of the bacterial phosphoenolpuyruvate-dependent phosphotransferase system belong to a superfamily of pore-forming proteins [68] because they are related by a low but statistically significant sequence similarity.

8.4 Modular Evolution of Complex I

8.4.1 Procaryotic Development

Complex I, like many multisubunit assemblies, has most probably developed by fusion of preexisting modules for electron transfer and proton transport [3–5]. Since neither the exact subunit arrangement nor the electron and proton pathways in complex I are known, such functional modules can only be defined by the relationship of parts of complex I to other enzymes (see Table 8.2).

From what has been discussed in Sections 8.2.1 and 8.2.3, it can be concluded that a proton-pumping quinone reductase existed already in the common ancestor of cyanobacteria and purple bacteria. This enzyme must have contained the homologues of the subunits Nuo A–D and H–N of today's *E. coli* complex I (Fig. 8.2). This ancestral enzyme might have employed hydrogen as the electron donor using the hydrogenase module made up of the subunits NuoB–D together with the ferredoxin-type subunit NuoI (Table 8.2.). A ferredoxin-type subunit is found in several hydrogenases [60]. The subunits NuoL, M, and N might have been occupied with proton translocation. Subunit NuoH served as the ubiquinone binding module. As long as we do not have more structural data, the role of the remaining subunits is obscure.

On further evolution, this ancestral hydrogen:quinone oxidoreductase diverged to form the NADH:ubiquinone oxidoreductase of purple bacteria and the (putative) ferredoxin:plastiquinone oxidoreductase in cyanobacteria. The former arose by addition of a NADH dehydrogenase module, constituted of the subunits NuoE–F, to the hydrogen-utilizing enzyme. Whether the putative cyanobacterial ferredoxin:plastoquinone oxidoreductase was equipped with a new electron input module or evolved by alterations of the already existing hydrogenase module cannot (yet) be decided.

This concept for the modular evolution of complex I is further supported by the conserved arrangement of the genes coding for the modules (Fig. 8.2) and by the fragmentation behavior of the complex I preparation from *E. coli* (Section 8.2.1 and Fig. 8.2).

8.4.2 Eucaryotic Refinement

A general phenomenon among respiratory complexes is that the bacterial and the mitochondrial enzymes have all subunits in common that constitute the redox-driven proton translocation machinery while the mitochondrial enzymes contain many additional subunits that do not directly participate in electron and proton transport. Only a few of these additional subunits appear to have particular functions, for example, the processing peptidase function of the "core-proteins" in ubiquinol:cytochrome *c* oxidoreductase of fungi [69] and plants [70] or the assembly function of a subunit in the ubiquinol:cytochrome *c* reductase of yeast [71].

The function of most of the additional subunits of the mitochondrial complex I is unknown. It may be speculated that they form a frame or stabilizer keeping the redox groups in the right position to prevent the electrons from escaping and forming reactive oxygen species. These subunits would therefore render energy conversion in complex I safer for the cells.

A 10-kDa additional subunit of the mitochondrial complex I proved to be an acyl carrier protein (ACP) with a phosphopantetheine as prosthetic group [31, 72]. This mitochondrial ACP closely resembles the bacterial or plastidial ACPs that participate in fatty acid synthesis. Disruption of the gene of the mitochondrial ACP in *N. crassa* causes two biochemical lesions that are not found in mutants lacking other peripheral subunits of complex I [73]. First, complex I is not assembled although most (if not all) other subunits are made. Second, the lysophospholipid content of the mitochondrial membranes is increased compared to the wild type [74]. The mitochondrial ACP also exists in *S. cerevisiae*, a yeast that does not have a complex I. In *S. cerevisiae*, the loss of the ACP leads to a pleiotropic respiratory-deficient phenotype unable to grow on nonfermentable substrates and lacking all respiratory enzymes [74].

The mitochondrial ACP is certainly a remnant of the fatty acid synthase in the procaryotic ancestor of this organelle. It is reasonable to assume that the ACP is involved in a synthetic pathway specific for mitochondria, but so far the role of this pathway is unknown. A 40-kDa additional subunit of the mitochondrial complex I might also belong to this pathway. The subunit contains a NAD(P)-binding motif and is related to other NAD(P)-dependent dehydrogenases/reductases [3, 4, 9]. It may function as a reductase in the putative synthetic pathway.

The plant mitochondrial complex I also contains many additional subunits for which, however, sequence data are available only in a few cases. With regard to the minimal subunits, the plant mitochondrial complex I resembles much more the animal or fungal mitochondrial complex I than its plastidial homologue (the putative ferredoxin:plastoquinone oxidoreductase) in the same plant. This is in agreement with the assumption that mitochondria of plants have the same endosymbiontic ancestor as mitochondria of other species and that chloroplasts have been brought about into plants by a later, separate endosymbiontic event [75].

8.5 Origin of Subunit Assemblies of Complex I

From what has been said in the previous sections, the electron transfer moiety of complex I can be traced back to at least two separate origins and the proton translocation moiety to another origin.

The three subunits NuoE, F, and G that are concerned with the electron subtraction from NADH share a common ancestor with the diaphorase moiety of the soluble NAD^+-reducing hydrogenase since the degree of sequence identity among the diaphorase subunits is in the same range as that of the corresponding complex I subunits. Therefore, it is rather unlikely that NuoE, F, and G descended directly from one NAD^+-reducing hydrogenase. The three subunits NuoB, C, and D that conduct the electrons into the poorly understood energy conversion machinery of complex I originate from a hydrogenase. As hydrogenases evolved from various ancestors [76], it can be speculated from sequence comparisons that they most likely share a common ancestor with the membrane-bound NiFe hydrogenases, the periplasmic or membrane-bound NiFe(Se) hydrogenases of sulfate-reducing bacteria, the F_{420} and methylviologen reducing hydrogenases of Archaea, and the hydrogenase part of the above mentioned NAD^+-reducing hydrogenases. Electron injection into the energy conversion machinery is also accomplished by the ferredoxin-type, peripheral NuoI subunit which is also present in several hydrogenases [60].

Two other membrane intrinsic subunits of complex I, NuoH and L, are probably directly involved in the electron-driven proton translocation. NuoH is suggested to contain the ubiquinone reduction site. NuoL functions most likely as a proton channel. Counterparts of these two subunits together with three typical hydrogenase subunits NuoB, C, and D and the ferredoxin-type subunit NuoI are also found in the formate hydrogenlyase of *E. coli*. For thermodynamic reasons, it is, however, unlikely that this enzyme operates as a proton pumping ubiquinone reductase.

Complex I and the formate hydrogenlyase might have emerged independently from preexisting subunit assemblies so that the two complexes are the result of convergent evolution. It is noteworthy that the orders of the corresponding genes in the *nuo* locus and the *hyc* locus are different. Phylogenetic trees with these proteins are still hard to construct because the number of examples is still too small to obtain reliable data.

The membrane intrinsic subunits NuoN, M, and L of complex I which are assumed to be engaged in proton translocation have most likely arisen from a common ancestor [77]. The three proteins are weakly related to one another;

NuoM and L are more closely related to each other than any one of them to NuoN. This implies that either NuoM is the common ancestor of or an evolutionary intermediate between NuoN and NuoL. The original form might have been a transporter protein. Other descendants of this transporter are found in the sugar permeases of the PTS system and in the K^+/H^+ and the Na^+/H^+ antiporters. NuoL and M show a higher degree of sequence identity to the antiporter proteins than to NuoN, suggesting that its divergence was a very early event. The rather high degree of sequence identity compared to other hydrophobic proteins that do not carry any redox group suggests an evolutionary late separation of the complex I and the antiporter genes. There is so far no homologue of NuoN in the database. The origin of the other membrane intrinsic subunits of complex I — NuoA, H, J, and K — is unknown.

Acknowledgment

This work was supported by the Deutsche Forschungsgemeinschaft and the Fonds der Chemischen Industrie. We thank Dr. H. Bothe, University of Köln, and Dr. P. Putnoky, Biological Research Center, Szeged, for providing data prior to publication. Expert linguistic help from B. Brors is gratefully acknowledged.

References

1. Hatefi, Y. *Annu. Rev. Biochem.* 1985; *54*, 1015–1069.

2. Matsushita, K., Ohnishi, T., Kaback, H. R. *Biochemistry* 1987; *26*, 7732–7737.

3. Friedrich, T., Steinmüller, K., Weiss, H. *FEBS Lett.* 1995; *367*, 107–111.

4. Weiss, H., Friedrich, T., Hofhaus, G., Preis, D. *Eur. J. Biochem.* 1991; *197*, 563–576.

5. Walker, J. E. *Quarterly Rev. Biophys.* 1992; *25*, 253–324.

6. Bendall, D. S., Manasse, R. S. *Biochim. Biophys. Acta* 1995; *1229*, 23–38.

7. Weiss, H., Friedrich, T. *J. Bioenerg. Biomembr.* 1991; *23*, 743–754.

8. Yagi, T., Xu, X., Matsuno-Yagi, A. *Biochim. Biophys. Acta* 1992; *1101*, 181–183.

9. Fearnley, I. M., Walker, J. E. *Biochim. Biophys. Acta.* 1992; *1140*, 105–134.

10. Schulte, U., Weiss, H. *Meth. Enzymol.* 1995; *260*, 3–14.

11. Sled, V. D., Friedrich, T., Leif, H., Weiss, H., Meinhardt, S. W., Fukumori, Y., Calhoun, M. W., Gennis, R. B., Ohnishi, T. *J. Bioenerg. Biomembr.* 1993; *25*, 347–356.

12. Dupuis, A. *FEBS Lett.* 1992; *301*, 215–218.

13. Strohdeicher, M., Neuß, B., Bringer-Meyer S., Sahm, H. H. *Arch. Microbiol.* 1990; *154*, 536–543.

14. Kömen, R., Zannoni, D., Ingledew, W. J., Schmidt, K. *Arch. Microbiol.* 1991; *155*, 382–390.

15. Weidner, U., Geier, S., Ptock, A., Friedrich, T., Leif, H., Weiss, H. *J. Mol. Biol.* 1993; *233*, 109–122.

16. Leif, H., Sled, V. D., Ohnishi, T., Weiss, H., Friedrich, T. *Eur. J. Biochem.* 1995; *230*, 538–548.

17. Yagi, T., Hon-nami, K., Ohnishi, T. *Biochemistry* 1988; *27*, 2008–2013.

18. Young, J. G., Rogers, B. L., Campbell, H. D., Jaworowski, A., Shaw, D. C. *Eur. J. Biochem.* 1981; *116*, 165–170.

19. Woese, C. R., Kandler, O., Wheelis, M. L. *Proc Natl. Acad. Sci.* USA 1990; *87*, 4576–4579.

20. Kunow, J., Linder, D., Stetter, K. O., Thauer, R. K. *Eur. J. Biochem.* 1994; *223*, 503–511.

21. Bongaerts, J., Zoske, S., Weidner, U., Unden, G. *Mol. Microbiol.* 1995; *16*, 521–534.

22. Wissenbach, U., Ternes, D., Unden, G. *Arch. Microbiol.* 1992; *158*, 68–73.

23. Anraku, Y., Gennis, R. B. *Trends Biochem. Sci.* 1987; *12*, 262–266.

24. Meijer, E. M., van Versenfeld, H. W., van der Beek, E. G., Stouthamer, A. H. *Arch. Microbiol.* 1977; *112*, 25–34.

25. Xu, X., Matsuno-Yagi, A., Yagi, T. *Biochemistry* 1993; *32*, 968–981.

26. Issartel, J. P., Lunardi, J., Cauvin, B., Chevallet, A., Peinnequin, A., Dupuis, A. *EBEC Short Reports* 1994; *8*, 55.

27. Friedrich, T., Strohdeicher, M., Hofhaus, G., Preis, D., Sahm, H., Weiss, H. *FEBS Lett.* 1990; *265*, 37–40.

28. Nosek, J., Fukuhara, H. *J. Bacteriol.* 1994; *176*, 5622–5630.

29. Douce, R., Neuberger, M. *Annu. Rev. Plant Physiol. Plant Mol. Biol.* 1989; *40*, 371–414.

30. de Vries, S., Grivell, L. A. *Eur. J. Biochem.* 1988; *176*, 377–384.

31. Sackmann, U., Zensen, R., Röhlen, D., Jahnke, U., Weiss, H. *Eur. J. Biochem.* 1991; *200*, 463–469.

32. Prömper, C., Schneider, R., Weiss, H. *Eur. J. Biochem.* 1993; *216*, 223–230.

33. Laterme, S., Boutry, M. *Plant Physiol.* 1993; *102*, 435–441.

34. Rasmussen, A. G., Mendel-Hartvig, J., Möller, I. Wiskich, J. T. *Physiol. Plant* 1994; *90*, 607–615.

35. Herz, U., Schröder, W., Liddel, A., Leaver, C. J., Brennicke, A., Grohmann, L. *J. Biol. Chem.* 1994; *269*, 2263–2269.

36. Van der Pas, J. C., Röhlen, D. A., Weidner, U., Weiss, H. *Biochim. Biophys. Acta* 1991; *1089*, 389–390.

37. Azevedo, J. E., Duarte, M., Belo, J. A., Werner, S., Videira, A. *Biochim. Biophys. Acta* 1994; *1188*, 159–161.

38. Attardi, G., Schatz, G. *Annu. Rev. Biochem.* 1988; *44*, 289–333.

39. Grohmann, L., Herz, U., Thieck, O., Heiser, V., Schmidt-Bleek, K., Lin, T., Brennicke, A. In *Progress in Cell Research—Symposium on "Thirty Years of Progress in Mitochondrial Bioenergetics and Molecular Biology,"* Palmieri, F., Papa, S., Saccone, C. Gadaleta, M. M. (eds.), Elsevier, Amsterdam, 1995; 189–193.

40. Pritchard, A. E., Seilhamer, J. J., Mahalingam, R., Sable, C. L., Venuti, S. E., Cummings, D. *Nuc. Acid Res.* 1990; *18*, 173–180.

41. Cole, R. A., Williams, K. L. *J. Mol. Evol.* 1994; *39*, 579–588.

42. Peterson, G. C., Souza, A. E., Parsons, M. *Mol. Biochem. Parasit.* 1993; *58*, 63–70.

43. Ohyama, K., et al. *Nature* 1986; *322*, 572–574.

44. Shimada, H., Sugiura, M. *Nuc. Acid Res.* 1991; *19*, 983–995.

45. Shinozaki, K., et al. *EMBO J.* 1986; *5*, 2043–2049.

46. Ellersiek, U., Steinmüller, K. *Plant Mol. Biol.* 1992; *20*, 1097–1110.

47. Berger, S., Ellersiek, U., Kinzelt, D., Steinmüller, K. *FEBS Lett.* 1993; *326*, 246–250.

48. Ogawa, T. *Proc. Natl. Acad. Sci.* USA 1991; *88*, 4275–4279.

49. Berger, S. Ellersiek, U., Westhoff, P., Steinmüller, K. *Planta* 1993; *190*, 25–31.

50. Mi, H., Endo, T., Schreiber, U., Asada, K. *Plant Cell Physiol.* 1992; *33*, 1099–1105.

51. Tran-Betcke, A., Warnecke, U., Böcker, C., Zaborosch, C., Friedrich, B. *J. Bacteriol.* 1990; *172*, 2920–2929.

52. Schmitz, O., Boison, G., Hilscher, R., Hundshagen, B., Zimmer, W., Lottspeich, F., Bothe, H. *Eur. J. Biochem.* 1995; *233*, 266.

53. Pilkington, S., Skehel, J. M., Gennis, R. B., Walker, J. E. *Biochemistry* 1991; *30*, 2166–2175.

54. Friedrich, B., Schwarz, E. *Annu. Rev. Microbiol.* 1993; *47*, 351–383.

55. Böhm, R., Sauter, M., Böck, A. *Mol. Microbiol.* 1990; *4*, 231–243.

56. Sauter, M., Böhm, R., Böck, A. *Mol. Microbiol.* 1992; *6*, 1523–1532.

57. Videira, A., Azevedo, J. E. *Int. J. Biochem.* 1994; *26*, 1391–1393.

58. Volbeda, A., Charon, M. H., Piras, C., Hatchikian, E. C., Fey, M., Fontecilla-Camps, J. C. *Nature* 1995; *373*, 580–587.

59. Fukuyama, K., Nakahara, Y., Tsukihara, T., Katsube, Y., Hase, T., Matsubara, H. *J. Mol. Biol.* 1988; *199*, 183–193.

60. Przybyla A. E., Robbins, J., Menon, N., Peck, H. D. *FEMS Microbiol. Rev.* 1992; *88*, 109–136.

61. Matsushita, K., Shinagawa, E., Adachi, O., Ameyama, M. *J. Biochem.* 1989; *105*, 633–637.

62. Friedrich, T., van Heek, P., Leif, H., Ohnishi, T., Forche, E., Kunze, B., Janssen, R., Trowitzsch-Kienast, W., Höfle, G., Reichenbach, H., Weiss, H. *Eur. J. Biochem.* 1994; *219*, 691–698.

63. Yamada, M., Sumi, K., Matsushita, K., Adachi, O., Yamada, Y. *J. Biol. Chem.* 1993; *268*, 12812–12817.

64. Pinner, E., Padan, E., Schuldiner, S. *J. Biol. Chem.* 1992; *267*, 11064–11068.

65. Gerchman, Y., Olami, Y., Rimon, A., Taglicht, D., Schuldiner, S., Padan, E. *Proc. Natl. Acad. Sci USA* 1993; *90*, 1212–1216.

66. Hamomoto, T., Hashimoto, M., Hino, M., Kidata, M., Seto, Y., Kudo, T., Horikoshi, K. *Mol. Microbiol.* 1994; *14*, 939–946.

67. Sakoda, H., Imanaka, T. *J. Ferment. Bioeng.* 1993; *75*, 454–456.

68. Reizer, A., Pao, G. M., Saier, M. H. *J. Mol. Evol.* 1991; *33*, 179–193.

69. Weiss, H., Leonard, K., Neupert, W. *Trends Biochem. Sci.* 1990; *15*, 178–180.

70. Braun, H. P., Schmitz, U. K. *Trends Biochem. Sci.* 1995; *20*, 171–175.

71. Schmitt, M. E., Trumpower, B. L. *J. Biol. Chem.* 1990; *266*, 14958–14963.

72. Runswick, M. J., Fearnley, I. M., Skehel, J. M., Walker, J. E. *FEBS Lett.* 1991; *286*, 121–124.

73. Nehls, U., Friedrich, T., Schmiede, A., Ohnishi, T., Weiss, H. *J. Mol. Biol.* 1992; *227*, 1032–1042.

74. Schneider, R., Massow, M., Lisowsky, T., Weiss, H. *Curr. Genet.* 1996; *29*, 10–17.

75. Yang, D., Oyaizu, Y., Olsen, G. J., Woese, C. R. *Proc. Natl. Acad. Sci. USA* 1985; *82*, 4443–4447.

76. Wu, L. F., Mandrand, M. A. *FEMS Microbiol. Rev.* 1993; *104*, 243–269.

77. Kikuno, R., Miyata, T. *FEBS Lett.* 1985; *189*, 85–88.

Evolution and Origins of the Cytochrome bc_1 and b_6f Complexes

P. N. Furbacher, G.-S. Tae, and W. A. Cramer

9.1 Introduction

The cytochrome bc_1 and b_6f complexes play a central role in the cellular bioenergetics of many organisms from bacteria to higher plants and animals. A clear picture has evolved over the past two decades for many aspects of bioenergetic mechanisms. In contrast, relatively little is known of the origin and evolution of these complexes, despite the fact that the cytochrome b subunit of the cytochrome bc_1 complex has provided a useful tool for phylogenetic analyses of a wide variety of taxonomic groups, primarily due to the functional necessity and hence ubiquity of the cytochrome and its associated complex [1–5]. Complete or partial sequences from more than 800 species [6] may make cytochromes b and b_6 the most sequenced integral membrane proteins.[1] Here, we focus on evolutionary considerations after presenting a brief review of bioenergetic and structural information regarding these complexes. For more thorough reviews of function, the reader is directed to Refs. [7] to [13].

9.2 Electron and Proton Transfer Mechanisms

The cytochrome bc_1/b_6f complex in the energy-transducing membranes of mitochondria, photosynthetic bacteria, and membranes of oxygenic photosyn-

---

[1] To avoid confusion, in this chapter cytochrome b refers only to the b cytochrome of the cytochrome bc_1 complex, and is thereby to be distinguished from cytochrome b_6 of the cytochrome b_6f complex.

thesis contains four redox centers: two b-hemes, one on each side of the membrane in the cytochrome b polypeptide, one high-potential (ca. $+0.3$ V) iron–sulfur [2Fe–2S*] center in the Rieske iron–sulfur polypeptide, and one covalently bound c-type heme in cytochrome c_1 ($E_m = +0.22$ V, mitochondria and photosynthetic bacteria) or cytochrome f ($E_m = +0.37$ V, oxygenic photosynthetic membranes, where E_m is the midpoint potential.

The cytochrome bc_1/b_6f complexes catalyze the oxidation of a quinol and the reduction of a mobile electron carrier, either a cytochrome c or plastocyanin. The quinone species varies, depending on the type of complex. Ubiquinone (UQ) is generally associated with cytochrome bc_1 complexes and plastoquinone (PQ) is invariantly associated with cytochrome b_6f complexes. The occurrence of other quinone species, such as menaquinone (MQ), found in the membranes of the purple photosynthetic bacterium *Rhodopseudomonas viridis*, *Escherichia coli*, and other bacteria is discussed below.

The oxidation of the quinol results in the deposition at pH values near neutrality of two H^+ into the bulk aqueous phase on the positive (p)-side of the membrane, thus contributing to the transmembrane proton electrochemical potential ($\Delta\mu_{H^+}$), and generates a low-potential semiquinone $Q_p^{\cdot-}$ that can reduce the b heme (b_p) on the p-side of the membrane, resulting in the phenomenon of "oxidant-induced reduction" [14]. In mitochondria and chromatophores, there is much evidence to support a 'Q cycle' mechanism for generation of a $\Delta\tilde{\mu}_{H^+}$ in the cytochrome bc_1 complex [7, 9]. In this mechanism, an electron is transferred across the membrane bilayer from heme b_p to a higher potential heme b_n on the electrochemically negative side of the membrane. In one version of the Q cycle, this electron is then transferred to a quinone (Q_n) bound near heme b_n, forming $Q_n^{\cdot-}$. Transfer of a second electron from $Q_p^{\cdot-} \rightarrow Q_n^{\cdot-}$ via hemes b_p and b_n, accompanied by the uptake of two protons from the negative (n) side, results in formation of Q_nH_2 that can diffuse to the quinone pool and complete the cycle [9]. In a modified Q cycle mechanism, electron transfer through the cytochrome b hemes to the quinone bound at the Q_n-site occurs cooperatively in a two-electron process after an electron has been transferred sequentially through heme b_p to heme b_n [15]. In either case, the interheme electron transfer between the hemes is believed to be electrogenic, and contributes significantly to the membrane electrical potential ($\Delta\psi$) component of the $\Delta\mu_{H^+}$. Electron transfer, from the Rieske iron–sulfur protein to cytochrome c_1 or f, and then to a cytochrome oxidase or photosynthetic reaction center complex via the soluble distributive electron carriers, cytochrome c, or plastocyanin, completes the oxidative reaction on the p-side of the membrane that contributes a negative free energy ($-\Delta G$) to the overall reaction.

It is sometimes argued that the quinone(ol) is solely responsible for carrying protons across the membrane in the Q cycle mechanism [9]. However, from the precedent of the involvement of protein groups of the photosynthetic reaction center in the H^+ uptake mechanism at the n-side membrane interface [16, 17], it is highly likely that the H^+ uptake and deposition reactions of the

bc_1/b_6f complexes also utilize protonatable amino acids of the quinone binding polypeptides of the complex.

9.3 Structural Considerations

The minimal oligomeric cytochrome bc_1 complexes able to accomplish the above redox and proton translocation reactions are the complexes from *Paracoccus denitrificans* and the purple bacterium, *Rhodobacter capsulatus* (Fig. 9.1C), composed of three polypeptides: cytochrome b ($M_r \approx 42{,}000$), cytochrome c_1 ($M_r \approx 34{,}000$), and the Rieske iron–sulfur protein ($M_r \approx 20{,}000$). Additional polypeptides found in the mitochondrial complex, a total of 10 in yeast (Fig. 9.1D) and 11 in beef heart (*Bos taurus*) complexes (not shown), may regulate assembly and function.

9.3.1 Cytochromes b and b_6, and Subunit IV

The cytochrome b polypeptide of the cytochrome bc_1 complex is an integral membrane protein of approximately 400 residues with eight transmembrane α-helices. The N-terminal half provides the binding sites for the two hemes, and contributes to the structure of at least a portion of the Q_p and Q_n quinone binding niches [11, 18–21]. The two hemes are coordinated by two pairs of histidine residues, two each in helices B and D, and are arranged parallel to the membrane normal, one heme on each side of the membrane with an edge-to-edge distance of ca. 12 Å [22, 23]. The C-terminal half of the cytochrome polypeptide is thought to contribute to the quinone binding sites on each side of the membrane, and is presumably involved in the packing interactions of the complex. The cytochrome c_1 polypeptide spans the membrane bilayer once, as does cytochrome f discussed below, with most of the mass and the heme projecting into the aqueous phase on the p-side of the membrane. The hydrophobic nature of the N-terminal region of the Rieske protein suggested that it may span the bilayer, but both in mitochondrial [24] and thylakoid [25] membranes, there is evidence that the protein is not tightly bound in the bilayer. The arguments about the Rieske protein in this respect are somewhat reminiscent of those that have taken place over a longer period of time about the membrane topography of cytochrome b_5.

The cytochrome b_6f complex, with four large polypeptides of $M_r > 15{,}000$ (Fig. 9.1A), may also be thought of as a minimal complex because cytochrome b_6 and subunit IV perform the functions of the larger cytochrome b polypeptide in the cytochrome bc_1 complex. The 215-residue cytochrome b_6 polypeptide corresponds to the N-terminal half, or heme-binding domain, of the larger cytochrome b and is characterized by its four transmembrane α-helical segments [26] and the invariable presence of 14 residues between the conserved heme-coordinating His residues on helix D (Fig. 9.2A). Subunit IV,

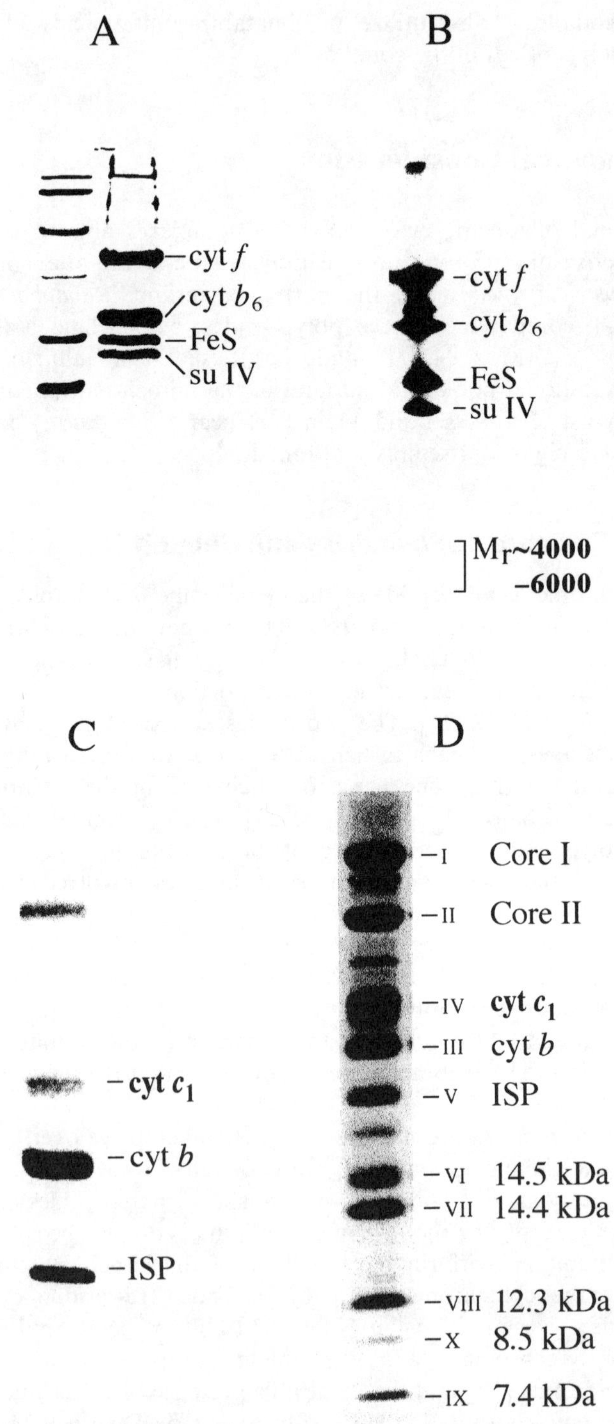
A
B
—cyt f
cyt b_6
—FeS
su IV
—cyt f
—cyt b_6
—FeS
—su IV
M_r~4000
—6000
C
D
—I   Core I
—II   Core II
—IV   cyt c_1
—III cyt b
—cyt c_1
—V   ISP
—cyt b
—VI   14.5 kDa
—VII 14.4 kDa
—ISP
—VIII 12.3 kDa
—X   8.5 kDa
—IX   7.4 kDa

a 160-residue polypeptide, corresponds to the C-terminal half of the ca. 400-residue cytochrome b of the bc_1 complex. It most likely spans the membrane three times [23] (Fig. 9.2B). Depending upon one's viewpoint, the larger cytochrome b could be a fusion product of, or has undergone fission to yield, the cytochrome b_6 and subunit IV polypeptides. In addition to the four Mr > 15,000 polypeptides, three or four small (M_r = 4 to 6 kDa) hydrophobic polypeptides (Fig. 9.1B; also Refs. [27] and [28]) of unknown function and uncertain stoichiometry have been identified.

9.3.2 Cytochrome f

The heme and most of the 285-residue cytochrome f is on the lumen (p) side of the membrane. It has one classic 20-residue transmembrane α-helix punctuated by a triple lysine on the n-side (residues 271 to 273 and Arg-250, conserved in 13 sequences [29]), and 250 of the 285 residues with the covalently bound heme in the p-side (lumen) aqueous phase, with a short C-terminus on the n-side of the membrane. The crystal structure of a 252-residue lumen side fragment of turnip (*Brassica campestris*) cytochrome f (Table 9.1, pp. 232–237) has been determined to a resolution of 2.3 Å [30]. Two general features of the structure, the existence of two distinct domains and a predominantly β secondary structure, are unprecedented for c-type cytochromes. Utilization of the α-amino group of the N-terminal amino acid as the axial sixth heme ligand is unprecedented for heme proteins. A C_α skeleton of cytochrome f and plastocyanin (*Populus nigra* var. *italica*) in a predocking state is shown in Figure 9.3 with putative key amino acid residues enumerated [30, 31].

9.4 The Earliest Complexes

What were the properties of the ancestor to the cytochrome bc_1/b_6f complex? What functions did it initially have, and which functions arose later in

<hr>

Figure 9.1 (A) Four large and (B) three to four small (M_r = 4000 to 6000) subunits in the b_6f complex from spinach chloroplast thylakoid membranes. (B) Second-dimension SDS-PAGE of cytochrome b_6f dimer scanned by J. B. Heymann using the Macintosh program *DeskScan II* at high contrast; the program NIH Image 1.53b7 was used to subtract the background and to amplify the density by a factor of two. (C) SDS-PAGE of cytochrome bc_1/b_6f complex consisting of three known subunits in the bc_1 complex derived from chromatophore membranes of the photosynthetic bacterium *Rb. capsulatus* (kindly provided by F. Daldal). (D) SDS-PAGE showing the 10 subunits of the mitochondrial cytochrome bc_1 complex purified from strain W303-1A by anion-exchange chromatography, omitting the second chromatography step [62]. The purified complex was denatured and subunits resolved by SDS-PAGE [63] (kindly provided by B. L. Trumpower).

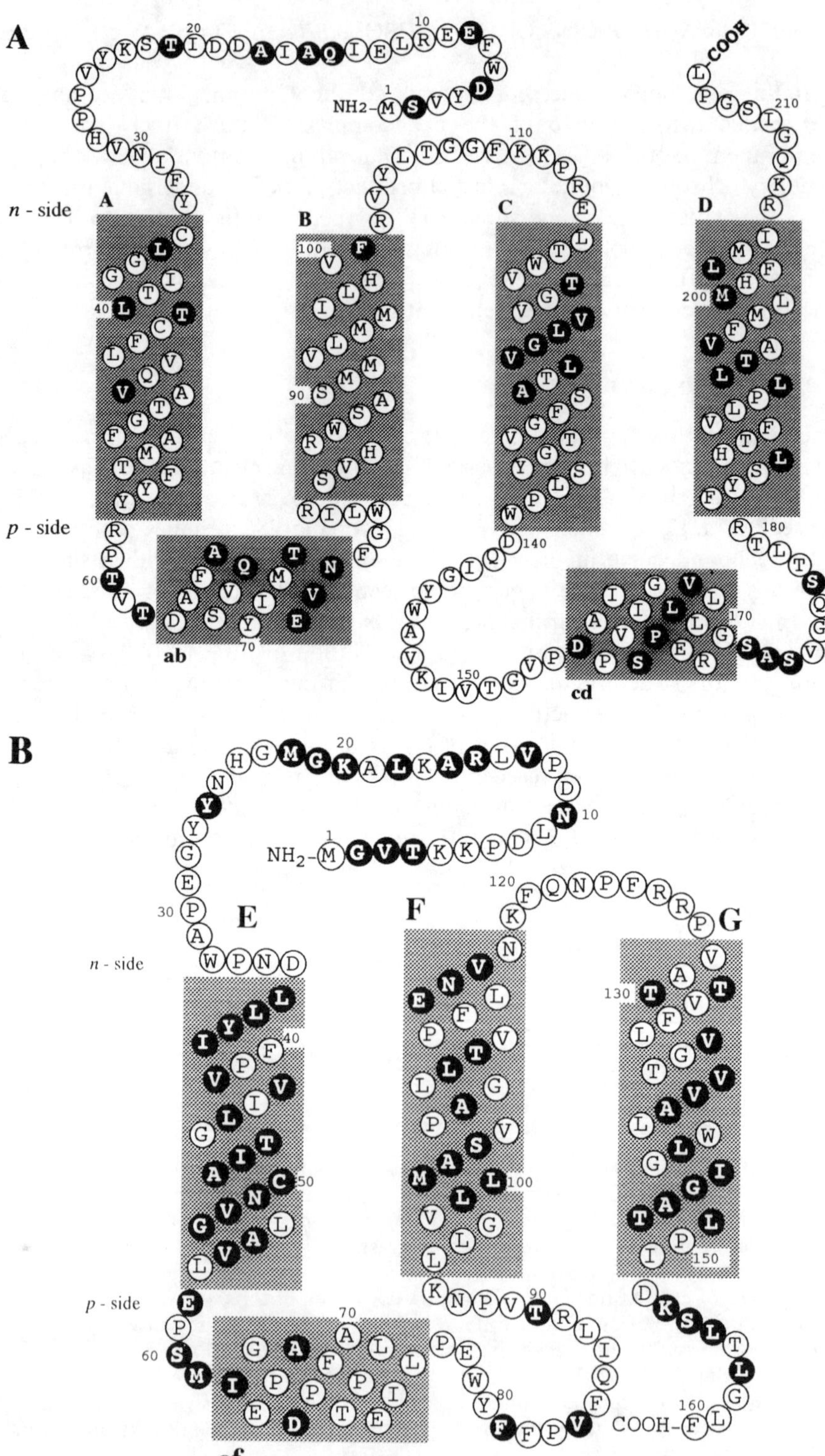
A
n - side
p - side
NH2-
COOH
ab
cd
B
n - side
p - side
NH2-
COOH-
ef

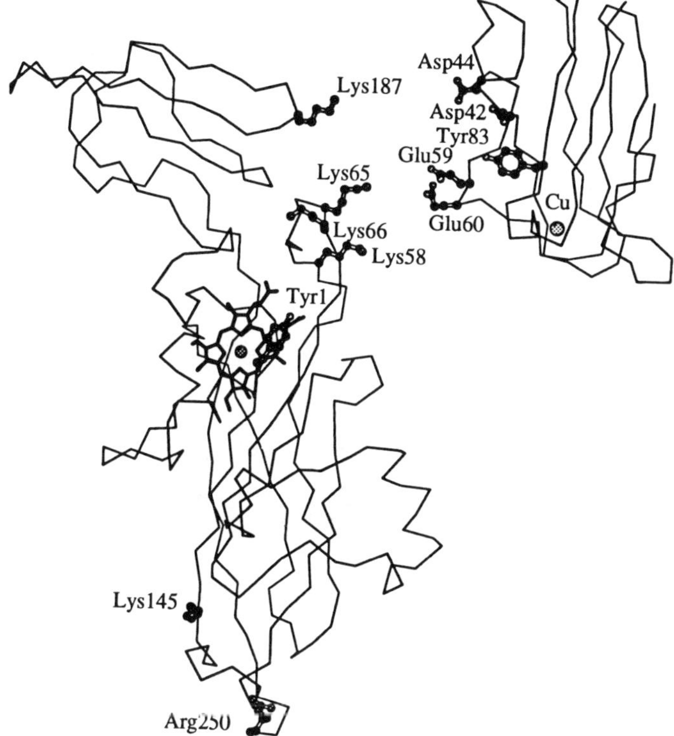

Figure 9.3 C_α traces of the 252-residue lumen-side domain of cytochrome f [30] and plastocyanin in a "predocking" state, showing some key residues, including those in an acidic patch of plastocyanin (Asp-44, Asp-59) and a complementary basic patch of cytochrome f (Lys-187, -66, -65, and -58), thought to be involved in the electrostatic docking. The dimensions of cytochrome f are 75 × 35 × 25 Å, with Arg-250 the lumen side boundary of the membrane-spanning α-helix. The heme ligands of cytochrome f are His-25 and the α-amino group of the N-terminal residue, Tyr-1.

evolution? Are there any clues about the ancestral state to be found in extant organisms? The general view is that the two complexes share a common ancestor. This hypothesis is based primarily on the relatively high degree of sequence identity between cytochrome b and b_6, and structural and functional homologies between the two complexes. When such an ancestor arose remains

Figure 9.2 Folding pattern across the thylakoid membrane of (A) cytochrome b_6 and (B) subunit IV predicted by the distribution and conservation in 13 and 16 sequences, respectively, of long (ca. 20-residue) hydrophobic segments. The hydrophobic membrane bilayer is defined by the rectangular boxes. Hatched circles represent residues that are not invariant or pseudo-invariant. (Reprinted with copyright permission of Plenum Press, New York.).

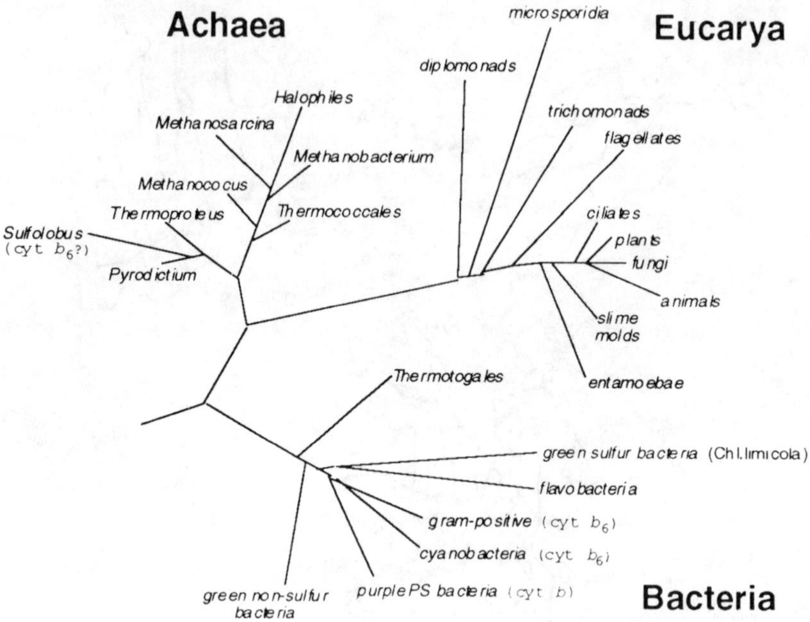

Figure 9.4 Universal phylogram illustrating the relationships of the three "domains" [34] and major groups within them. The diagram is redrawn from Ref. [32]. The groups in which cytochrome bc_1/b_6f complexes have been identified are appropriately labeled.

an unresolved question. Today, cytochrome bc_1/b_6f complexes have been found to be present in a number of groups of the Bacteria (PS3, Gram-positive; *Chlorobium limicola*, green sulfur; cyanobacteria; and the purple photosynthetic bacteria), the Archaea[2] (*Sulfolobus acidocaldarius*), and in eukaryotic organelles (see Fig. 9.4). What seems clear is that the prototype of the hydrophobic cytochrome *b*, with an N-terminal heme binding four transmembrane helix domain and the downstream four helix domain on the same polypeptide involved in quinone binding and structural interaction, appeared very early in evolution. The identification of a cytochrome bc_1-like complex in one member of the Archaea, *S. acidocaldarius* [33], suggests that the ancestor to the cytochrome bc_1/b_6f complex may have been present in organisms before the split that yielded the Bacteria and the "Urcarya" (the progenitors of the Archaea and Eucarya [34]).

Cytochrome bc_1 and b_6f complexes are present in the cells of modern eukaryotes only as a result of the endosymbiotic events from which the

---

[2] In the remainder of the text, we use the designations proposed by Wheelis et al. [32] for the three main branches of evolution, or "domains": Archaea, Bacteria, and Eucarya.

mitochondrion and chloroplast originated [35, 36]. If the ancestor to the cytochrome bc_1 and b_6f complexes existed at the time of divergence between the Bacteria and Urcarya, the earliest urkaryotes would most likely have contained a cytoplasmic cytochrome bc_1 complex prior to the endosymbiotic events. The absence of such a complex in the cytoplasm of modern eukaryotes raises the question: Was the cytoplasmic complex lost as a result of the endosymbiosis? If the complex was not present at the time of divergence, several important implications become apparent: (1) the distribution of cytochrome bc_1-like complexes in the Archaea will be found to be paraphyletic (i.e., not sharing common ancestry), and would imply gene transfer events from bacterial species; and (2) the endosymbiosis events in the eukaryotes were advantageous because they provided the urkaryotic cell with a bioenergetic pathway it did not possess.

It may be speculated that the cytochrome bc_1 complex of the Anoxyphotobacteria (purple photosynthetic bacteria) most accurately reflects the ancestral complex. This view holds that the immediate ancestor to the two complexes was composed of three major polypeptides, cytochromes b and c_1, and the Rieske iron–sulfur protein. Alternatively, the operon coding the respiratory electron transport proteins in *S. acidocaldarius* provides some support for a model of the evolution of electron transport chains from fermentation in which fumarate reductase evolved from a soluble enzyme to a membrane-bound complex containing an iron–sulfur complex, ubiquinone-10, and a cytochrome b [37]. The operon encodes three genes (*sox*ABC), two of which are related to the cytochrome oxidase subunits I and II, and one, *sox*C, that appears to be related to the cytochrome b of the bc_1 complex [33]. The latter gene codes for a 563-residue polypeptide with the eight putative transmembrane helices ascribed to the cytochrome b of the bc_1 complex and a C-terminal extension of approximately 150 residues with four additional putative transmembrane domains which may provide the quinone-binding functions. In addition, there are 13 residues between the two heme-coordinating His residues in helix D, a diagnostic character of cytochrome b compared to cytochrome b_6 [23]. The complex contains four hemes a, and no heme b. Alignment against sets of cytochromes b or b_6 resulted in low levels of identity for either set (Table 9.1).

There is at least one high-potential [2Fe–2S*] protein in *S. acidocalarius* [38]. However, no c-type cytochrome has been found in the respiratory chain of these organisms [39] or any other aerobic Archaea [40]. The apparent absence of a soluble cytochrome c in the Archaea may represent an adaptation in these organisms that avoids the use of a soluble or extrinsic redox protein; the alternative adaptation, toward a thermotolerant cytochrome c, appears not to have occurred. As more acidothermophiles are investigated, it will be interesting to see if this is a unique occurrence or an adaptation imposed by harsh environments. In any case, the presence of a cytochrome b-like domain and a high-potential iron–sulfur protein in *S. acidocaldarius* suggests that the primitive complex may not have had a cytochrome c_1/f subunit (see the discussion of *Chlorobium limicola* below).

9.5 Subunit Evolution

Evolution of the structure and function of the three primary subunits within the complex has also occurred. The "long" cytochrome b of the cytochrome bc_1 complex appears to have undergone two important structural changes that characterize cytochrome b_6 and subunit IV of the cytochrome b_6f complex: (1) the bifurcation of the longer cytochrome into two subunits, and (2) the insertion of an additional residue between the heme-ligating His residues in helix D. Cytochrome b_6f complexes are found in the cyanobacteria, chloroplasts, and in at least some members of the Gram-positive bacteria. It is generally accepted that the chloroplasts of all higher plants represent a monophyletic group, having arisen from a single endosymbiotic event involving an ancestor of the modern cyanobacteria (cf. [36]). The cytochrome b of *Bacillus stearothermophilus* [41], a Gram-positive thermophilic bacterium, is cytochrome b_6-like in length and has the 14-residue span between the two heme-liganding His residues in helix D, a diagnostic character associated with cytochrome b_6. The polypeptide equivalent of subunit IV is also present.

A cytochrome b with mixed properties has been discovered in *Chlorobium limicola*, a green sulfur bacterium. The *pet*CB operon of *C. limicola* encodes a Rieske iron–sulfur protein (*pet*C) of 181 residues and a cytochrome b (*pet*B) of 428 residues [42]. The four Cys and two His residues of the Rieske iron–sulfur protein are conserved, although the E_m of the menaquinol-oxidizing Rieske center in the *Chlorobium* complex is 150 mV more negative than that oxidizing ubiquinol or plastoquinol. The cytochrome b sequence shows hybrid features, with some key features resembling more closely cytochrome b_6. Such features include 14 residues between the two histidines of transmembrane helix D, and the lack of an eighth transmembrane helix, both of which are diagnostic of cytochrome b_6–subunit IV [23]. Interestingly, the *pet*CB operon of *C. limicola* lacks the *pet*A gene encoding cytochrome c_1. The function of the cytochrome c_1 may be provided by a 24-kDa cytochrome c-551 component that copurified with the B840 reaction center [42]. Although the cytochrome b of *C. limicola* may provide a glimpse of the transition from cytochrome b to b_6, it also may provide evidence that the ancestral complex was comprised of only a cytochrome b and a Rieske iron–sulfur protein. Further study is needed to verify that there is no c-type cytochrome associated with the complex.

Evidence for the evolution of function in the Rieske iron–sulfur protein is provided by the range of quinone specificities that exist. The midpoint potentials of the iron–sulfur centers have evolved, although the selective advantages for each species of quinone are not clear. *Chlorobium* [43] is one of seven examples (*Bacillus alkalophilus* [44], *Chloroflexus aurantiacus* [45], *Thermus thermophilus* [46], *Heliobacterium chlorum* [47], *Bacillus* PS3 [48], and *Bacillus firmus* [49]) in which the Rieske proteins have E_m values approximately 150 mV more negative than those of chloroplasts, mitochondria, and purple bacteria; these low-potential Rieske iron–sulfur proteins oxidize menaquinol whose E_{m7} of $-60\,\text{mV}$ is approximately $150\,\text{mV}$ more negative than those of PQ and UQ. In considering the evolutionary niche of MQ, it is of interest (1) that *E. coli*, which does not have a cytochrome bc_1 complex, but instead contains b heme-quinol oxidases, also utilizes MQ when grown under microaerophilic conditions, and (2) that the reaction center of *Rhodopseudomonas viridis*, but not *Rb. sphaeroides*, contains menaquinone (n.b., a phylogenetic analysis of reaction center polypeptides concluded that *Rps. viridis* is more primitive than *Rb. sphaeroides* [50]). In addition to the more negative E_m ($=100$–$160\,\text{mV}$) of the Rieske iron–sulfur center, and the absence of sensitivity to stigmatellin, the EPR spectrum of the Rieske protein shows features that are characteristic of a $MQ^{\doteq}$, as opposed to $UQ^{\doteq}$ or $PQ^{\doteq}$ [49]. The Rieske iron–sulfur proteins of *H. chlorum*, *Bacillus* PS3, and *B. firmus* OF4 exhibit these same features. Nitschke et al. [51] suggest that separate transitions occurred from the lower E_m MQ to higher E_m PQ (from the Gram-positive bacilli, or firmicutes, to cyanobacteria) and UQ (between branches of the proteobacteria, including the purple bacteria). The E_m shift of the quinone pool, accompanied by a similar shift in E_m of the Rieske center, may have been a response to increased O_2 in the environment. The lower E_m system would have been more O_2-sensitive [51].

The membrane-bound c-type cytochromes of the bc_1/b_6f complexes perform equivalent functions: (1) transferring electrons from the Rieske iron–sulfur proteins to a mobile, soluble electron carrier (cytochrome c or plastocyanin, respectively), and (2) possibly anchoring the Rieske protein to the membrane. It is increasingly evident that they share little in common at the structural level [30, 31]. Their differences at the gene level are so striking that it can be questioned whether they share a genetic ancestry or represent an evolutionary convergence of function. The evolutionary relationship of cytochromes c_1 and f is analyzed and discussed below.

9.6 Sequence Comparisons: Function and Evolution

Compilations of cytochrome b, cytochrome b_6, and subunit IV sequences are shown in Tables 9.1 to 9.3. As previously noted [52], the overall sequence identity of cytochrome b_6 (69% and 79% for identical and pseudo-identical residues) and subunit IV (53% and 63%, respectively) is much greater than that of cytochrome b. It has been noted that the sequence identity is greater on the p-side of cytochrome b [2, 6], and greater on the n-side of cytochrome b_6, where it is 78% and 87% for invariance and pseudo-invariance [52]. The high degree of invariance on the n-side of cytochrome b_6 is consistent with a role in a cyclic pathway in which a low-potential photoreductant can dock to the n-side of the membrane and transfer electrons to the cytochrome b_6f complex. The invariance of the four histidine residues on helices B and D is clearly related to conservation of the active site. As discussed previously, there is one

Table 9.1 Amino Acid Sequences of Cytochrome *b* of the Cytochrome *bc*$_1$ Complex from 53 Species. The Eight Transmembrane Helices (A to H) are Underscored.

```
A.american  MI----NIR----------- -------------------- ----------------YQP- -------------------- LGICLILQILTGLFLAMHYT AD-TTTAFS-----SVTHIC
A.francisc  K--ML-SLP----------- -------------------- -------------------- -------------------- LGLCLLIQIVIGLFLAMHYT AS-VELAFS-----SVANIC
A.nidulans  M-------R----------- -------------------- -------------------- -------------------- LALCLGIQIVTGVTLAMHYT PS-VSEAFN-----SVEHIM
A.suum      MK------------------ -------------------- -------------------- -------------------- LGMVLGFQILTGTFLAFYYS ND-GALAFL-----SVQYIM
A.transmon  M---A-NIR----------- -------------------- -------------------- -------------------- LGLCLITQILTGLFLAMHYT AD-ISTAFS-----SVAHIC
B.japonicu  M--SG-PSD----------- -------------------- ----------------YQP- --------------SNPALQ LSFMLGMQILTGVILAMHYT PH-ADLAFK-----SVELIV
B.taurus    M---T-NIR----------- -------------------- -------------------- -------------------- LGICLILQILTGLFLAMHYT SD-TTTAFS-----SVTHIC
C.dromedar  MT----NIR----------- -------------------- -------------------- -------------------- LGVCLIMQILTGLFLAMHYT SD-TTTAFS-----SVAHIC
C.elegans   LK------------------ -------------------- -------------------- -------------------- LGMVLIFQILTGTFLAFYYT PD-SLMAFS-----TVQYIM
C.reinhard  M--RM-HN------------ -------------------- -------------------- -------------------K AGMMLASQMLTGILLAMHYV GH-VDYAFA-----SVQHLM
C.smithii   M-----RMH----------- -------------------- -------------------- -------------------- AGMMLASQMLTGILLAMHYV GH-VDYAFA-----SVQHLM
carp        MA--S--LR----------- -------------------- -------------------- -------------------- LGLCLITQILTGLFLAMHYT SD-ISTAFS-----SVTHIC
chicken     MA--P-NIR----------- -------------------- -------------------- -------------------- LAVCLMTQILTGLLLAMHYT AD-TSLAFS-----SVAHTC
Chlorobium  MAENT-PKPAAGTAPAKPKP AAPGAAKPAAPKAARPGAAK PAAKPAAPRAAAPSGVYKKP PVDRPDPNPFKDSKRDAVAG GLFFFVIQILTGLLLLQYYK PT-ETDAFA-----SFLFIQ
D.yakuba    MH--K-PLR----------- -------------------- -------------------- -------------------- LGLCLIIQILTGLFLAMHYT AD-VNLAFY-----SVNHIC
deer        MI----NIR----------- -------------------- -------------------- -------------------- LGICLILQILTGLFLAMHYT SD-TMTAFS-----SVTHIC
Dolphin     M---T-NIR----------- -------------------- -------------------- -------------------- LGLCLIMQILTGLFLAMHYT PD-TTTAFS-----SVAHIC
Drosophila  MN--K-PLR----------- -------------------- -------------------- -------------------- LGLCLIIQILTGLFLAMHYT AD-INLAFY-----SVNHIC
frog        MA--P-NIR----------- -------------------- -------------------- -------------------- LGVCLIAQIITGLFLAMHYT AD-TSMAFS-----SVAHIC
goat        MT----NIR----------- -------------------- -------------------- -------------------- LGICLILQILTGLFLAMHYT SD-TMTAFS-----SVTHIC
H.africaea  M---T-NIR----------- -------------------- -------------------- -------------------- LGACLIIQILTGLFLAMHYT AY-TTTAFS-----SVAHIC
Human       MT----PMR----------- -------------------- -------------------- -------------------- LGACLILQITTGLFLAMHYS PD-ASTAFS-----SIAHIT
L.tarentol  MF--F-RVR----------- -------------------- -------------------- -------------------- LGFFICMQIICGVCLAWLFF SC-FICTNW-----YFVLFL
M.polymorp  -------AR----------- -------------------- -------------------- -------------------R AGLCLVIQILTGVFLAMHYT PH-VDLAFL-----SVEHIM
Maize       M--TI-RNQ----------- -------------------- -------------------- -------------------R AGICLVIQIVTGVFLAMHYT PH-VDLAFN-----SVEHIM
mouse       MT----NMR----------- -------------------- -------------------- -------------------- LGVCLMVQIITGLFLAMHYT SD-TMTAFS-----SVTHIC
N.crass     M-------R----------- -------------------- -------------------- -------------------- LACCLIMQIVTGVTLAMHYS PN-VLEAFN-----SMEHIM
O.bertiana  MA--T-IRN----------- -------------------- -------------------- -------------------- AGICLVIQIVTGVFLAMHYT PH-VDLAFN-----SVEHIM
opossum     MT----NLR----------- -------------------- -------------------- -------------------- LGMCLIIQILTGLFLAMHYT SD-TLTAFS-----SVAHIC
P.anserina  M-----RIL----------- -------------------- -------------------- -------------------- LLLCLIIQIITGVTLAMHYS PN-VLEAFN-----SIEHIM
P.aurelia   IL---NIF-N---------- -------------------- ----------------Y--- ----------------FKN- TFMTIIVQLVSGTMLAFSSV PEPMLIPTV-----RDEEDI
P.denitrif  MAGIP-HDH----------- -------------------- ----------------YEP- --------------KTGFER LAFCLVLQIATGIVLVMHYT PH-VDLAFA-----SVEHIM
P.falcipar  M-----N------------- -------------------- ----------------FYS- -------------------- LGMMFFIQIMTGVFLASRYT PD-VSYAYY-----SMQHIL
P.tetraure  ML--N--------------- -------------------- -------------------- -------------------- TFMTIIVQLVSGTMLAFSSV PE-PMLIPTVRDEEDIEDLY
P.yoelii    M-------N----------- -------------------- ----------------YNS- -------------------- LGIIFFIQILTGVFLASRYS PE-ISYAYY-----SIQHIL
Rat         M---T-NIR----------- -------------------- -------------------- -------------------- LGVCLMVQILTGLFLAMHYT SD-TMTAFS-----SVTHIC
Rb.capsula  MSGIP-HDH----------- -------------------- ----------------YEP- --------------KTGIEK LAFTLVLQIVTGIVLAMHYT PH-VDLAFA-----SVEHIM
Rb.sphaero  MSGIP-HDH----------- -------------------- ----------------YEP- --------------RTGIEK LAFCLVLQIVTGIVLAMHYT PH-VDLAFA-----SVEHIM
rice        M---T-IRN----------- -------------------- -------------------- -------------------- AGICLVIQIVTGVFLAMNHT PH-VDLAFN-----SVEHIM
Rs.rubrum   M--YT-PPR----------- -------------------- ----------------WN-- ---------------NKALK AGIAMIIMIATGIFLAMSYT AH-VDHAFD-----SVERIM
S.acidocal  -------------------- -------------------- -------------------- ---------LVEEKKSGIID VAASFAYTIITGLFLLLYYQ P---AFAYQ-----STQTII
S.cerevisi  M--A---------------- -------------------- -------------------- -------------------- LGLCLVIQIVTGIFMAMHYS SN-IELAFS-----SVEHIM
S.pombe     M-------K----------- -------------------- -------------------- -------------------- LACVLVIQIVIGMTLACFYI PN-MDLAFT-----SVERIV
S.scrofa    MT----NIR----------- -------------------- -------------------- -------------------- LGICLILQILTGLFLAMHYT SD-TTTAFS-----SVTHIC
S.tuberosu  ----T-IRN----------- -------------------- -------------------- -------------------- AGICLVIQIVTGVFLAMHYT PH-VDLAFN-----SVEHIM
seal        MT----NIR----------- -------------------- -------------------- -------------------- LGICLILQILTGLFLAMHYT SD-TTTAFS-----SVTHIC
T.annulata  -------------------- -------------------- -------------------- -------------------- LGILLVLQIISGLMLSFFYV PA-KGMAFE-----STLAVM
T.brucei    FR------------------ -------------------- -------------------- ------------------CR LGFFIALQIICGVCLAWLFF SC-FICSNW-----YFVLFL
T.napu      MI----NIR----------- -------------------- -------------------- -------------------- LGICLILQILTGLFLAMHYT SD-TSTAFS-----SVTHIC
T.tajacu    MT----NIR----------- -------------------- -------------------- -------------------- LGICLLLQILTGLFLAMHYT PD-TTTAFS-----SVTHIC
urchin      ML--G-PLR----------- -------------------- -------------------- -------------------- LGLCLITQILTGLFLAMHYT AD-ISLAFS-----SASHIC
V.faba      M---T-IRNQRLS------- -------------------- -------------------- -------------------- ---KSHPLLKQPISSTLNQH LIDYPTPSN-LSYWWGFGSL
whale       MT----NIR----------- -------------------- -------------------- -------------------- ---KSHPLMK-----IVNDA FVDLPTPSN-ISSWWNFGSL
```

A.american   ---QRFSLMK-----IVNNA FIDLPAPSN-ISSWWNFGSL LGICLILQILTGLFLAMHYT AD-TTTAFS-----SVTHIC RDVNYGWIIRYMHANGASMF FICLFMHVGRGLYYGSYMFL ET--WNIGVILLFTVMATAF
A.francisc   ---EQQPTLK-----IINSA LVDLPVPAN-ISIWWNFGSL LGLCLLIQIVIGLFLAMHYT AS-VELAFS-----SVANIC RDVNYGWLLRTVHANGASFF FICIYFHIGRGMYYGSFHYF ET--WMTGIALLFLVMAAAF
A.nidulans   -ILKSHPLLK-----IVNSY IIDSPQPAN-LSYLWNFGSL LALCLGIQIVTGVTLAMHYT PS-VSEAFN-----SVEHIM RDVNNGWLVRYLHSNTASAF FFLVYLHIGRGLYYGSYKTP RTLTWAIGTVILIVMMATAF
A.suum   ---KSHP-LD-----FVNSM VVRLPSSKV-LTYGWNFGSM LGMVLGFQILTGTFLAFYYS ND-GALAFL-----SVQYIM YEVNFGWIFRVLHFNGASLF FIFLYLHLFKGLFFMSYRLK K--VWVSGIVITTLVMMEAF
A.transmon   ---KTHPLLK-----IINGA FIDLPTPSN-ISVWWNFGSL LGLCLITQILTGLFLAMHYT AD-ISTAFS-----SVAHIC RDVNYGWLIRNIHANGASFF FICLYLHVARGMYYGSYLQK ET--WNIGVILLLLTMMTAF
B.japonicu   WIERRLPILG-----LMHSS FVAYPTPRN-LNYWWTFGAI LSFMLGMQILTGVILAMHYT PH-ADLAFK-----SVELIV RDVNYGWLLRNMHACGASMF FFAVYVHMLRGLYYGSYKEP REVLWILGVIIYLLMMATGF
B.taurus   ---KSHPLMK-----IVNNA FIDLPAPSN-ISSWWNFGSL LGICLILQILTGLFLAMHYT SD-TTTAFS-----SVTHIC RDVNYGWIIRYMHANGASMF FICLYMHVGRGLYYGSYTFL ET--WNIGVILLLTVMATAF
C.dromedar   ---QRFSLLK-----IMNDA FIDLPAPSN-ISSWWNFGSL LGVCLIMQILTGLFLAMHYT SD-TTTAFS-----SVAHIC RDVNYGWIIRYLHANGASMF FICLYIHVGRGLYYGSYTFS ET--WNVGMVLLFTVMATAF
C.elegans   ---INNSLLN-----FVNGM LVTLPSSKT-LTLSWNFGSM LGMVLIFQILTGTFLAFYYT PD-SLMAFS-----TVQYIM YEVNFGWVFRIFHFNGASLF FIFLYLHIFKGLFFMSYRLK K--VWMSGLTIYLLVMMEAF
C.reinhard   IQLL-----S-----VLNTH LVAYPTPMN-LNYSWNGGSL AGMMLASQMLTGILLAMHYV GH-VDYAFA-----SVQHLM TDVPSGMILRYAHANGASLF FIVVYLHVLRGMYYGSGAQP REIVWISGVVILLVMIITAF
C.smithii   ---NSHPLL-----SVLNTH LVAYPTPMN-LNYSWNGGSL AGMMLASQMLTGILLAMHYV GH-VDYAFA-----SVQHLM TDVPSGMILRYAHANGASLF FIVVYLHVLRGMYYGSGAQP REIVWISGVVILLVMIITAF
carp   ---NSHPLIK-----IANDA LVDLPTPSN-ISAWWNFGSL LGLCLITQILTGLFLAMHYT SD-ISTAFS-----SVTHIC RDVNYGWLIRNVHANGASFF FICIYMHIARGLYYGSYLYK ET--WNTGVILLLTLMATAF
chicken   ---KTHPLLK-----MINNS LIDLPAPSN-ISAWWNFGSL LAVCLMTQILTGLLLAMHYT AD-TSLAFS-----SVAHTC RNVQYGWLIRNLHANGASFF FICIFLHIGRGLYYGSYLYK ET--WNTGVILLLTLMATAF
Chlorobium   WFQERFYVLN-----PIIDY LKHKEVPKHALSFWYYFGGL GLFFFVIQILTGLLLLQYYK PT-ETDAFA-----SFLFIQ GEVPFGWLLRQIHAWSANLM IMMLFIHMFSTFFMKSYRKP RELMWVSGFVLLLLSLGFGF
D.yakuba   ---KSHPLFK-----IANNA LVDLPAPIN-ISSWWNFGSL LGLCLIIQILTGLFLAMHYT AD-VNLAFY-----SVNHIC RDVNYGWLLRTLHANGASFF FICIYLHIGRGIYYGSYLFT PT--WLVGVIILFLVMGTAF
deer   ---KTHPLMK-----IVNNA FIDLPAPSN-ISSWWNFGSL LGICLILQILTGLFLAMHYT SD-TMTAFS-----SVTHIC RDVNYGWIIRYMHANGASMF FICLFMHVGRGLYYGSYMFL ET--WNIGVILLFTVMATAF
Dolphin   ---KTHPLMK-----ILNDA FIDLPTPSN-ISSWWNFGSL LGLCLIMQILTGLFLAMHYT PD-TTTAFS-----SVAHIC RDVNYGWFIRYLHANGASMF FICLYAHMGRGLYYGSYMFQ ET--WNIGVLLLFTVMATAF
Drosophila   ---KTHPLFK-----IANNA LVDLPAPIN-ISSWWNFGSL LGLCLIIQILTGLFLAMHYT AD-INLAFY-----SVNHIC RDVNYGWLLRTLHANGASFF FICIYLHVGRGIYYGSYKFT PT--WLIGVIILFLVMGTAF
frog   ---KTHPLIK-----IINNS FIDLPTPSN-ISSLWNFGSL LGVCLIAQIITGLFLAMHYT AD-TSMAFS-----SVAHIC RDVNYGWLIRNLHANGASFF FICIYLHIGRGLYYGSFLYK ET--WNIGILLLFTVMATAF
goat   ---KINPLMK-----IVNNA FIDLPTPSN-ISSWWNFGSL LGICLILQILTGLFLAMHYT SD-TMTAFS-----SVTHIC RDVNYGWIIRYMHANGASMF FICLFMHIGRGLYYGSYTFL ET--WNIGVLLLATMATAF
H.africaea   ---KSHPLLK-----IINHS FIDLPTPSN-ISTWWNFGSL LGACLIIQILTGLFLAMHYT AY-TTTAFS-----SVAHIC RDVNYGWLIRYLHANGASMF FICLYLHVGRGLYYGSYMFT ET--WNIGILLLFTVMATAF
Human   ---KTHPLMK-----LINHS FIDLPTPSN-ISAWWNFGSL LGACLILQITTGLFLAMHYS PD-ASTAFS-----SIAHIT RDVNYGWIIRYLHANGASMF FICLFLHIGRGLYYGSFLYS ET--WNIGILLLATMATAF
L.tarentol   -------FL---LFFLLFRN LCCLLTSGC-LLRVYGVGFS LGFFICMQIICGVCLAWLFF SC-FICTNW-----YFVLFL WDFDLGFVIRSTHICFTSLL FFLLYVHIFKCIVLIILFDT HILVWAVGFIIYIFIVVIGF
M.polymorp   LSILKQPIFS-----TFNNH LIDYPTPSN-ISYWWGFGSL AGLCLVIQILTGVFLAMHYT PH-VDLAFL-----SVEHIM RDVKGGWLLRYMHANGASMF FIVVYLHFFRGLYYGSYASP KELVWCLGVILLLMIVTAF
Maize   FSLLKQPIYS-----TLNQH LIDYPTPSN-LSYWWGFGCL AGICLVIQIVTGVFLAMHYT PH-VDLAFN-----SVEHIM RDVEGGWLLRYMHANGASMF LIVVHLHIFRGLYHASYSSP REFVWCLGVVIFLLMIVTAF
mouse   ---KSHPLFK-----IINHS FIDLPAPSN-ISSWWNFGSL LGVCLMVQIITGLFLAMHYT SD-TMTAFS-----SVTHIC RDVNYGWIIRYMHANGASMF FICLFLHVGRGLYYGSYTFM ET--WNIGVLLLFAVMATAF
N.crass   -LLKSHPLLK-----LVNSY LIDASQPSN-MSYLWNFGSL LACCLIMQIVTGVTLAMHYS PN-VLEAFN-----SMEHIM RDVNNGWLVRYLHSNTASAF FFLVYLHMGRGMYYGSYRAP RTLVWAIGTVILMLMMATAF
O.bertiana   ---KEHPLLKQPISSILNQH LIDYPTPSN-LSYWWGFGSL AGICLVIQIVTGVFLAMHYT PH-VDLAFN-----SVEHIM RDVEGGWLLRYMHANGASMF FIVVYLHTFRGLYYASYSSP REFVWCLGVVIFLLMIVTAF
opossum   -------LMK-----IINHS FIDLPAPSN-ISAWWNFGSL LGMCLIIQILTGLFLAMHYT SD-TLTAFS-----SVAHIC RDVNYGWLIRNLHANGASMF FMCLFLHVGRGIYYGSYLYK ET--WNIGVILMLTVMATAF
P.anserina   ---KSHPLLK-----LVNSY LIDASQPSN-ISYLWNFGSL LLLCLIIQIITGVTLAMHYS PN-VLEAFN-----SIEHIM RDVNNGWLVRYLHSNTASAF FFLVYLHIGRGMYYGSYRAP RTLVWAIGTVILIVMIVTAF
P.aurelia   --------LR-----VSFHE VFSL-----------FGFF TFMTIIVQLVSGTMLAFSSV PEPMLIPTV-----RDEEDI EDLYTGDFFW-LHTRGASLI FIFSYFHLLRKLYLNVFDLE TEASWKSGVFSFLVFQVVAF
P.denitrif   WLHRRLPIVS-----LVYDT LM-IPTPKN-LNWWWIWGIV LAFCLVLQIATGIVLVMHYT PH-VDLAFA-----SVEHIM RDVNGGYMLRYLHANGASLF FLAVYIHIFRGLYYGSYKAP REVTWIVGMLIYLMMMGTAF
P.falcipar   --------IN-----LVKAH LMNYPCPLN-MNFLWNYGFL LGMMFFIQIMTGVFLASRYT PD-VSYAYY-----SMQHIL RELWSGWCFRYMHATGASLV FLLTYLHILRGLNYSYMYLP --LSWMSGLILFMMFIVTAF
P.tetraure   ------------IFNYFKNL RVSFHEVFS-L-----FGFF TFMTIIVQLVSGTMLAFSSV PE-PMLIPTVRDEEDIEDLY TD-DFFW----LHERGVDLI FIFSYFHLLRKLYLNVFDLE TEASWKSGVFSFLVFQVVVF
P.yoelii   --------IN-----LVKTH LINYPCPLN-INFLWNYGFL LGIIFFIQILTGVFLASRYS PE-ISYAYY-----SIQHIL RELWSGWCFRYMHATGASLV FFLTYLHILRGLNYSYLYLP --LSWISGLIIFALFIVTAF
Rat   ---KSHPLFK-----IINHS FIDLPAPSN-ISSWWNFGSL LGVCLMVQILTGLFLAMHYT SD-TMTAFS-----SVTHIC RDVNYGWLIRYLQANGASMF FICLFLHVGRGLYYGSYTFL ET--WNIGIILLFAVMATAF
Rb.capsula   WLHDRLPIVG-----LVYDT IM-IPTPKN-LNWWWIWGIV LAFTLVLQIVTGIVLAMHYT PH-VDLAFA-----SVEHIM RDVNGGWAMRYIHANGASLF FLAVYIHIFRGLYYGSYKAP REITWIVGMVIYLLMMGTAF
Rb.sphaero   WLHSRLPIVA-----LAYDT IM-IPTPRN-LNWMWIWGVV LAFCLVLQIVTGIVLAMHYT PH-VDLAFA-----SVEHIM RNVNGGFMLRYLHANGASLF FIAVYLHIFRGLYYGSYKAP REVTWIVGMLIYLAMMATAF
rice   ---KSHPLLKQPIYSTLNQH LIDYPLPSI-LSYWWGFGSL AGICLVIQIVTGVFLAMHHT PH-VDLAFN-----SVEHIM RDVEGGWLLRYMHANGASMF LIVVHLHIFRGLYHASYSSP REFVWCLGVVIFLLMIVTAF
Rs.rubrum   WFDERLPVLT-----VAHKE LVVYPAPRN-LNYFWNFGSL AGIAMIIMIATGIFLAMSYT AH-VDHAFD-----SVERIM RDVNYGWLMRYMHANGASMF FIVVYVHMFRGLYYGSYKPP REVLWWLGLVILLLMMATAF
S.acidocal   SILER---LG-----INEAP LFRTPDYMYNISYW--LGAM VAASFAYTIITGLFLLLTYQ P---AFAYQ-----STQTII NSVPYGSVLLFSHLYGSYIM ILLAYIHMFRNFYKGAYKKP RELQWVTGVLLLALTLGASF
S.cerevisi   -FRKSNVYLS-----LVNSY IIDSPQPSS-INYWWNMGSL LGLCLVIQIVTGIFMAMHYS SN-IELAFS-----SVEHIM RDVHNGYILRYLHANGASFF FMVMFMHMAKGLYYGSYRSP RVTLWNVGVIIFLTIATAF
S.pombe   -ILKSNPFLA-----LANNY MIDAPEPSN-ISYFWNFGST LACVLVIQIVIGMTLACFYI PN-MDLAFT-----SVERIV RDVNYGFLLRAFHANGASFF FIFLYLHMGRGLYYGSYKYP RTMTWNIGVIIFLLTIITAF
S.scrofa   ---KSHPLMK-----IINNA FIDLPAPSN-ISSWWNFGSL LGICLILQILTGLFLAMHYT SD-TTTAFS-----SVTHIC RDVNYGWVIRYLHANGASMF FICLFIHVGRGLYYGSYMFL ET--WNIGVVLLFTVMATAF
S.tuberosu   ---KSHPLLKQPISSTLNQH LIDYPTPSN-LSYWWGFGSL AGICLVIQIVTGVFLAMHYT PH-VDLAFN-----SVEHIM RDVEGGWLLRYMHANGASMF FIVVHLHIFRGLYHASYSSP REFVRCLGVVIFLLMIVTAF
seal   ---KNYPLMK-----IINNS FIDLPTPSN-ISAWWNFGSL LGICLILQILTGLFLAMHYT SD-TTTAFS-----SVTHIC RDVNYGWIIRYLHANGASMF FICLYMHVGRGLYYGSYTFT ET--WNIGIILLFTVMATAF
T.annulata   ------------------H LLSYMVPKN-LNLNWNFGFI LGILLVLQIISGLMLSFFYV PA-KGMAFE-----STLAVM LNICFGWFVRLYHSFGVSFY FFFMFLHIMKGMWYSSNHLP --WSYSGVVIFVLSIATAF
T.brucei   FL-----LFF-----LLFRN LCCLLMSGC-LYRIYGVGFS LGFFIALQIICGVCLAWLFF SC-FICSNW-----YFVLFL WDFDLGFVIRSVHICFTSLL YLLLYIHIFKSITLIILFDT HILVWFIGFILFVFIIIIAF
T.napu   -------LMK-----IVNNA FIDLPAPSN-ISSWWNFGSL LGICLILQILTGLFLAMHYT SD-TSTAFS-----SVTHIC RDVNYGWIIRYMHANGASMF FICLYMHVGRGLYYGSYTFL ET--WNIGVILLLTVMATAF
T.tajacu   -------LMK-----IINNT FIDLPTPSN-ISSWWNFGSL LGICLLLQILTGLFLAMHYT PD-TTTAFS-----SVTHIC RDVNYGWIIRYLHANGASMF FICLFIHVGRGLYYGSYLFL ET--WNIGILLLTVMATAF
urchin   ---QRLSIFR-----ILNST FVDLPLPSN-LSIWWNFGSL LGLCLITQILTGLFLAMHYT AD-ISLAFS-----SASHIC RDVNYGWLLRNVHANGASLF FICMYCHIGRGLYYGGSNKI ET--WNVGVILFLVTVLTAF
V.faba   AGICLVIQIVTGVFLAMHYT PH-VDLAFN-----SVEHVM AGICLVIQIVTGVFLAMHYT PH-VDLAFN-----SVEHVM RDVEGGWLLRYMHANGASMF LIVVHLHIFRGLYHASYSSP REFVWCLGVVIFLLMIVTAF
whale   LGLCLIMQILTGLFLAMHYT PD-TTTAFS-----SVTHIC LGLCLIMQILTGLFLAMHYT PD-TTTAFS-----SVTHIC RDVNYGWIIRYLHANGASMF FICLYAHMGRGLYYGSYAFR ET--WNIGVILLFTVMATAF

                                           helix A   helix B   helix C

233

Continued

Table 9.1 *(Continued)*


```
A.american  MGYVLPWGQMSFWGATVITN LLSAIPYIGTN-LVEW-IWG GFS---------VDKATLTR FFAFHFILPFIIAALAMV-H LLFLHET-GSNNPTG-IPS- -------DADKIPFHPYYTI KDIL---GALLMILALMMLV
A.francisc  LGYVLPWGQMSFWGATVITN LVSAVPYIGND-VVQW-LWG GFA---------NDNPTLTR FFTFHFLIPFLVAGLTMI-H LLFLHQS-GSNNPLG-INA- -------NLDKLPFHPYFTI KDTVGF---MVLIFFLVTLS
A.nidulans  LGYVLPYGQMSLWGATVITN LMSAIPWIGQD-IVEF-IWG GFS---------VNNATLNR FFALHFLLPFVLAALALM-H LIAMHDTVGSGNPLG-ISA- -------NYDRLPFAPYFIF KDLITI---FIFFIVLSIFV
A.suum      MGYVLVWAQMRFWASVVITS LLSVIPVWGFA-IVTW-IWS GFT---------VSSATLKF FFVLHFLVPWGLLLLVLL-H LVFLHET-GRTSKLY-CHG- -------DYDKVCFYPEYWV KDFL---NVVVW-FVFIFFS
A.transmon  VGYVLPWGQMSFWGATVITN LLSAFPDIGDT-LVQW-IWG GFS---------VDNATLTR FFAFHFLLPFVIAGASMI-H LLFLHQT-GSNNPTG-LNS- -------DADKVTFHPYFSY KDLFGF---TLMLVGLTSVA
B.japonicu  MGYVLPWGQMSFWGATVITN LFSAIPYFGES-IVTL-LWG GYS---------VGNPTLNR FFSLHYLLPFLIAGVVVL-H VWALHVA-GQNNPEG-VEPK ------SEKDTVPFTPHATI KDMFGV---ACFLLLYAWFI
B.taurus    MGYVLPWGQMSFWGATVITN LLSAIPYIGTN-LVEW-IWG GFS---------VDKATLTR FFAFHFILPFIIMAIAMV-H LLFLHET-GSNNPTG-ISS- -------DVDKIPFHPYYTI KDILGA---LLLILALMLLV
C.dromedar  MGYVLPWGQMSFWGATVITN LLSAIPYIGTT-LVEW-IWG GFS---------VDKATLTR FFAFHFILPFIITALVAV-H LLFLHET-GSNNPTG-ISS- -------DMDKIPFHPYYTI KDIL---GALLLMLALLILV
C.elegans   MGYVLVWAQMSFWAAVVITS LLSVIPIWGPT-IVTW-IWS GFG---------VTGATLKF FFVLHFLLPWAILVIVLG-H LIFLHST-GSTSSLY-CH-- ----GDYDRKVCFSPEYLG KDAYNIVIWLLFIV----LS
C.reinhard  IGYVLPWGQMSFWGATVITS LATAIPVVGKH-IMYW-LWG GFS---------VDNPTLNR FYSFHYTLPFILAGLSVF-H IAALHQY-GSTNPLG-VN-- ------SQSSLISFGSYFGA KDLVVLCSWLLCSAFL---V
C.smithii   IGYVLPWGQMSFWGATVITS LATAIPVVGKH-IMYW-LWG GFS---------VDNPTLNR FYSFHYTLPFILAGLSVF-H IAALHQY-GSTNPLG-VNS- -------QSSLISFGSYFGA KDLV---GALFLALVFSILV
carp        VGYVLPWGQMSFWGATVITN LLSAVPYMGDM-LVQW-IWG GFS---------VDNATLTR FFAFHFLLPFVIAAATII-H LLFLHET-GSNNPLG-LNS- -------DADKVSFHPYFSY KDLL---GFVIMLLALTLLA
chicken     VGYVLPWGQMSFWGATVITN LFSAIPYIGHT-LVEW-AWG GFS---------VDNPTLTR FFALHFLLPFAIAGITII-H LTFLHES-GSNNPLG-ISS- -------DSDKIPFHPYYSF KDIL---GLTLMLTPFLTLA
Chlorobium  TGYLLPWNELAFFATQVGTE VPKVAPG-GAF-LVEI-LRG GPE---------VGGETLTR MFSLHVVLLPGLVMLVLAAH LTLVQIL-GTSAPIG-YKE- --AGLIKGYDK--FFPTFLA KDGIG---WLIGFALLIYLA
D.yakuba    MGYVLPWGQMSFWGATVITN LLSAIPYLGMD-LVQW-LWG GFA---------VDNATLTR FFTFHFILPFIVLAMTMI-H LLFLHQT-GSNNPIG-LNS- -------NIDKIPFHPYFTF KDIV---GFIVMIFILISLV
deer        MGYVLPWGQMSFWGATVITN LLSAIPYIGTN-LVEW-IWG GFS---------VDKATLTR FFAFHFILPFIIAALAMV-H LLFLHET-GSNNPTG-IPS- -------DADKIPFHPYYTI KDIL---GALLMILVLMMLV
Dolphin     VGYVLPWGQMSFWGATVITN LLSAIPYIGTT-LVEW-IWG GFS---------VDKATLTR FFAFHFILPFIITALAAV-H LLFLHET-GSNNPTG-IPS- -------NMDMIPFHPYYTI KDILGG---LLLILTLLALT
Drosophila  MGYVLPWGQMSFWVATVITN LLYAIPYLGMD-LVQW-LWG GFA---------VDNATLTR FFTFHFILPFIVLAMTMI-H LLFLHQT-GSNNPIG-LNS- -------NIDKIPFHPYFTF KDIV---GFIVMIFILISLV
frog        VGYVLPWGQMSFWGATVITN LLSAIPYIGNV-LVQW-IWG GFS---------VDNATLTR FFAFHFLLPFIIAGASIL-H LLFLHET-GSTNPTG-LNS- -------DPDKVPFHPYFSY KDLL---GFLIMLTALTLLA
goat        MGYVLPWGQMSFWGATVITN LLSAIPYIGTN-LVEW-IWG GFS---------VDKATLTR FFAFHFILPFIITALAMV-H LLFLHET-GSNNPTG-IPS- -------DTDKIPFHPYYTI KDIL---GAMLLILVLMLLV
H.africaea  MGYVLPWGQMSFWGATVITN LLSAIPYIGTT-LVEW-IWG GFS---------VDKATLTR FFAFHFSLPFIITALVLV-H LLFLHET-GSNNPSG-IDS- -------NSDKIPFHPYYTI KDILGL---LLMLTALLILV
Human       MGYVLPWGQMSFWGATVITN LLSAIPYIGTD-LVQW-IWG GYS---------VDSPTLTR FPTFHFILPFIIAALATL-H LLFLHET-GSNNPLG-ITS- -------HSDKITFHPYYTI KDAL---GLLLFLLSLMTLT
L.tarentol  IGYVLPCTMMSYWGLTVFSN ILATVPVIGTW-LCYW-IWG SEY---------INDFTLLK LHVLHVLLPFVLILVIFM-H LFCLHYFMSSDGFCDRFAF- -------YCERLCFCMWFYL RDMF---LAFLILFYVVFI
M.polymorp  IGYVLPWGQMSFWGATVITS LASAIPVVGDT-IVTW-LWG GFS---------VDNATLNR FFSLHYLLPFIIAGASIL-H LAALHQY-GSNNPLG-IN-- ------SSVDKIAFYPYIYV KDLV---GWVAFAIFFSIFV
Maize       IGYVPPWGQMSFWGATVITS LASAIPVVGDT-IVTW-LWG GFS---------VDNATLNR FFSLHHLLPLILAGASLL-H LAALHQY-GSNNPLG-VH-- ------SEMDKIASYPYFYV KDLV---GRVASAIFFSIWI
mouse       MGYVLPWGQMSFWGATVITN LLSAIPYIGTT-LVEW-IWG GFS---------VDKATLTR FFAFHFILPFIIAALAIV-H LLFLHET-GSNNPTG-LNS- -------DADKIPFHPYYTI KDIL---GILIMFLILMTLV
N.crass     LGYVLPYGQMSLWGATVITN LMSAIPWIGQD-IVEL-HLG GFS---------VNNATLNR FFALYFVLLFILVLVLM-Y LIVLYDMVGLSNPLG-ALG- -------NYDRMMFAPYYLF KDLMTI---FIFMYVLSSFV
O.bertiana  IGYVLPWGQMSFWGATVITS LASAIPVVGDT-IVTW-LWG GFS---------VDNATLNR FFSLYHLLPFILVGASLL-H LAALHQY-GSSNPLG-VHS- -------EMDKISFYPYFYV KDLV---GWVAFAIFFSIWI
opossum     VGYVLPWGQMSFWGATVITN LLSAIPYIGNT-LVEW-IWG GFS---------VDKATLTR FFAFHFILPFIIALVIV-H LLFLHET-GSNNPTG-INP- -------NSDKIPFHPYYTI KDAL---GLILMLLILMSLA
P.anserina  LGYVLPYGQMSLWGATVITN LVSAIPWIGQD-IVEF-IWG GFS---------VNNATLNR FFALHYLLPFILVALVLM-H LIALHDTAGSSNPLG-ISG- -------NYDRITFAPYYLF KDLI---TIFIFIFVLSSFV
P.aurelia   FGLVLCCTHLSEITLTIAAN IFHTFFMFKGK-AYWF-LFT DKQ---------LNTDTLIR LAYAHYVSAFYLSFLGLL-H GIDMH-----------YDWK NEPFYDGLSSEMLWWDEALS NDLTNFFVLLVFITLAFFLL
P.denitrif  MGYVLPWGQMSFWGATVITG LFGAIPGVGEA-IQTW-LLG GPA---------VDNPTLNR FFSLHYLLPFVIAALVVV-H IWAFHTT-GNNNPTG-VEVR RGSKEEAKKDTLPFWPYFVI KDLFAL---AVVLVVFFAIV
P.falcipar  VGYVLPWGQMSYWGATVITN LLSSIP----V-AVIW-MCG GYT---------VSDPTMKR FFVLHFILPFIGLCIVFM-H MFFLHL-HGSTNPLG-Y--- -----DTALKM-PFYPNLLS LDVKGFNNVMILFL-MQSLF
P.tetraure  FGLVLCCTHLSEITLTIAAN IFHTF-FMFKG-KAYWFLFT DKQ---------LNTDTLIR LAYAHYVSAFYLSFLGLL-H GIDIHYDWKNEPFYDGLSS- -------EMLWWDEALSNEL TNFF---VLLVFITLAFFLL
P.yoelii    IGYVLPWGQMSYWGATVITN LLSGIP----A-LVIW-LCG GYT---------VSDPTIKR FFVLHFILPFVALCIVFM-H IFFLHL-HGSTNPLG-Y--- -----DTALKI-PFYPNLLS LDVKGFNNILILFL-IQSIF
Rat         MGYVLPWGQMSFWGATVITN LLSAIPYIGTT-LVEW-IWG GFS---------VDKATLTR FFAFHFILPFIIAALAIV-H LLFLHET-GSNNPTG-LNS- -------DADKIPFHPYYTI KDLLGV---FMLLLFLMTLV
Rb.capsula  MGYVLPWGQMSFWGATVITG LFGAIPGIGPS-IQAW-LLG GPA---------VDNATLNR FFSLHYLLPFVIAALVAI-H IWAFHTT-GNNNPTG-VEVR RTSKADAEKDTLPFWPYFVI KDLFAL---ALVLLGFFAVV
Rb.sphaero  MGYVLPWGQMSFWGATVITG LFGAIPGIGHS-IQTW-LLG GPA---------VDNATLNR FFSLHYLLPFVIAALVAI-H IWAFHST-GNNNPTG-VEVR RTSKAEAQKDTVPFWPYFII KDVFAL---AVVLLVFFAIV
rice        IGYVPPWGQMSFWGATVITS LASAIPVVGDT-IVTW-LWG GFS---------VDNATLNR FFSLHHLLPLILVGASLL-H LAALHQY-GSNNPLG-VHS- -------EMDKIASYPYFYV KDLV---GRVASAIFFSIWI
Rs.rubrum   MGYVLPWGQMSFWGATVITN LFSAIPVVGDD-IVTL-LWG GFS---------VDNPTLNR FFSLHYLFPMLLFAVVFL-H MWALHVK-KSNNPLG-IDAK ------GPFDTIPFHPYYTV KDAFGL---GIFLMVFCFFV
S.acidocal  FGYSLVSDVLGVNAIDIGDQ LLVGTGIPGATAIVGW-LFG PGGSAALSSNPLVRSELFDR LLGWHIIMVFLLGVLFLF-H FMLSERY-G-MTPAT-REKP KVPSYYTKEEQEKFNPWWPR NFVYMLSIVLITWGIILFVP
S.cerevisi  LGYCCVYGQMSHWGATVITN LFSAIPFVGND-IVSW-LWG GFS---------VSNPTIQR FFALHYLVPFIIAAMVIM-H LMALH-IHGSSNPLG-ITG- -------NLDRIPMHSYFIF KDLVTV---FLFMLILALFV
S.pombe     LGYCLPANQMSFWGATVITN LLSAVPFIGDD-LVHT-LWG GFS---------VSNPTLNR FFSTHYLMPFVIAALSVM-H LIATH-TNGSSNPLG-VTA- -------NMDRIPMNPYYTM KDLMTM---FIFLIGMNYMA
S.scrofa    MGYVLPWGQMSFWGATVITN LLSAIPYIGTD-LVEW-IWG GFS---------VDKATLTR FFAFHFILPFIITALAAV-H LMFLHET-GSNNPTG-ISS- -------DMDKIPFHPYYTI KDIL---GALFMMLILLILV
S.tuberosu  IGYVLPWGQMSFWGATVITS LASAIPVVGDT-IVTW-LWG GFS---------VDNATLNR FFSLHHLLPFILVGASLL-H LAALHQY-GSNNPLG-VHS- -------EMDKIASYPYFYV KDLV---GWVAFAIFFSIWI
seal        MGYVLPWGQMSFWGATVITN LLSAIPYVGTD-LVQW-IWG GFS---------VDKATLTR FFAFHFILPFVVLALDAV-H LLFLHET-GSNNPSG-IMS- -------DSDKIPFHPYYTI KDIL---GALLLILVLTLLV
T.annulata  VGYVLPDGQMSFWGATVIGG LLKFFG----K-ANVL-IFG GQT---------VGPETLER FFSIHVILPVIILLVVIF-H LYVLHR-DGSSNPLA-V--- -----IDMLAIFRFHPVVLF SDIR-FIVIVILLIGVQSGY
T.brucei    IGYVLPCTMMSYWGLTVFSN IIATVPILGIW-LCYW-IWG SEF---------INDFTLLK LHVLHVLLPFILLIILIL-H LFCLHYFMSSDAF------- -CDRFAFYCERLSFCMWFYL RDMFLA---FSILLCMMYVI
T.napu      MGYVLPWGQMSFWGATVITN LLSAIPYIGTE-LVEW-IWG GFS---------VDKATLTR FFAFHFILPFVITALALV-H LLFLHET-GSNNPTG-IPS- -------DADKIPFHPYYTI KDVL---GALVLMLVLLLLV
T.tajacu    MGYVLPWGQMSFWAATVITN LLSAIPYIGTD-LVEW-IWG GFS---------VDKATLTR FFAFHFILPFIITALVIV-H LLFLHET-GSNNPTG-IPS- -------NMDKIPFHPYYTI KDIL---GATLMILILLLLV
urchin      VGYVLVWGRMSFWAATVIAN LVTAVPCVGTT-IVQW-LWG GFS---------VDNATLTR FFAFHFLFPFIIAALAII-D LVFLHNS-GANNPVG-LNS- -------NYDKAPFHIYTT KDTV---GFIALIAALFVLA
whale       VGYVLPWGQMSFWGATVITN LLSAIPYIGTT-LVEW-IWG GFS---------VDKATLTR FFAFHFILPFIILALAIV-H LIFLHET-GSNNPTG-IPS- -------DMDKIPFHPYHTI KDIL---GALLLILLMLT
                                                                       helix D                                                      helix E
```

```
A.american  LFSPDLLGDPDNYTP-AN-P LNTPPH---IKP--EWYFLF AYAILRS----------------------IPNKLGGVLA-- -------------LVLSIL ILIFMP---LLHTSKQRSMM
A.francisc  LTSPYLLGDPDNFIP-AN-P LVTPAH---IQP--EWYFLF AYAILRS----------------------IPNKLGGVIA-- -------------LVSSIL ILVSLP---FTFKPKFRGLE
A.nidulans  FFMPNALGDSENYVM-AN-P MQTPPA---IVP--EWYLLP FYAILRS----------------------IPNKLLGVIA-- -------------MFAAIL ALMVMP---ITDLSKLRGVQ
A.suum      LGYPFTLGDPEMFIE-SD-P MMRPVH---IVP--EWYFLF AYAILRA----------------------IPNKVLGVVS-- -------------LFASIL VTVVFV---LVNNY---VSV
A.transmon  LFSPNLLGDPDNFTP-AN-P LVTPPH---IKP--EWYFLF AYAILRS----------------------IPNKLGGVLA-- -------------LLFSIL VLMLVP---MLHTSKQRGNT
B.japonicu  FYMPNYLGDADNYIP-AN-P GVTPPH---IVP--EWYYLP FYAILRS----------------------IPNKLAGVIG-- -------------MFSAII ILCFLP---WLDAAKTRSSK
B.taurus    LFAPDLLGDPDNYTP-AN-P LNTPPH---IKP--EWYFLF AYAILRS----------------------IPNKLGGVLA-- -------------LAFSIL ILALIP---LLHTSKQRSMM
C.dromedar  LFSPDLLGDPDNYTP-AN-P LNTPPH---IKP--EWYFLF AYAILRS----------------------IPNKLGGVLA-- -------------LVLSIL ILAFIP---ALHTSKQRSMT
C.elegans   LIYPFNLGDAEMFIE-AD-P MMSPVH---IVP--EWYFLF AYAILRA----------------------IPNKVLGVIA-- -------------LLMSIV TFYFFA---LVNNY---TSC
C.reinhard  FFYPDLLGHPDNLIP-AN-P YSTPQH---IVP--EWYFLW VYAILRS----------------------IPNKAMGVLA-- -------------IGLVFA SLFAMP---FIG---LGGGK
C.smithii   FFYPDLLGHPDNLIP-AN-P YSTPQH---IVP--EWYFLW VYAILRS----------------------IPNKAMGVLA-- -------------IGLVFA SLFAMP---FIG---LGGGK
carp        LFSPNLLGDPENFTP-AN-P LVTPPH---IKP--EWYFLF AYAILRS----------------------IPNKLGGVLA-- -------------LLFSIL VLMVVP---LLHTSKQRGLT
chicken     LFSPNLLGDPENFTP-AN-P LVTPPH---IKP--EWYFLF AYAILRS----------------------IPNKLGGVLA-- -------------LAASVL ILFLIP---FLHKSKQRTMT
Chlorobium  VMFPWEIGVKANPLS-PA-P LG-------IKP--EWYFWA QFQLLKDF--------------------KFEGGEL-- -------------LAIILF TIGGV--------------
D.yakuba    LISPNLLGDPDNFIP-AN-P LVTPAH---IQP--EWYFLF AYAILRS----------------------IPNKLGGVIA-- -------------LVLSIA ILMILP---FYNLSKFRGIQ
deer        LFSPDVLGDPDNYTP-AN-P LNTPPL---IKP--EWYFLF AYAILRS----------------------IPNKLGGVLA-- -------------LVLSIL ILIFMP---LLHTSKQRSMM
Dolphin     LFTPDLLGDPDNYTP-AN-P LSTPPH---IKP--EWYFLF AYAILRS----------------------IPNKLGGVLA-- -------------LLLSIL ILIFIP---MLQTSKQRSMM
Drosophila  LISPNLLGDPDNFIP-AT-P LVTPAH---IQP--EWYFLF AYAILRS----------------------IPNKLGGVIA-- -------------LVLSIA ILMILP---FYNLSKFRGIQ
frog        MFSPNLLGDPDNFTP-AN-P LITPPH---IKP--EWYFLF AYAILRS----------------------IPNKLGGVLA-- -------------LVLSIL ILALMP---LLHTSKQRSLM
goat        LFTPDLLGDPDNYTP-AN-P LNTPPH---IKP--EWYFLF AYAILRS----------------------IPNKLGGVLA-- -------------LVLSIL ILVLVP---FLHTSKQRSMM
H.africaea  LFSPDLLGDPDNYTP-AN-P LNTPPH---IKP--EWYFLF AYAILRS----------------------IPNKLGGVLA-- -------------LIFSIL ILAIIP---LLHTSKQRSML
Human       LFSPDLLGDPDNYTL-AN-P LNTPPH---IKP--EWYFLF AYTILRS----------------------VPNKLGGVLA-- -------------LLLSIL ILAMIP---ILHMSKQQSMM
L.tarentol  FINWYFVFHEESWVI-VD-T LKTSDK---ILP--EWFFLF LFGFLKA----------------------VPDKFTGLLL-- -------------MVILLF SLFLFI---L----------
M.polymorp  FYAPNVLGHPDNYIP-AN-P MSTPAH---IVP--EWYFLP VYAILRS----------------------IPNKLGGVAA-- -------------IGLVFV SLLALP---FINTSYVRSSS
Maize       FFAPNVLGHPDNYIP-AN-P MPTPPH---IVP--EWYFLP IHAILRS----------------------IPDKAGGVAA-- -------------IAPVFI SLLALP---FFKEMYVRSSS
mouse       LFFPDMLGDPDNYMP-AN-P LNTPPH---IKP--EWYFLF AYAILRS----------------------IPNKLGGVLA-- -------------LILSIL ILALMP---FLHTSKQRSLM
N.crass     FFMPNVLGDSENYMM-AN-P MQTPPA---IVP--EWYLLP FDAMLRS----------------------MPNKLLGVIA-- -------------MFSAML AMMLLP---MTDLGRSKGLQ
O.bertiana  FYAPNVLGHPDNYIP-AN-P MSTPPH---IVP--EWYFLP IYAILRS----------------------IPDKAGGVAA-- -------------IALVFI CLLALP---FFKDMYVRSSS
opossum     MFSPDMLGNPDNFTP-AN-P LNTPPH---IKP--EWYFLF AYAILRS----------------------IPNKLGGVLA-- -------------LLASLL ILLIIP---LLHTSKQRSLM
P.anserina  FFMSNVLGDSENYIM-AN-P MQTPAA---IVP--EWYLLP FYAILRS----------------------IPNKLLGVIA-- -------------MFAAIL AIMLLP---ITDLGRSKGLQ
P.aurelia   FEEPEALSYEIFMWG-DI-G LSTDVRFYGVAP--EWYF-------RPFMA-----WLIA---------CPFHKTGIFG-- -------------LLFFFV TLYYQPNLH----GVSDQNS
P.denitrif  GFMPNYLGHPDNYIE-AN-P LVTPAH---IVP--EWYFLP FYAILRAFTAD---VWVVML VNWLSFGIIDAKFFGVIA-- -------------MFGAIL VMALVP---WLDTSRVRSGQ
P.falcipar  GIMPL--SHPDNAIV-VN-T YVTPSQ---IVP--EWYFLP FYAMLK----------------------TVPSKPAGLVI-- -------------VLLSLQ LLFLLAEQRSLTTMIQFKMI
P.tetraure  FEEPEALSY-EIFMW-GDIG LSTDVRFYGVAP--HWYFRP FMAWLIA----------------------CPFHKTGIFG-- -------------LLFFFV TLYYQP---NLHGVSDQNS-
P.yoelii    GVIPL--SHPDNAII-VN-T YVTPLQ---IVP--EWYFLP FYAMLK----------------------TIPSKNAGLVI-- -------------VVASLQ LLFLLAEQRNLTTIIQFKMV
Rat         LFFPDLLGDPDNYTP-AN-P LNTPPH---IKP--EWYFLF AYAILRS----------------------IPNKLGGVVA-- -------------LILSIL ILAFLP---FLHTSKQRSLT
Rb.capsula  AYMPNYLGHPDNYVQ-AN-P LSTPAH---IVP--EWYFLP FYAILRAFAAD---VWVVIL VDGLTFGIVDAKFFGVIA-- -------------MFGAIA VMALAP---WLDTSKVRSGA
Rb.sphaero  GFMPNYLGHPDNYIE-AN-P LSTPAH---IVP--EWYFLP FYAILRAFTAD---VWVVQI ANFISFGIIDAKFFGVLA-- -------------MFGAIL VMALVP---WLDTSPVRSGR
rice        FFAPNVLGHPDNYIP-AN-P MPTPPH---IVP--EWYFLP IHAILRS----------------------IPDKAGGVAA-- -------------IAPVFI SLLALP---FFKEMYVRSSS
Rs.rubrum   FFAPNFFGEPDNYIP-AN-P MVTPTH---IVP--EWYFLP FYAILRA----------------------VPDKLGGVLA-- -------------MFGAIL ILFVLP---WLDTSKVRSAT
S.acidocal  NLLANINGLPIVINPYPA-P QAGSPQAVSVQPYPPWFFLF LFKLVDFLLPNGIPITPILT IALLVVGLVILMLLPFLDPS DSLYVTRRKFWTWIMTTLAV YLVELSVWGYLEPGVPEPTS
S.cerevisi  FYSPNTLGHPDNYIP-GN-P LVTPAS---IVP--EWYLLP FYAILRS----------------------IPDKLLGVIT-- -------------MFAAIL VLLVLP---FTDRSVVRGNT
S.pombe     FYNPYGFMEPDCALP-AD-P TKTPMS---IVP--EWYLLP FYAILRA----------------------MPNFQLGVMA-- -------------MLLSIL VLTLLP---LLDFSAIRGNS
S.scrofa    LFSPDLLGDPDNYTP-AN-P LNTPPH---IKP--EWYFLF AYAILRS----------------------IPNKLGGVLA-- -------------LVASIL ILILMP---MLHTSKQRGMM
S.tuberosu  FYAPNVLGHPDNYIP-AN-P MSTPPH---IVP--EWYFLP IHAILRS----------------------IPDKVGGVAA-- -------------IAPVFI CLLALP---FFKSMYVRSSS
seal        LFSPDLLGDPDNYIP-PN-P LSTPPH---IKP--EWYFLF AYAILRS----------------------IPNKLGGVLA-- -------------LVLSIL VLAIMP---LLHTSKQRGMM
T.annulata  GFISIFQADPDNSIL-SD-P LNTPAH---IIP--EWYLLL FYATLK----------------------VFPTKVAGLLA-- -------------MAGMLE LLVLLVESRY------FKQT
T.brucei    FINWYFYFHEESWVI-VD-T LKTSDK---ILP--EWFFLY LFGFLKA----------------------IPDKFMGLFL-- -------------MVILLF SLFLF--------------
T.napu      LFSPDLLGDPDNYTP-AN-P LNTPPH---IKP--EWYFLF AYAILRS----------------------IPNKLGGVLA-- -------------LIASIL ILQLMP---LLHTSKQRSMM
T.tajacu    LFSPDLLGDPDNYTP-AN-P LNTPSH---IKP--EWYFLF AYAILRS----------------------IPNKLGGVLA-- -------------LALSIL ILALVP---ALHTSKQRSMM
urchin      LLFPCALNDPENFIP-AN-P LSHPPH---IQP--EWYFLF AYAILRS----------------------IPNKLGGVIA-- -------------LVAAIL VLFLMP---LLNTSKNESNS
V.faba      FYAPNVLGHPDNYIP-AN-P MPTPPH---IVP--EWYFLP IHAILRS----------------------IPDKSGGVAA-- -------------IAPVFI CLLALP---FFKSMYVRSSS
whale       LFAPDLLGDPDNYTP-AN-P LSTPAH---IKP--EWYFLF AYAILRS----------------------IPNKLGGVLA-- -------------LLLSIL ILAFIP---MLHTSNQRSMM

                                                                                              helix F
```


Continued

Table 9.1 *(Continued)*

```
A.american   ----F--RPFSQC----LFWI ------LLVADLLTLTWIGG QPVEHPFIIIGQLASI---M YFLIILVL-MPVTSTIEN-N -------------------- --------------------
A.francisc   ----F--YSVAQP----LFWS ------WVSVFLLL-TWIGA RPVEDPYNFLGQILTC---A YFS-YFVF-TPIVINLND-K
A.nidulans   ----F--RPLSKV----VFYI ------FVANFLIL-MQIGA KHVETPFIEFGQISTI---I YFAYFFVI-VPVVSLIEN--
A.suum       ----M--SKLNKF----TVFV ------FFIFVLVVLSWTGQ CLVEDPFVFLSMVFSF---L YFFVIFLL-F---------
A.transmon   ----F--RPLSQI----LFWA ------LVADMLVL-TWIGG QPVEHPFVLIGQVAST---V YFALFLIA-LPLTGWLEN-K
B.japonicu   ----Y--RPLAKQ----FFWI ------FVAVCILL-GYLGA QPPEGIYVIAGRVLTV---C YFAYFLIV-LPLLSRIET-P RPVPNSI-------SEAILA KGGKAVASVA----------
B.taurus     ----F--RPLSQC----LFWA ------LVADLLTL-TWIGG QPVEHPYITIGQLASV---L YFLLILVL-MPTAGTIEN-K
C.dromedar   ----F--RPISQC----LFWV ------LLVADLLTLTWIGG QPVEPPFIMIGQVASI---L YFSLILIL-MPVAGIIEN-R
C.elegans    ----L--TKLNKF----LVFM ------FIISSTIL-SWLGQ CTVEDPFTILSPLFSF---I YFGLAYLM-L---------
C.reinhard   ----F--RIITEW----LYWT ------FLADVLLL-TWLGG NEITPITSFVGQCCTA---Y LFFYLLVC-QPLVGYLET-Q ----------------F- -------AHG---------
C.smithii    ----F--RTITEW----LYWT ------LFLADVLLLTWLGG NEITPITSFVGQCCTA---Y LFFYLLVC-QPLVGYLE---
carp         ----F--RPITQF----LFWT -------LVADMIILTWIGG MPVEHPFIIIGQIASV---L YFALFLIF-MPLAGWLEN-K
chicken      ----F--RPLSQT----LFWL ------LLVANLLILTWIGS QPVEHPFIIIGQMASL---S YFTILLIL-FPTIGTLEN-K
Chlorobium   ----------------VW- ----------LLV-PFIDR QASEEK-------KSP---I FTIFGILV-LAFL------
D.yakuba     ----F--YPINQI----LFWS ------LMLVTVILLTWIGA RPVEEPYVLIGQILTI---I YFLYYLI--NPLVTKWWD-N
deer         ----F--RPFSQC----LFWI ------LLVADLLTLTWIGG QPVEHPFIIIGQLASI---L YFLIILVL-MPATSTIQN-N
Dolphin      ----F--RPFSQL----LFWV ------LIADLLTL-TWIGG QPVEHPYIIVGQLASI---L YFLLILVL-MPTAGLIEN-K
Drosophila   ----F--YPINQV----MFWS ------FMLVTVILLTWIGA RPVEEPYVLIGQILTV---V YFLYYLV--NPLITKWWD-N
frog         ----F--RPFTQI----MFWA -------LVADTLILTWIGG QPVEDPYTMIGQLASV---I YFSIFIIM-FPLMGWVEN-K
goat         ----F--RPISQC----MFWI ------FLVADLLTLTWIGG QPVEHPYIIIGQLASI---M YFLIILVM-MPVASTIEN-N
H.africaea   ----F--RPFSQC----LFWI ------LAANLLIL-TWIGG QPVEHPYITIGQLASI---S YFSILLII-MPLTSIMEN-K
Human        ----F--RPLSQS----LYWL ------LLAADLLILTWIGG QPVSYPFTIIGQVASV---L YFTTILIL-MPTISLIEN-K
L.tarentol   ----------NC----ILWF ------LVYCRSSLL-WF-- ---TYSLILFYSIFMS---G FLALYVILAYPIWMELQF-W
M.polymorp   ----F--RPIHQK----FFWL ------LVADCLLL-GWIGC QPVEAPYVTIGQIASV---G FFFYFAI--TPILGKCDA-R -----------------LI KNSNACEARS----------
Maize        ----F--RPIHQG----IFWL ------LLADCLLL-GWIGC QPVEAPFVTIGQISSF---F FFLFFAI--TPIPGRVG--- -------------------- --------RG----------
mouse        ----F--RPITQI----LYWI -------LVANLLILTWIGG QPVEHPFIIIGQLASI---S YFSIILIL-MPISGIIED-K
N.crass      ----F--RPLSKF----AFWA ------FVVNFLIL-MKLGA CHVESPFIELGQFSTI---F YLSYFMFM-VPVLSLIEN--
O.bertiana   ----F--RPIYQG----IFWL ------LLLADCLLLGWIGC QPVEAPFVTIGQI-SS---I VFFLFFAI-TPILGRVGR-G
opossum      ----F--RPISQI----MFWL ------LLVANLLTLTWIGG QPVEQPFIIIGQLAST---L YFSLIIIF-MPLAGMYED-H
P.anserina   ----F--RPLSKF----AFWV ------LFVVNFLILMKLGA CHVETPFIELGQLSTA---L YFGHFIII-VPIISLIEN-T
P.aurelia    ----YGKKTLT---------I SST-V-----------LAK KNTATPFSISIDSNLY---H QITYFFFI-MCCLY------
P.denitrif   ----Y--RPLFKW----WFWL ------LAVDFVVL-MWVGA MPAEGIYPYIALAGSA---Y WFAYFLII-LPLLGIIEK-P DAMPQTI-------EEDF--
P.falcipar   ---FGARDYSVP----IMW- ------FMCAFYAL-LWIGC QLPQDMFILYGR------- LFIVLFFC-SGLFVLVHY-R RT----------------
P.tetraure   ----Y--GKKTLT----ISST ------LVLAKKNTATPFSI SIDSNLYHQITYFFFIMCCL YTPSFLPYGRFFNQIGGNWG
P.yoelii     ----FSAREYSVP----IIW- ------FMCSFYAL-LWIGC QLPQDIFILYGR------- LFIILFFS-SGLFSLVQY-K KT---------------
Rat          ----F--RPITQI----LYWI ------LVANLLVL-TWIGG QPVEHPFIIIGQLASI---S YFSIILIL-MPISG------
Rb.capsula   ----Y--RPKFRM----WFWF ------LVLDFVVL-TWVGA MPTEYPYDWISLIAST---Y WFAYFLVI-LPLLGATEK-P EPIPASI-------EEDF--
Rb.sphaero   ----Y--RPMFKI----YFWL ------LAADFVIL-TWVGA QQTTFPYDWISLIASA---Y WFAYFLVI-LPILGAIEK-P VAPPATI-------EEDF--
rice         ----F--RPIHQG----IFWL -------LLADCLLLGWIGC QPVEAPFVTIGQI-PS---F FFFLFFAI-TPIPGRVGR-G
Rs.rubrum    ----F--RPVFKG----FFWV ------FLADCLLL-GYLGA MPAEEPYVTITQLATI---Y YFLHFLVI-TPLVGWFEK-P KPLPVSI-------SSPVTT QA----------------
S.acidocal   AQIEFLGPPLVIIGIIVYLWP TERKTKTVSTTATD-SRVIK MNITPMEILLGAVGTL---S FAATLFNF-IQFPTLING-I ILVPLGLFAIYALRRISFYV LGGKPVASVGNTSSRISLRK
S.cerevisi   ----F--KVLSKF----FFFI ------FVFNFVLL-GQIGA CHVEVPYVLMGQIATF---I YFAYFLII-VPVISTIEN--
S.pombe      ----F--NPFGKF----FFWT ------FVADFVIT-AWIGG SHPENVFITIGAIATI---F YFSYFFIL-MPVYTMLGN--
S.scrofa     ----F--RPLSQC----LFWM ------LLVADLITLTWIGG QPVEHPFIIIGQLASI---L YFLIILVL-MPITSIIEN-N
S.tuberosu   ----F--RPIHQG----IFWL ------LLLADCLLLGWIGC QPVEAPFVTIGQI-SP---L VFFLFFAI-TPILGRVGR-G
seal         ----F--RPISQC----LFWF ------LLVADLLTLTWIGG QPVEHPYITVGQLASI---L YFTILLVL-MPIASIIEN-N
T.annulata   ----VSAMNY--H----RVWT TSS-VPLVPVLFML-GSIGK MVVHVDLIAIGT--CV---V LSVVLF-----IYKLLD---
T.brucei     --------ILNC----ILW- ------FVYCRSSL-LWLT- ------YSLI-----L---F YSIWMSGF-LALYVVLAY-- ---PIWM-------ELQY--
T.napu       ----F--RPISQC----LFWL ------FLAADLLTLTWIGG QPVEHPYVVIGQLASI---L YFSIILVL-MPVAGVIEN-K
T.tajacu     ----F--RPLSQL----LFWM -------LVADFLTLTWIGS QPVEHPFIIIGQLASI---L YFLIILVL-MPVANIIEN-N
urchin       ----F--RPLSQA----TFWI ------FLVATFFVLTWIGS QPVEQPFVLIGQIASL---L YFSLFIFG-FPLVSSLEN-K
V.faba       ----F--RPIHQG----IFWL ------FLLADRLLLGWIGC QPVEAPFVTIGQI-PP---F VFFLFFAI-TPIPGRVGR-G
whale        ----F--RPFSQF----LFWV ------LLVADLLTLTWIGG QPVEHPYMIVGQLASI---L YFLLILVL-MPVTSLIEN-K
                           helix G                                   helix H
```

```
A.american   --------------------  --------------------  ------LLLK---------W---
A.francisc   --------------------  --------------------  -------------------IV-
A.nidulans   --TLVE--------------  --------------------  -------------LG-TKK-NF-
A.suum       --------------------  --------------------  ------LLVYYFAGRV----FM-
A.transmon   --AL----------------  --------------------  --N-WN-----------------
B.japonicu   -IALVAAGALFL--------  ----GSLQDARANEGSDKPP  GNK-WSFAGPFGKFD-RGALQR-
B.taurus     --LL----------------  --------------------  --K-W------------------
C.dromedar   --------------------  --------------------  ------LILK---------W---
C.elegans    --------------------  --------------------  ----------FIFMS-SKLLFK-
C.reinhard   -TQ-----TN----------  --------------------  -----------------------
C.smithii    --------------------  --------------------  --------TQFAHGTQT----N-
carp         --------------------  --------------------  ------ALK----------WAC-
chicken      --------------------  --------------------  -----MLN----------Y---
Chlorobium   ---LINTYRVYA--------  ----E---------------  ---------------YSMLK-
D.yakuba     --------------------  --------------------  ------LLN-------------
deer         --------------------  --------------------  ------LLK---------W---
Dolphin      --LL----------------  --------------------  --K-W------------------
Drosophila   --------------------  --------------------  ------LLN-------------
frog         --------------------  --------------------  ------LLN---------W---
goat         --------------------  --------------------  -----ILLK---------W---
H.africaea   --LL----------------  --------------------  --K-WS-----------VLVV-
Human        --------------------  --------------------  ------MMLK--------W-A-
L.tarentol   --------------------  --------------------  ------VLLLF---MLVVCRLD-
M.polymorp   -VL-----ASFL--------  ----TSI-------------  GWI-W--------------W-
Maize        -IP-----KYYT--------  ----E---------------  -----------------------
mouse        --------------------  --------------------  ------AMLK--------LYP-
N.crass      --TLVD--------------  --------------------  ------------LN-YLK----
O.bertiana   --------------------  --------------------  ------LIPNSYTDETDHT----
opossum      --------------------  --------------------  ------LLE--------PKFP-
P.anserina   --------------------  --------------------  -----LVDL------NTMSKG-
P.aurelia    -TPFSLPYGRFF--------  ----NQIGC-----------  --N-WGFLFSYFYVFCYLAFTD-
P.denitrif   -------NAHYG--------  ----PETHE-----------  -----------------AE-
P.falcipar   ---------HY---------  --------------------  -----DYSSQ---------ANM-
P.tetraure   --------------------  --------------------  ------FLFSYFYVFCYLAFTD-
P.yoelii     ---------HY---------  --------------------  -----DYSSQ---------ANI-
Rat          --------------------  --------------------  -------------------IV-
Rb.capsula   -------NSHYG--------  ----NP--------------  ----------------AE-
Rb.sphaero   -------NAHYS--------  ----PATGCTKT--------  ----------------VVAE-
rice         --------------------  --------------------  -----LIPKYYTDETHRTGSFS
Rs.rubrum    --------------------  --------------------  -----------------------
S.acidocal   KIAFFGIIALFVVSLVLLGL  MWTLPSVGFQATYAGMDLGV  ILLLWGVAIQLYHYEIF--VKE-
S.cerevisi   --VLFY--------------  --------------------  -------------IG-RVN-K--
S.pombe      --TLID--------------  --------------------  -------------LN-LSSIKR-
S.scrofa     --------------------  --------------------  ------LLK---------W---
S.tuberosu   --------------------  --------------------  ------IPNSYTDETDHT-----
seal         --------------------  --------------------  ------FILK---------W---
T.annulata   --------------------  --------------------  -------SAR---------VRA-
T.brucei     --------------------  --------------------  ----WVL---LLFLL-IVCRLD-
T.napu       --------------------  --------------------  ------MLK---------W---
T.tajacu     --------------------  --------------------  ------LLK---------W---
urchin       --------------------  --------------------  ------MI-----------FS-
V.faba       --------------------  --------------------  ------IPNSYTDETDQ------
whale        --------------------  --------------------  ------LMK---------W---
```

Table 9.2 Amino Acid Sequences of Cytochrome b_6 of the Cytochrome b_6f Complex from 13 Species. The Four Transmembrane Helices (A to D) are Underscored.

```
Agmen        -MFTKEVTDSKLYKWFNERL  EIQAISDDISSKYVPPHVNI  FYCLGGITLTCFIIQFATG-  MFAMTFYYKPTVAEAFTSVQ
Barley       ----------KVYDWFEERL  EIQAIADDITSKYVPPHVNI  FYCLGGITLTCFLVQVATG-  -FAMTFYYRPTVTEAFSSVQ
E.gracilis   -------MS-RVYDWFEERL  EIQAIADDVSSKYVPPHVNI  FYCLGGITFTCFIIQVATG-  -FAMTFYYRPTVTEAFLSVK
P.hollandica HMFTKQVQESGVYKWFNDRL  EIEAISDDISSKYVPPHVNI  FYCLGGITLVCFIIQFATGH  MFAMTFYYKPSVTEAFTSVQ
Lvwort       -------MVV--YDWFEERL  EIQAIADDITSKYVPPHVNI  FYCLGGITLTCFLVQVATG-  -FAMTFYYRPTVTEAFSSVQ
Maize        ----------KVYDWFEERL  EIQAIADDITSKYVPPHVNI  FYCLGGITLTCFLVQVATG-  -FAMTFYYRPTVTEAFSSVQ
Nostoc       -------MA-NVYDWFEERL  EIQAIAEDVTSKYVPPHVNI  FYCLGGITLTCFLIQFATG-  -FAMTFYYKPTVAEAFSSVE
C.Protothec  -------MS-KIYDWFEERL  EIQSIADDISSKYVPPHVNI  FYCFGGITFTCFLVQVATG-  -FAMTFYYRPTVAEAFTSVQ
C.reinhard   -------MS-KVYDWFEERL  EIQAIADDITSKYVPPHVNI  FYCIGGITFTCFLVQVATG-  -FAMTFYYRPTVAEAFASVQ
Rice         -------MS-KVYDWFEERL  EIQAIADDITSKYVPPHVNI  FYCLGGITLTCFLVQVATG-  -FAMTFYYRPTVTEAFSSVQ
Spinach      -------MS-KVYDWFEERL  EIQAIADDITSKYVPPHVNI  FYCLGGITLTCFLVQVATG-  -FAMTFYYRPTVTDAFASVQ
Tobacco      -------MSVKVYDWFEERL  EIPAIADDITSKYVPPHVNI  FYCLGGITLTCFLVPVATG-  -FAMTFYYRPTVTEAFASVP
Wheat        -------MS-KVYDWFEERL  EIQAIADDITSKYVPPHVNI  FYCLGGITLTCFLVQVATG-  -FAMTFYYRPTVTEAFSSVQ
                                                             helix A
```

```
Agmen        YIMNEVNFGWLIRSIHRWSA  SMMVLMMILHIFRVYLTGGF  KR-PRELTWITGVIMATITV  SFGVTGYSLPWDQVGYWAVK
Barley       YIMTEANFGWLIRSVHRWSA  SMMVLMMILHVFRVYLTGGF  KK-PRELTWVTGVVLAVLTA  SFGVTGYSLPWDQIGYWAVK
E.gracilis   YIMNEVNFGWLIRSIHRWSA  SMMVLMMILHVCRVYLTGGF  KK-PRELTWVTGIILAILTV  SFGVTGYSLPWDQVGYWAVK
P.hollandica YLMNEVSFGWLIRSIHRWSA  SMMVLMMILHVFRVYLTGGF  KHNPRELTWITGVILAVITV  SFGVTGYSLPWDQVGYWAVK
Lvwort       YIMTEVNFGWLIRSVHRWSA  SMMVLMMILHIFRVYLTGGF  KK-PRELTWVTGVILAVLTV  SFGVTGYSLPWDQIGYWAVK
Maize        YIMTEANFGWLIRSVHRWSA  SMMVLMMILHVFRVYLTGGF  KK-PRELTWVTGVVLAVLTA  SFGVTGYSLPWDQIGYWAVK
Nostoc       YIMNEVNFGWLIRSIHRWSA  SMMVLMMILHVFRVYLTGGF  KK-PRELTWVSGVILAVITV  SFGVTGYSLPWDQVGYWAVK
C.Protothec  YLMTQVNFGWLIRSIHRWSA  SMMVLMMILHIFRVYLTGGF  KK-PRELTWVTGVLMAVCTV  SFGVTGYSLPWDQIGYWAVK
C.reinhard   YIMTDVNFGWLIRSIHRWSA  SMMVLMMVLHVFRVYLTGGF  KR-PRELTWVTGVIMAVCTV  SFGVTGYSLPWDQVGYWAVK
Rice         YIMTEANFGWLIRSVHRWSA  SMMVLMMILHVFRVYLTGGF  KK-PRELTWVTGVVLAVLTA  SFGVTGYSLPWDQIGYWAVK
Spinach      YIMTEVNFGWLIRSVHRWSA  SMMVLMMILHVFRVYLTGGF  KK-PRELTWVTGVVLGVLTA  SFGVTGYSLPWDQIGYWAVK
Tobacco      YIMTEANFGWLIRSVHRWSA  SMMVLMMILHVFRVYLTGGF  KK-PRELTWVTGVVLAVLTA  SFGVTGYSLPWDPVGYWAVK
Wheat        YIMTEANFGWLIRSVHRWSA  SMMVLMMILHVFRVYLTGGF  KK-PRELTWVTGVVLAVLTA  SFGVTGYSLPWDQIGYWAVK
                       helix B                                      helix C
```

```
Agmen        IVSGVPAAIPVVGDQMVELL  RG-GASVGQATLTRFYSLHT  FVLPWLIAVFMLAHFLMIRK  QGISGPL
Barley       IVTGVPDAIPVIGSPLVELL  RG-SASVGQSTLTRFYSLHT  FVLPLLTAVFMLMHFPMIRK  QGISGPL
E.gracilis   IVTGVPEAIPLIGNFIVELL  RG-SVSVGQSTLTRFYSLHT  FVLPLLTATFMLGHFLMIRK  QGISGPL
P.hollandica IVSGVPEAIPLVGPLMVELI  RGHSASVGQATLTRFYSLHT  FVLPWFIAVFMLMHFLMIRK  QGISGPL
Lvwort       IVTGVPEAIPIIGSPLVELL  RG-SVSVGQSTLTRFYSLHT  FVLPLLTAIFMLMHFLMIRK  QGISGPL
Maize        IVTGVPEAIPVIGSPLVELL  RG-SASVGQSTLTRFYSLHT  FVLPLLTAVFMLMHFPMIRK  QGISGPL
Nostoc       IVSGVPEAIPVVGVLISDLL  RG-GSSVGQATLTRYYSAHT  FVLPWLIAVFMLFHFLMIRK  QGISGPL
C.Protothec  IVTGVPDAIPVIGQVLLELL  RG-GVAVGQSTLTRFYSLHT  FVLPLFTAVFMLMHFLMIRK  QGISGPL
C.reinhard   IVTGVPDAIPGVGGFIVELL  RG-GVGVGQATLTRFYSLHT  FVLPLLTAVFMLMHFLMIRK  QGISGPL
Rice         IVTGVPDAIPVIGSPLVELL  RG-SASVGQSTLTRFYSLHT  FVLPLLTAVFMLMHFLMIRK  QGISGPL
Spinach      IVTGVPDAIPVIGSPLVELL  RG-SASVGQSTLTRFYSLHT  FVLPLLTAVFMLMHFLMIRK  QGISGPL
Tobacco      IVTGVPDAIPVIGSPLVELL  RG-SASVGQSTLTRFYSLHT  FVLPLLTAVFMLMHFPMIRK  QGISGPL
Wheat        IVTGVPDAIPVIGSPLVELL  RG-SASVGPSTLTRFYSLHT  FVLPLLTAVFMLMHFLMIRK  PGISGPL
                                                         helix D
```

change in helix D, an increase from 13 to 14 in the number of residues between its two histidine residues [23], which appears to be an accurate indicator of the evolutionary transition between the cytochrome bc_1 and b_6f complexes.

The higher degree of sequence identity is also found in cytochrome f sequences of higher plant chloroplasts (82% pseudo-identity). When the comparison is broadened to include sequences from four cyanobacteria and *Chlamydomonas reinhardtii*, the level of identity drops to 28% (Table 9.4). In contrast, cytochrome c_1 exhibits only 7% identity in the 11 sequences com-

Table 9.3 Amino Acid Sequences of Cytochrome Subunit IV of the Cytochrome b_6f Complex from 16 Species. The Three Transmembrane Helices (E to G) are Underscored.

```
Spinach       MGVTKKPDLNDPVLRAKLAK GMGHNYYGEPAWPNDLLYIF PVVILGTIACNVGLAVLEPS MIGEPADPFATPLEILPEWY
Barley        MGVTKKPDLNDPVLRAKLAK GMGHNYYGEPAWPNDLLYIF PVVILGTIACNVGLAVLEPS MIGEPADPFATPLEILPEWY
Maize         MGVTKKPDLNDPVLRAKLAK GMGHNYYGEPAWPNDLLYIF PVVILGTIACNVGLAVLEPS MIGEPAD-------------
RICE          MGVTKKPDLNDPVLRAKLAK GMGHNYYGEPAWPNDLLYIF PVVILGTIACNVGLAVLEPS MIGEPADPFATPLEILPEWY
Tobacco       MGVTKKPDLNDPVLRAKLAK GMGHNYYGEPAWPNDLLYIF PVVILGTIACNVGLAVLEPS MIGEPADPFATPLEILPEWY
Pea           MG------------------ ---HNYYGEPAWPNDLLYIF PVVILGTIACNVGLAVLEPS MIGEPADPFATPLEILPEWY
Wheat         MGVTKKPDLNDPVLRAKLAK GMGHNYYGEPAWPNDLLYIF PVVILGTIACNVGLAVLEPS MIGEPADPFATPLEILPEWY
Lvwort        MGVTKKPDLSDPILRAKLAK GMGHNYYGEPAWPNDLLYIF PVVILGTIACTVGLAVLEPS MIGEPANPFATPLEILPEWY
C.Reinhard    MSVTKKPDLSDPVLKAKLAK GMGHNTYGEPAWPNDLLYMF PVVILGTFACVIGLSVLDPA AMGEPANPFATPLEILPEWY
C.Eugametos   MSVTKKPDLNDPVLRAKLAK GFGHNTYGEPAWPNDLLYIF PVVIFGTFACCIGLAVLDPA AMGEPANPFATPLEILPEWY
C.Protothec   MAVTKKPDLSDPQLRAKLAK GMGHNYYGEPAWPNDIFYMF PVVIFGTFAGVIGLAVLDPA AIGEPANPFATPLEILPEWY
Agmenellum    MSIMKKPDLSDPKLRAKLAQ NMGHNYYGEPAWPNDILFTF PICIAGTIGLITGLAILDPA MIGEPGNPFATPLEILPEWY
S.6803        MSIIKKPDLSDPDLRAKLAK GMGHNYYGEPAWPNDILYMF PICILGALGLIAGLAILDPA MIGEPADPFATPLEILPEWY
Nostoc        MATQKKPDLSDPQLRAKLAK GMGHNYYGEPAWPNDLLYVF PIVIMGSFAAIVALAVLDPA MTGEPANPFATPLEILPEWY
P.hollandica  MSVLKKPDLTDPVLLEKLAQ NMGHNYYGEPAWPNDLLYTF PVVILGTLACVVGLAVLDPA MVGEPANPFATPLEILPEWY
S.obliquus    MSVTKKPDLTDPVLKEKFAK GMGHNYYGEPAWPNDLLYIF PVVILGTFACVIGLSVLDPA AIGEPANPFATPLEILPEWY
                                                         helix E

Spinach       FFPVFQILRTVPNKLLGVLL MASVPAGLLTVPFLENVNKF QNPFRRPVATTVFLVGTVVA LWLGIGATLPIDKSLTLGLF*
Barley        FFPVFQILRTV--------- -------------------- -------------------- --------------------*
Maize         -------------------- -------------------- -------------------- --------------------*
RICE          FFPVFQILRTVPNKLLGVLL MVSVPTGLLTVPFLENVNKF QNPFRRPVATTVFLIGTAVA LWLGIGATLPIEKSLTLGLF*
Tobacco       FFPVFQILRTVPNKLLGVLL MVSVPAGLLTVPFLENVNKF QNPFRRPVATTVFLIGTAVA LWLGIGATLPIDKSLTLGLF*
Pea           FFPVFQILRTVPNKLLGVLL MVSVPAGLLTVPFLENVNKF QNPFRRPVATTVFLIGTVVA LWLGIGATLPIEKSLTLGLF*
Wheat         FFPVFQILRTVPNKLLGVLL MVSVPTGLFTVPFLENVNKF QNPFRRPVATTVFLIGTVVA LWLGIGATLPIDKSLTLGLF*
Lvwort        FFPVFQILRTVPNKLLGVLL MAAVPAGLLTVPFLENVNKF QNPFRRPVATTVFLIGTVVA LWLGIGAALPIDKSLTLGLF*
C.Reinhard    FYPVFQILRVVPNKLLGVLL MAAVPAGLITVPFIESINKF QNPYRRPIATILFLLGTLVA VWLGIGSTFPIDISLTLGLF*
C.Eugametos   FYPVFQILRTVPNKLLGVLA MAAVPVGLLTVPFIESINKF QNPYRRPIATILFLVGTLVA VWLGIGATFPIDISLTLGLF*
C.Protothec   FYPVFQLLRTVPNKLLGVLL MAAVPAGLITVPFIKIYNKF QNPFRRPVATTVFLVGTVAA IWLGIGAALPIDISLTLGLF*
Agmenellum    LYPVFQILRVLPNKLLGIAC QGAIPLGLMMVPFIESVNKF QNPFRRPVAMAVFLFGTAVT LWLGAGACFPIDESLTLGLF*
S.6803        LYPTFQILRILPNKLLGIAG MAAIPLGLMLVPFIESVNKF QNPFRRPIAMTVFLFGTAAA LWLGAGATFPIDKSLTLGLF*
Nostoc        LYPVTQILRSLPNKLLGVLA MASVPLGLILVPFIENVNKF QNPFRRPVATTVFLFGTLVT LWLGIGAALPLDKSLTLGLF*
P.hollandica  LYPAFQILRVVPNKLLGILL QTAIPLGLMLVPFIENINKF QNPFRRPIAMAVFLFGTLVT LWMGVAATLPIDKFFTLGLF*
S.obliquus    FYPVFQLLRTVPNKLLGVLL MAAVPAGLITVPFIENINKF QNPYRRPIATTLFLVGTLVA VWLGIGATLPIEISLTFGLF*
                             helix F                                          helix G
```

pared in Table 9.5. Whereas the alignment of cytochrome f sequences from higher plants had gaps only in the leaders, that is, none after Tyr-1, the alignment of cytochrome c_1 sequences exhibits numerous gaps throughout. Tyr-1 is conserved in all but the cytochrome f from *Marchantia polymorpha*, which appears to be Phe-1 instead. If this proves not to be a sequencing error, yet another precedent will have been set for heme ligation.

Within the Rieske iron–sulfur protein, there appear to be five conserved structural domains [53]. Within these domains, there is considerable sequence variation, as illustrated in the alignment in Table 9.6 and Ref. [53]. However, when only chloroplast sequences are compared, the degree of identity is 48% (alignment not shown). This higher degree of identity is somewhat surprising because the Rieske iron–sulfur protein is nuclear encoded. In general, sequence evolution is slower in the chloroplast than in the nucleus [36, 54, 55]. Both cytochrome b_6 and cytochrome f are chloroplast encoded. The high levels of identity among the cytochrome b_6 sequences, and to some extent the degree to

Table 9.4 Amino Acid Sequences of Cytochrome f of the Cytochrome b_6f Complex from 13 Species. The Following Features are Indicated: N-terminal Amino Acid, Tyr-1 (Bold), the Heme-Binding Domain (Italics), and the Single Transmembrane Helix Region (Underscored).

```
S.6803      MRN--PDTLGLWTKTMVA--  -----------LRRFTVLAI  ATVSVFLITDLGLPQAASAY  PFWAQETAPLTPREATGRIV
A.quadrupl  MKT--PELMAIWQR------  -----------LKTACLVAI  ATFGLFFASDVLFPQAAAAY  PFWAQQTAPETPREATGRIV
Nostoc.790  MRN--ASVTARLTRSVRA--  -----------IVKTLLIAI  ATVTFYFSCDLALPQSAAAY  PFWAQQTYPETPREPTGRIV
C.reinhard  MSNQV--FTTLRAATL----  --------------AVILGM  AGGLAV--------SPAQAY  PVFAQQNY-ANPREANGRIV
M.polymorp  MQN--RNFNNLIIK------  ------WA----IRLI      SIMIIINT--IFWSSISEAY  PIYAQQGY-ENPREATGRIV
B.campestr  ------------------    ------------------    -------------------Y  PIFAQQNY-ENPREATGRIV
S.oleracea  MQ----------------    ----TINTFSWIKEQITRSI  SISLILYI--ITRSSIANAY  PIFAQQGY-ENPREATGRIV
T.aestivum  ME----------------    ----NRNTFSWVKEQITRSI  SVSIMIYV--ITRTSISNAY  PIFAQQGY-ENPREATGRIV
O.sativa    ME----------------    ----NRNTFSWVKEQMTRSI  SVSIMIYV--ITRTSISNAY  PIFAQQGY-ENPREATGRIV
N.tabacum   MQ----------------    ----TRNAFSWLKKQITRSI  SVSLMIYI--LTRTSISSAY  PIFAQQGY-ENPREATGRIV
O.hookerii  M-----------------    -----KNTFSWIKKEITRSI  SLSLMIYI--ITRTSISNAY  PIFAQQGY-ENPREATGRIV
P.sativum   MDRELSNLPNLIVEIFRIKD  CTMQTRNAFSWIKKEITRSI  SVLLMIYI--ITRAPISNAY  PIFAQQGY-ENPREATGWIV
V.faba      M-----------------    ---QTRNAFSWIKKEITRSI  SVLLMIYI--ITRAPISNAY  PIFAQQGY-ENPREATGRIV

S.6803      CANCHLAQKAAEVEIPQAVL  PDTVFEAVVKIPYDLDSQQV  LGDGSKGGLNVGAVLMLPEG  FKIAPPDRLSEGLKEKVGGT
A.quadrupl  CANCHLAAKEAEVEIPQSVL  PDQVFEAVVKIPYDHSQQQV  LGDGSKGGLNVGAVLMLPDG  FKIAPADRLSDELKEKTEGL
Nostoc.790  CANCHLAAKPTEVEVPQSVL  PDTVFKAVVKIPYDTSAQQV  GADGSKVGLNVGAVLMLPEG  FKIAPEDRISEELQEEIGDT
C.reinhard  CANCHLAQKAVEIEVPQAVL  PDTVFEAVIELPYDKQVKQV  LANGKKGDLNVGMVLILPEG  FELAPPDRVPAEIKEKVGNL
M.polymorp  CANCHLAKKPVDIEVPQSVL  PNTVFEAVVKIPYDMQIKQV  LANGKKGSLNVGAVLILPEG  FELAPSDRIPPEMKEKIGNL
B.campestr  CANCHLASKPVDIEVPQAVL  PDTVFEAVVKIPYDMQLKQV  LANGKKGALNVGAVLILPEG  FELAPPDRISPEMKEKIGNL
S.oleracea  CANCHLANKPVDIEVPQAVL  PDTVFEAVVRIPYDMQLKQV  LANGKKGGLNVGAVLILPEG  FELAPPDRISPEMKEKMGNL
T.aestivum  CANCHLASKPVDIEVPQAVL  PDTVFEAVLRIPYDMQLKQV  LANGKKGGLNVGAVLILPEG  FELAPPDRISPELKEKIGNL
O.sativa    CANCHLANKPVDIEVPQAVL  PDTVFEAVLRIPYDMQLKQV  LANGKKGGLNVGAVLILPEG  FELAPPDRISPELKEKIGNL
N.tabacum   CANCHLANKPVEIEVPQAVL  PDTVFEAVVRIPYDMQLKQV  LANGKRGGLNVGAVLILPEG  FELAPPDRISPEMKEKIGNL
O.hookerii  CANCHLANKPVDIEVPQAVL  PDTVFEAVVRIPYDRQVKQV  LANGKKGGLNVGAVLILPEG  FELAPPARISPEMKERIGNP
P.sativum   CANCHLANKPVDIEVPQAVL  PDTVFEAVVRIPYDMQVKQV  LANGKKGALNVGAVLILPEG  FELAPPHRLSPQIKEKIGNL
V.faba      CANCHLANKPVDIEVPQAIL  PDTVFEAVVRIPYDMQVKQV  LANGKKGALNVGAVLILPEG  FELAPPDRLSPEIKEKIGNL

S.6803      YFQPYREDMENVVIVGPLPG  EQYQEIVFPVLSPDPAKDKS  INYGKFAVHLGANRGRGQIY  PTGLLSNNNAFKAPNAGTIS
A.quadrupl  YFQSYAPDQENVVIIGPISG  DQYEEIVFPVLSPDPKTDKN  INYGKYAVHLGANRGRGQVY  PTGELSNNNQFKASATGTIT
Nostoc.790  YFQPYSEDKENIVIVGPLPG  EQYQEIVFPVLSPNPATDKN  IHFGKYSVHVGGNRGRGQVY  PTGEKSNNNLYNASATGTIA
C.reinhard  YYQPYSPEQKNILVVGPVPG  KKYSEMVVPILSPDPAKNKN  VSYLKYPIYFGGNRGRGQVY  PDGKKSNNTIYNASAAGKIV
M.polymorp  FFQPYSNDKKNILVIGPVPG  KKYSEMVFPILSPDPATNKE  AHFLKYPIYVGGNRGRGQIY  PDGSKSNNTVYNASITGKVS
B.campestr  SFQNYRPNKKNILVIGPVPG  QKYSEITFPILAPDPATNKD  VHFLKYPIYVGGNRGRGQIY  PDGSKSNNTVYNATAGGIIS
S.oleracea  SFQSYRPNKQNILVIGPVPG  QKYSEITFPILAPDPATKKD  VHFLKYPIYVGGNRGRGQIY  PDGSKSNNTVYNSTATGIVK
T.aestivum  AFQSYRPDKKNILVIGPVPG  KKYSEIVFPILSPDPATKKD  AHFLKYPIYVGGNRGRGQIY  PDGSKSNNTVYNATSTGIVR
O.sativa    SFQSYRPNKKNILVIGPVPG  KKYSEIVFPILSPDPAMKKD  VHFLKYPIYVGGNRGRGQIY  PDGSKSNNTVYNATSTGVVR
N.tabacum   SFQSYRPNKKNILVIGPVPG  QKYSEITFPILSPDPATKKD  VHFLKYPIYVGGNRGRGQIY  PDGSKSNNTVYNATAAGIVS
O.hookerii  SFQSYRPTKKNILVIGPVPG  QKYSEITFPILSPDPATNKD  VHFLKYPIYVGGNRGRGQIY  PDGSKSNNTVYNATAAGIVS
P.sativum   SFQSYRPTKKNILVIGPVPG  KKYSEITFPILSPDPATKRD  VYFLKYPLYVGGNRGRGQIY  PDGSKSNNNVSNATATGVVK
V.faba      SFQSYRPTKKNIIVIGPVPG  KKYSEITFPILSPDPATKRD  VYFLKYPIYVGGTRGRGQIY  PDGSKSNNNVYNATATGVVN

S.6803      EVNALEAGG----YQLIL-T  TADGTETVD-IPAGPELIVS  AGQTVEAGEFLTNNPNVGGF  GQKDTEVVLQNPTRIKFLVL
A.quadrupl  NIAVNEAAG----TDITI-S  TEAGEVIDT-IPAGPEVIVS  EGQAIAAGEALTNNPNVGGF  GQKDTEVVLQNPARIYGYMA
Nostoc.790  KIAKEEDEDGNVKYQVNI-Q  PESGDVVVDTVPAGPELIVS  EGQAVKAGDALTNNPNVGGF  GQRDAEIVLQDAGRVKGLIA
C.reinhard  AITALSEKKGG--FEVSI-E  KANGEVVVDKIPAGPDLIVK  EGQTVQADQPLTNNPNVGGF  GQAETEIVLQNPARIQGLLV
M.polymorp  KI--FRKEKGG--YEITIDD  ISDGHKVVDISAAGPELIIS  EGELVKVDQPLTNNPNVGGF  GQGDAEVVLQDPLRIQGLLL
B.campestr  KI--LRKEKGG--YEITIVD  ASNERQVIDIIPRGLELLVS  EGESIKLDQPLTSNPNVGGF  GQGDAEIVLQDPLRVQGLLF
S.oleracea  KI--VRKEKGG--YEINIAD  ASDRREVVDIIPRGPELLVS  EGESIKLDQPLTSNPNVGGF  GQGDAEVVLQDPLRIQGLLF
T.aestivum  KI--LRKEKGG--YEISIVD  ASDGRQVIDIIPPGPELLVS  EGESIKLDQPLTSNPNVGGF  GQGDAEIVLQDPLRVQGLLF
O.sativa    KI--LRKEKGG--YEISIVD  ASDGRQVIDLIPPGPELLVS  EGESIKLDQPLTSNPNVGGF  GQGDAEIVLQDPLRVQGLLF
N.tabacum   KI--IRKEKGG--YEITITD  ASDGRQVVDIIPPGPELLVS  EGESIKFDQPLTSNPNVGGF  GQGDAEIVLQDPLRVQGLLF
O.hookerii  KI--IRKEKGG--YEITITD  ASDGRQVVDIIPSGPELLVS  EGESTKLDQPLTSNPNVGGF  GQGDAEVVLQDPLRVQGLLF
P.sativum   QI--IRKEKGG--YEITIVD  ASDGSEVIDIIPPGPELLVS  EGESIKLDQPLTSNPNVGGF  GQGDAEIVLQDPLRVQGLLL
V.faba      KK--IRKEKGG--YEITIVD  GSDGREVIDIIPPGPELLVS  EGESIKLDQPLTSNPNVGGF  GQGDAEIVLQDPLRVQGLLL

S.6803      FLAGIMLSQILLVLKKKQIE  KVQAAEL-NF
A.quadrupl  FVAGIMLTQIFLVLKKKQVE  RVQAAGNCDF
Nostoc.790  FVALVMLAQVMLVLKKKQVE  RVQAAEM-NF
C.reinhard  FFSFVLLTQVLLVLKKKQFE  KVQLAEM-NF
M.polymorp  FFGSVILAQIFLVLKKKQFE  KVQLAEM-NF
B.campestr  FLGSVVLAQIFLVLKKKQFE  KVQLSEM-NF
S.oleracea  FFASVILAQIFLVLKKKQFE  KVQLSEM-NF
T.aestivum  FFASVILAQVFLVLKKKQFE  KVQLYEM-NF
O.sativa    FFASVILAQIFLVLKKKQFE  KVQLYEM-NF
N.tabacum   FLASVILAQIFLVLKKKQFE  KVQLAEM-NF
O.hookerii  FLQSVILAQIFLVLKKKQFE  KVQLSEM-NF
P.sativum   FLASIILAQILLVLKKKQFE  KVQLSEM-NF
V.faba      FLASIILAQIFLVLKKKQFE  KVQLSEM-NF
```

Table 9.5 Amino Acid Sequences of Cytochrome c_1 of the Cytochrome bc_1 Complex. The Following Features are Indicated: N-Terminal Amino Acid of *Rb. capsulatus* (Bold; arrow), the Heme-binding Domain (Italics), and the Single Transmembrane Region (Underscored).

```
B.taurus     --------------------  ----SDLELHPPSYPWSHRG LLSSLDHTSIRRGFQVYKQV CSSCHSMDYVAYRHLV---G
Br.japonic   --------------------  ----NEGSDKPPGNKWSFAG PFGKFDRGALQRGLKVYKEV CASCHGLSYIAFRNLAEAGG
E.gracilis   --------------------  -----GVDSHPPALPWPHFQ WFQGLDWRSVRRGKEVYEQV FAPCHSLSFIKYRHF----E
Human        --------------------  ----SDLEVHPPSYPWSHRG LLSSLDHTSIRRGFQVYKQV CASCHSMDFVAYRHLV---G
N.crassa     MTP-----------------  ----AEEGLHATKYPWVHEQ WLKTFDHQALRRGFQVYREV CASCHSLSRVPYRALV---G
Rb.capsula   M----KKLLISAVSALVLGS GAAFANSNVPDHAF--SFEG IFGKYDQAQLRRGFQVYNEV CSACHGMKFVPIRTLADDGG
S.cerevisi   MTA-----------------  ----AEHGLHAPAYAWSHNG PFETFDHASIRRGYQVYREV CAACHSLDRVAWRTLV---G
Potato(M)    YSD-----------------  ---EAEHGLECPNYPWPHEG ILSSYDHASIRRGHQVYQQV CASCHSMSLISYRDLV---G
P.denitrif   ---------DHGDAAAQEA GDSHAAAHIEDISF--SFEG PFGKFDQHQLQRGLQVYTEV CSACHGLRYVPLRTLADEGG
R.rubrum     MTTIVKRALVAAGMVLAIGG AAQANEGGVSLHKQDWSWKG IFGRYDQPQLQRGFQVFHEV CSTCHGMKRVAYRNLSALG-
Rb.sphaero   M----KKLLISAVSALVLGS GAALANSNVQDHAF--SFEG IFGKFDQAQLRRGFQVYSEV CSTCHGMKFVPIRTLSDDGG
                                              ↑

B.taurus     VCYTEDEAKALAEEVEVQDG PNEDGEMFMRPGKLSDYFPK PYPNPEAARAANNGALPPDL SYIVRAR-------------
Br.japonic   PSYSVAQVAAFASDYKIKDG PNDAGDMFERPGRPADYFPS PFPNEQAARAANGGAAPPDL SLITKAR-----SYGRGFPW
E.gracilis   AFMSKEEVKNMAASFEVDDD PDEKGEARKRPGKRFDTVVQ PYKNEQEARYANNGALPPDL SVITNAR-------------
Human        VCYTEDEAKELAAEVEVQDG PNEDGEMFMRPGKLFDYFPK PYPNSEAARAANNGALPPDL SYIVRAR-------------
N.crassa     TILTVDEAKALAEENEYDTE PNDQGEIEKRPGKLSDYLPD PYKNDEAARFANNGALPPDL SLIVKAR-------------
Rb.capsula   PQLDPTFVREYAAGLDTIID KDSGEE---RDRKETDMFPT R---------VGDGMGPDL SVMAKARAGFSGPAGSGMN-
S.cerevisi   VSHTNEEVRNMAEEFEYDDE PDEQGNPKKRPGKLSDYIPG PYPNEQAARAANQGALPPDL SLIVKAR-------------
Potato(M)    VAYTEEETKAMAAEIEVVDG PNDEGEMFTRPGKLSDRFPQ PYANEAAARFANGGAYPPDL SLITKAR-------------
P.denitrif   PQLPEDQVRAYAANFD-ITD PETEED---RPRVPTDHFPT V---------SGEGMGPDL SLMAKARAGFHGPYGTGLS-
R.rubrum     --FSEDGIKELAAEKEFPAG PDDNGDMFTRPGTPADHIPS PFANDKAAAAANGGAAPPDL SLLAKAR-----P-------
Rb.sphaero   PQLDPTFVREYAAGLDTIID KDSGEE---RDRKETDMFPT R---------VGDGMGPDL SVMAKARAGFSGPAGSGMN-

B.taurus     ----------HGGEDYVFSL LTGYCE---------PPTGV SLREGLYFNPYFPGQAIG-- ----------------MAPP
Br.japonic   FIFDFFTQYQEQGPDYVSAV LQGFEEK--------VPEGV TIPEGSYYNKYFPGHAIK-- ----------------MPKP
E.gracilis   ----------HGGVDYIYAL LTGYGRP--------VPGGV QLSTTQWYNPYFHGGIIG-- ----------------MPPP
Human        ----------HGGEDYVFSL LTGYCE---------PPTGV SLREGLYFNPYFPGQAIA-- ----------------MAPP
N.crassa     ----------HGGCDYIFSL LTGYPDE--------PPAGA SVGAGLNFNPYFPGTGIA-- ----------------MARV
Rb.capsula   ---QLFKGM--GGPEYIYNY VIGF-EENPEC----APEGI ---DGYYNKTFQIGGVPDT CKDAAGVKITHGSWARMPPP
S.cerevisi   ----------HGGCDYIFSL LTGYPDE--------PPAGV ALPPGSNYNPYFPGGSIA-- ----------------MARV
Potato(M)    ----------HNGQNYVFAL LTGYRD---------PPAGV SIREGLHYNPYFPGGAIA-- ----------------MPKM
P.denitrif   ---QLFNGI--GGPEYIHAV LTGYDGEEKE-------EAG ---AVLYHNAAFA------- ----------GNWIQMAAP
R.rubrum     ----------GGPNYIYSL LEGYASMTTIVKRALVAAGM VLAIGGAAQANEGGVSLH-- ----------------KQDW
Rb.sphaero   ---QLFKGI--GGPEYIYRY VTGFPEENPAC----APEGI ---DGYYNEVFQVGGVPDT CKDAAGIKTTHGSWAQMPPA

B.taurus     IYNEVLE-FDDGTPATMSQV AKDVCTFLRWAAEPEHDHRK RMGLKMLLMMGL-------- ---------------LLPL
Br.japonic   LSDGQVT-YDDGSPATVAQY SKDVTTFLMWTAEPHMEARK RLGFQVFVFL-I-------- ---------------IFAGL
E.gracilis   LTDDMIE-YEDGTPASVPQM AKDVTCFLEWCSNPWWDERK LLGYKTIATLAV-------- ---------------IAVSS
Human        IYTDVLE-FDDGTPATMSQI AKDVCTFLRWASEPEHDHRK RMGLKMLMMMAL-------- ---------------LVPL
N.crassa     LYDGLVD-YEDGTPASTSQM AKDVVEFLNWAAEPEMDDRK RMGMKVLVVTS--------- --------------VLFAL
Rb.capsula   LVDDQVT-YEDGTPATVDQM AQDVSAFLMWAAEPKLVARK QMGLVAMVMLGL-------- --------------LSVM
S.cerevisi   LFDDMVE-YEDGTPATTSQM AKDVTTFLNWCAEPEHDERK RLGLKTVIILS--------- -------------SLYLL
Potato(M)    LNDGAVE-YEDGIPATEAQM GKDVVSFLSWAAEPEMEERK LMGFKWIFVLSL-------- -------------ALLQA
P.denitrif   LSDDQVT-YEDGTPATVDQM ATDVAAFLMWTAEPKMMDRK QVGFVSVIFLIV-------- -------------LAAL
R.rubrum     SWKGIFGRYDQ------PQL QRGFQVFHEVCSTCH--GMK RVAYRNLSALGFSEDGIKEL AAEKEFPAGPDDNGDMFTRP
Rb.sphaero   LFDDLVT-YEDGTPATVDQM GQDVASFLMWAAEPKLVARK QMGLVAVVMLGL-------- ----------------LSVM

B.taurus     VYAMKRHKWSVLKSRKLAYR PPK----------------- ---------- ---
Br.japonic   MYFTKKKVWA---------- -------------------D-- ---------- -SH
E.gracilis   GYY-NRFLSGLWRSRRLAFR PFN----------------- ---------- ---
Human        VYTIKRHKWSVLKSRKLAYR PPK----------------- ---------- ---
N.crassa     SVYVKRYKWAWLKSRKIVYD PPKSPPPATNLALPQQRA-- ---------- -KS
Rb.capsula   LYLTNKRLWAPYKGHK---- -------------------- ---------- --A
S.cerevisi   SIWVKKFKWAGIKTRKFVFN PPKP---------------- ---------- -RK
Potato(M)    AY-YRRLRWSVLKSRKLVLD VVN----------------- ---------- ---
P.denitrif   LYLTNKKLWQPIKHPRKP-- -------------------- ---------- --E
R.rubrum     GTPADHIPSPFANDKAAAAA NGGAAPPDLSL-LAKARPGG PNYIYSLLEG YAS
Rb.sphaero   LYLTNKRLWAPYKRQK---- -------------------- ---------- --A
```

Table 9.6 Amino Acid Sequences of Rieske Iron–Sulfur Protein of Cytochrome bc_1 and b_6f. Mitochondrial Sequences are Indicated with "(M)" to the Right of the Species Name. A Putative Transmembrane Helix of ~22 Residues is Indicated by an Underscore. Data on Presequence Cleavage Sites are Largely Missing; However, the First Residue After the Site for the Tobacco Chloroplast Sequence is Shown in Bold.

Species				
B.taurus	MLSVAARSRHS-----RPSY	----RPRPAGWRALRPWYSR	SSRESPVLDLKRSVLCRESL	RGQAAA-ALVASVSLNVPAS
rat	---------------------	---SRGVAGALRPLLQSAVP	ATSEPPVLDVKRPFLCRESL	SGQAATRPLVATVGLNVPAS
yeast	MLGIRSSVKTC-----FKPM	SLTSK----RLISQSLLASK	STYRTPNFD-------DVL	KENNDAD-------------
N.crassa	MAPVSIVSRAA-----MRAA	AAPAR----AVRALTTSTAL	QGSSSSTFE-------SPF	KGESKAA-------------
P.denitri	M-------------------	--------------------	--------------------	--------------------
Rb.capsula	M-------------------	--------------------	--------------------	--------------------
Rps.rubrum	A-------------------	------------EAEHTAS	T-------------------	--------------------
tobacco(M)	-------------WPVRSA	APSSSAFISANHFSSDDDSS	SPRSISP-SLASVFLHHT--	RGFSS-NSVSPAHDMGLVPD
Z.may(M)	MLRVAGRRLSSSLSWR-PAA	AVARGPLAGAGVPDRDDDSA	RGRSQPRFSIDSPFFVAS--	RGFSSTETVVPRNQDAGLAD
Nostoc	M-----------A---Q--	--------------------	--------------------	--------------------
tobacco	--------------------	--MASSTLSPVT--QLCSSK	SGLSSVSQCLLVKPMKINS-	HGLGKDK------RMKVKCM
pea	--------------------	--MSSTTLSPTTPSQLCSGK	SGISCPSIALLVKPTRTQM-	TGRG-NK------GMKITCQ
spinach	SMI-ISIFNQLHLT---ENS	SLMASFTLSSATPSQLCSSK	NGMFAPSLALAKAGRSVNVL	ISKERIR------GMKLTCQ
S.6803	MLVKILKFRRFIMT---Q--	--------------------	--------------------	--------------------
Chlorobium	-------------MA---QTG	NFKSPARMSSLGQGAAPASS	GAVTG--------------	-GKPREG------GLK----
B.japonic	M-------------------	----------------TTA	S-------------------	--------------------

Species				
B.taurus	VRYSHTDIKVPDFSDYRRPE	VLDSTKSSKESSE--ARKGF	SYLVTATTTVGVAYAAKNVV	SQFVSSMSASADVLAM-SKI
rat	VRYSHTDIKVPDFSDYRRAE	VLDSTKSSKESSE--ARKGF	SYLVTATTTVGVAYAAKNAV	SQFVSSMSASADVLAM-SKI
yeast	--------KGRSYA-----	--------------------	-YFMVGAMGLLSSAGAKSTV	ETFISSMTATADVLAM-AKV
N.crassa	--------KVPDFGKYM---	-------SKAPPS--TNMLF	SYFMVGTMGAITAAGAKSTI	QEFLKNMSASADVLAM-AKV
P.denitri	--------------------	---SHADEHAGDHGATRRDF	LYYATAGAGTVAAGAA---A	WTLVNQMNPSADVQAL-ASI
Rb.capsula	--------------------	---SHAEDNAG----TRRDF	LYHATAATGVVVTGAA---V	WPLINQMNASADVKAM-ASI
Rps.rubrum	--------------------	----------PGGESSRRDF	LIYGTTAVGAVGVALA---V	WPFIDFMNPAADTLAL-AST
tobacco(M)	LPPTVAAIKNPT-----SKI	VYDEHNHERYPPGDPSKRAF	AYFVLTGGRFVYASLMRLLI	LKFVLSMSASKDVLAL-ASL
Z.may(M)	LPATVAAVKNPN-----PKV	VYDEYNHERYPPGDPSKRAF	AYFVLSGGRFIYASLLRLLV	LKFVLSMSASKDVLAL-ASL
Nostoc	---FSESADVPDMGR-----	----------------RQFM	NLLTFGTVT-GVALGALYPV	VKYFIPPASGGAGGGTTAKD
tobacco	**A**TSIPADDRVPDMEK-----	----------------RNLM	NLLLLGALSLPTA-GMLVSY	GTFFVPPGSGGGSGGTPAKD
pea	ATSIPAD-RVPDMSK-----	----------------RKTL	NLLLLGALSLPTA-GMLPY	GSFLVPPGSGSSTGGTVAKD
spinach	ATSIPADN-VPDMQK-----	---------------RETL	NLLLLGALSLPTGY-MLLPY	ASFFVPPGGGAGTGGTIAKD
S.6803	---ISGSPDVPDLGR-----	----------------RQFM	NLLTFGTIT-GVAAGALYPA	VKYLIPPSSGGSGGGVTAKD
Chlorobium	---------GVDFER-----	----------------RGFL	HKIV-GGVGAVVAVSTLYPV	VKYIIPPAR---------KI
B.japonic	--------------------	----------SADHPTRRDF	<u>LFVATGAAAAVGGAAA---L</u>	WPFISQMNPDASTIAAGAPI

Species				
B.taurus	EIKLSDIPEGKNMAFKWRGK	PLFVRHRTKKEIDQEAAVEV	SQLRD-----------PQHD	LERVK-------KPEWVILI
rat	EIKLSDIPEGKNMAFKWRGK	PLFVRHRTKKEIDQEAAVEV	SQLRD-----------PQHD	LERVK-------KPEWVILI
yeast	EVNLAAIPLGKNVVVKWQGK	PVFIRHRTPHEIQEANSVDM	SALKD-----------PQTD	ADRVK-------DPQWLIML
N.crassa	EVDLNAIPEGKNVIIKWRGK	PVFIRHRTPAEIEEANKVNV	ATLRD-----------PETD	ADRVK-------KPEWLVML
P.denitri	QVDVSGVETGTQLTVKWLGK	PVFIRRRTEDEIQAGREVDL	GQLIDRSAQNSNKPDAPATD	ENRTM-----DEAGEWLVMI
Rb.capsula	FVDVSAVEVGTQLTVKWRGK	PVFIRRRDEKDIELARSVPL	GALRDTSAENANKPGAEATD	ENRTLPAFDGTNTGEWLVML
Rps.rubrum	EVDVSAIAEGQAITVTWRGK	PVFVRHRTQKEIVVARAVDP	ASLRD-----------PQTD	EARVQ-------QAQWLVMV
tobacco(M)	EVDLSSIEPGTTVTVKWRGK	PVFIRRRTEDDISLANSVDL	GSLRD-----------PQQD	AERVK-------NPEWLVVI
Z.may(M)	EVDLSSIEPGTTVTVKWRGK	PVFIRRRTEDDIKLANSVDV	ASLRH-----------PEQD	AERVK-------NPEWLVVI
Nostoc	EL---------------GN	DVSLSKFLENRNAGDRALVQ	G-LKGD--------PTYIVVE	NKQAIKDYG---------IN
tobacco	AL---------------GN	DVIASEWLKTHPPGNRTLTQ	G-LKGD--------PTYLVVE	NDGTVATYG---------IN
pea	AV---------------GN	DVVATEWLKTHAPGDRTLTQ	G-LKGD--------PTYLVVE	KDRTLATFA---------IN
spinach	AL---------------GN	DVIAAEWLKTHAPGDRTLTQ	G-LKGD--------PTYLVVE	SDKTLATFG---------IN
S.6803	AL---------------GN	DVKVTEFLASHNAGDRVLAQ	G-LKGD--------PTYIVVQ	GDDTIANYG---------IN
Chlorobium	KN---------------VD	ELTVGK-ASEVPDGKSKIFQ	FN---E--------DKVIVVN	KGGALT--A---------VS
B.japonic	EVDLSPIAEGQDIKVFWRGK	PIYISHRTKKQIDEARAVNV	ASLPD-----------PQSD	EARVKSG-----HEQWLVVI

Species					
B.taurus	GVCTHLGCVPIANA-----G	DFGGYYCPCHGSHY-DASGR	IRKGPAPLNLEVPSYEF-TS	DDMVIV-------------	---G
rat	GVCTHLGCVPIANA-----G	DFGGYYCPCHGSHY-DASGR	IRKGPAPLNLEVPTYEF-TS	GDVVVV-------------	---G
yeast	GICTHLGCVPIGEA-----G	DFGGWFCPCHGSHY-DISGR	IRKGPAPLNLEIPAYEF--D	GDKVIV-------------	---G
N.crassa	GVCTHLGCVPIGEA-----G	DYGGWFCPCHGSHY-DISGR	IRKGPAPLNLEIPLYEF-PE	EGKLVI-------------	---G
P.denitri	GVCTHLGCVPI----GDGAG	DFGGWFCPCHGSHY-DTSGR	IRRGPAPQNLHIPVAEF-LD	DTTIKL-------------	---G
Rb.capsula	GVCTHLGCVPM----GDKSG	DFGGWFCPCHGSHY-DSAGR	IRKGPAPRNLDIPVAAF-VD	ETTIKL-------------	---G
Rps.rubrum	GVCTHLGCIPLGQKAGDPKG	DFDGWFCPCHGSHY-DSAGR	IRKGPAPLNLPVPPYAF-TD	DTTVLI-------------	---G
tobacco(M)	GVCTHLGCIPLPNA-----G	DFGGWFCPCHGSHY-DISGR	IRKGPAPYNLEVPTYSF-LE	ENKLLI-------------	---G
Z.may(M)	GVCTHLGCIPLPNA-----G	DFGGWFCPCHGSHY-DISGR	IRKGPAPFNLEVPTYSF-LE	ENKLLV-------------	---G
Nostoc	AICTHLGCVVPWNVAENK--	----FKCPCHGSQY-DETGK	VVRGPAPLSLALAHAN-TVD	D-KIILSPWTETDFRTGDAP	WWA-
tobacco	AVCTHLGCVVPFNAAENK--	----FICPCHGSQY-NNQGR	VVRGPAPLSLALAHADI--D	DGKVVFVPWVETDFRTGEDP	WWA-
pea	AVCTHLGCVVPFNQAENK--	----FICPCHGSQY-NDQGR	VVRGPAPLSLALAHCDVGVE	DGKVVFVPWVETDFRTGDAP	WWS-
spinach	AVCTHLGCVVPFNAAENK--	----FICPCHGSQY-NNQGR	VVRGPAPLSLALAHCDV--D	DGKVVFVPWTETDFRTGEAP	WWSA
S.6803	AVCTHLGCVVPWNASENK--	----FMCPCHGSQY-NAEGK	VVRGPAPLSLALAHATVTDD	D-KLVLSTWTETDFRTDEDP	WWA-
Chlorobium	AVCTHLGCLVNWVDADNQ--	----YFCPCHGAKY-KLTGE	IISGPQPLPL--KQYKARIE	GDSIIISK-----------	---A
B.japonic	GICTHLGCIPIAHE-----G	NYDGFFCPCHGSQY-DSSGR	IRQGPAPANLPVPPYQF-VS	DTKIQI-------------	---G

which cytochrome f is conserved further, demonstrate the slower rate of sequence change in the chloroplast genes. In contrast, cytochrome c_1 is nuclear encoded and exhibits a notable lack of sequence conservation.

9.7 Phylogenetic Analyses

Amino acid sequences for cytochromes b (53 species) and b_6 (13), subunit IV (16), cytochromes f (13) and c_1 (11), and Rieske iron–sulfur proteins (16) were obtained from GenBank and the SWISS-PROT protein databases using TurboGopher (copyright, University of Minnesota) and are shown in Tables 9.1 to 9.6, respectively. The partial sequences available for avian species [56] were not used. Alignments were performed on a Macintosh Quadra 650 with *SeqApp* [D. G. Gilbert, Indiana University] and *ClustalV* [D. Higgins, EMBL-Heidelberg], both of which are available from *ftp.iubio.edu*. The N-terminal portion of cytochrome b was aligned with cytochrome b_6. Some manual adjustments to computer alignments were necessary. Subunit IV of the cytochrome b_6f complex was aligned with the C-terminal portion of cytochrome b. In both cases, due to the presence of large gaps at the termini of the respective alignments, relatively gapless segments of the aligned polypeptides were used for phylogenetic analysis. The well-reasoned corrections to obvious sequencing errors proposed by Degli Esposti et al. [6] were not made.

Phylogenetic trees were inferred by using the Heuristic Search Procedure in *PAUP: Phylogenetic Analysis Using Parsimony*, version 3.0d for the Macintosh [57]. To evaluate the reliability of the topologies, bootstrap probabilities [58] were calculated using *PAUP*. The trees in Figures 9.5 to 9.8 are bootstrap majority-rule consensus topologies. The number of repetitions is indicated in the each figure legend. Percentage bootstrap probabilities are indicated on the branches between the nodes of the trees. Outgroups, if any, are indicated in the figure legends. Branch lengths are not indicated.

9.7.1 Cytochromes b and b_6

Cytochrome b has been used to construct phylogenies for a number of groups: for example, the flock species of cichlid fishes that inhabit in the Great Lakes of East Africa [1, 5], mammals [2], sharks [4], and unisexual salamanders [3]. The ubiquity and relatively high degree of sequence conservation of the mitochondrial sequences make it a useful tool for phylogenetic analyses. However, it can be seen that cytochrome b_6 may not be as useful for investigating the phylogenetic relationships among closely related species of higher plants. The bootstrap probabilities for all of the higher plant branches

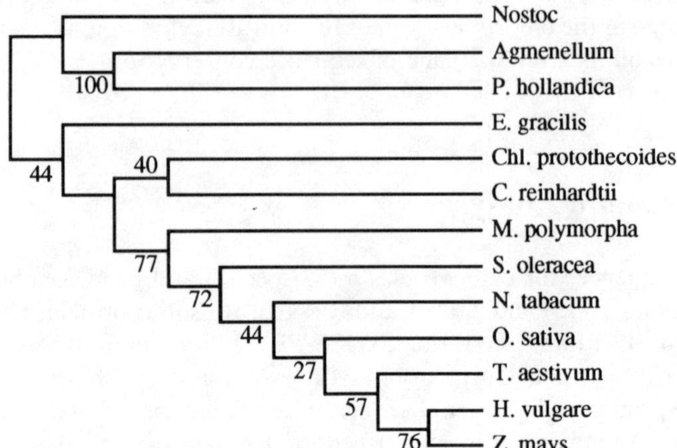

Figure 9.5   Majority-rule bootstrap tree of cytochrome b_6 sequences from 13 species. Frequencies (%) of occurrence among the bootstrap trees are shown on the internal branches (95% is considered significant). The three cyanobacterial species were assigned to the outgroup. The analysis is based on 1000 replications using the Heuristic Search algorithm in PAUP [57], set to the following parameters: (1) only minimal trees kept, (2) starting tree derived by stepwise addition, (3) swapping on minimal trees only, (4) random addition, and (5) tree bisection-reconnection (TBR) branch swapping.

are non-significant (Fig. 9.5). In addition, none of the internal branch probabilities[3] prove to be significant. The majority-rule consensus tree shows a very curious result: the dicot *Nicotiana tabacum* (tobacco) is clustered with the monocots, away from the other dicot, *Spinacia oleracea*. This inability to reliably resolve known relationships is due to the inordinately high degree of sequence identity evident in Table 9.2.

The cladistic analysis of cytochrome b_6 and the N-terminal portion of cytochrome b reveals a distinct division between the two types of b-cytochrome (Fig. 9.6). The analysis was performed with *C. limicola* arbitrarily assigned as the outgroup. The PS3 sequence clustered with the cytochrome b_6 sequences. Thus, the specific similarities (i.e, length equivalent to that of cytochrome b_6, and 14 residues between the heme-coordinating His residues of helix D, and the degree of sequence identity) place it closer to the cyanobacterial and higher plant sequences. When PS3 was assigned as the outgroup (analysis not shown), *C. limicola* essentially switched places with only a 3% (from 0.28 to 0.31) change in bootstrap probability. In both analyses, the sequence of *S. acidocal-*

---

[3] Internal branching probabilities (i.b.p.) can be approximated using the following equation: i.b.p. $= 1 - \Pi_{i=1}^{n}(1 - b_i)$ [2], where b_i is the probability at branch i. For example, between whale and seal in Figure 9.6 there are two nodes for which the i.b.p. $\approx [1 - (0.18)*(0.54)] \approx 0.90$; in other words, $p \nless 0.05$, and the branching pattern is not significant.

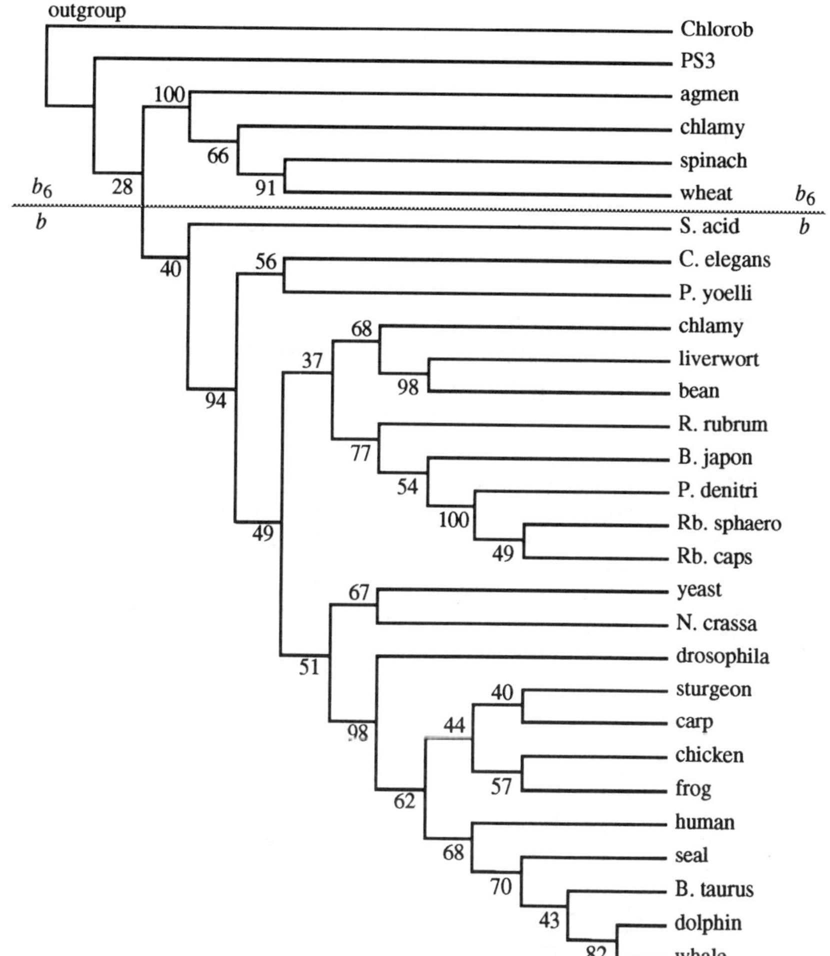

Figure 9.6 Phylogenetic analysis of a selected subset of cytochromes b and b_6 from the compilation presented in Tables 9.1 and 9.2. A subset was used because analysis of the complete set was not possible with the available computer resources. Tree representations and construction methodology are described in Figure 9.5.

darius clustered with the cytochromes b, diverging before all of the others, as illustrated in Figure 9.6. It is interesting to note that the sequence of *B. japonicum* occupies an intermediate position between that of *R. rubrum* and those of *Rb. sphaeroides* and *Rb. capsulatus*.

9.7.2 Cytochomes c_1 and f

The evolutionary relationship of cytochromes c_1 and f was previously considered [59]. From a comparison of the few sequences available at that time, five of cytochrome c_1 and one of cytochrome f, it was inferred that the eukaryotic cytochromes c_1 form a group distinct from those of bacteria, and that due to the rather low level of identity ($<40\%$) between the two groups, the time of divergence between them was considered to be remote. The amount of divergence between all cytochromes c_1 and the single cytochrome f was the greatest ($<20\%$ identity), from which it was inferred that the point of divergence was either extremely remote or the cytochrome f lineage had undergone more rapid sequence evolution. In an attempt to better understand the relationship between the two cytochromes for which more sequences are now known, we performed a phylogenetic analysis, the results of which are illustrated in Figure 9.7.

The majority-rule consensus cladogram of Figure 9.7 shows a strict bifurcation of the two c-type cytochromes with significant bootstrap probabilities for each of the two cytochrome branches (0.95 cytochrome f and 1.00 cytochrome c_1). This was expected owing to the low degree of sequence identity between the two cytochromes. Alignment of the two groups required the introduction of numerous gaps in the cytochrome c_1 sequences. In addition, it proved difficult to obtain consistent alignment of the major structural feature common to c-type cytochromes, that of the C–X–X–C–H, heme-liganding signature pentapeptide region. The bootstrap confidence limits are significant only for the major groupings and those distinguishing *B. taurus* from human cyto-

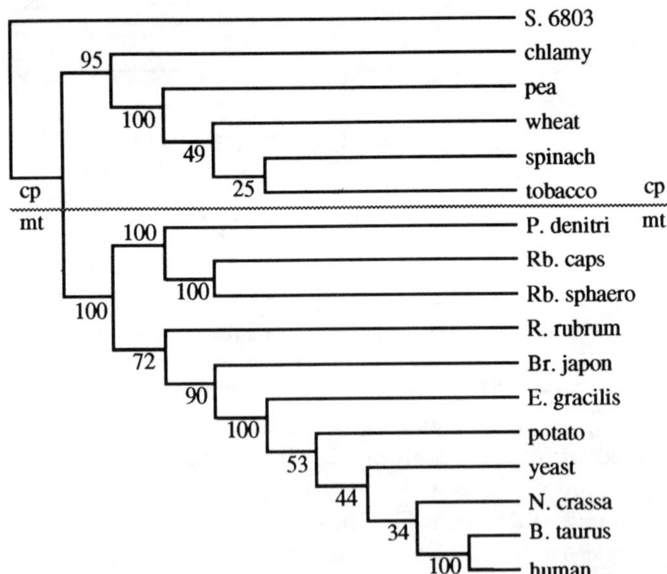

Figure 9.7 Majority-rule bootstrap tree of cytochromes c_1 (mt) and selected cytochromes f (cp) from the sequences presented in Tables 9.4 and 9.5. Tree representations and construction methodology are described in Figure 9.5. mt, mitochondria; cp, chloroplasts.

chrome c_1. However, there are some anomalies, such as the segregation of the monocot *Triticum aestivum* and two dicots, *S. oleracea* and *N. tabacum*, from another dicot, *P. sativum*. Among the cytochromes f of higher plant chloroplasts, the relatively high sequence identity ($>70\%$) lowers both the number of character states available for phylogenetic analysis and the likelihood for definitive clustering.

The extent to which cytochromes c_1 and f are dissimilar, and the possible absence of a c-type cytochrome from the quinol oxidase complex of *S. acidocaldarius* [33], supports the concept discussed previously that the primitive complex was composed only of a cytochrome b-like protein and a Rieske iron–sulfur protein. According to this view, recruitment of a c-type cytochrome into quinol oxidase would have occurred in at least two separate instances, so that the functionally analogous cytochromes c_1 and f could be seen as having appeared via convergent, rather than divergent, evolution.

The dearth of sequence information for cytochromes c_1 and f allows us to simply raise several possibilities without offering definitive evidence for any one hypothesis. Too few sequences are available from the Archaea, as well as from other poorly investigated groups of Bacteria (e.g., Oxyphotobacteria). The availability of the high-resolution crystal structure of cytochrome f [30], in which it is seen that the axial sixth ligand is shown to be the N-terminal amino group, indicates that the structure of cytochrome c_1, where this ligand has been inferred to be an internal methionine [60, 61], may be markedly different from that of cytochrome f. Thus, given the very low extent (approximately random) of amino acid sequence identity between cytochromes c_1 and f, different cellular origins (nuclear versus organellar encoded genes) and apparently different mechanisms of heme ligation and assembly, a radically different view of their evolutionary relationship is that they may be rather exquisite examples of convergent evolution [30]. Hence, we propose that the convergence would be limited to that of function and not necessarily be a convergence of structure. The resolution of a cytochrome c_1 structure is clearly needed to begin to solve this and other issues [31].

9.7.3 The Rieske Protein

A phylogenetic analysis was performed based on the amino acid sequences of the Rieske iron–sulfur protein from 16 species, including those from mammals, fungi, oxygenic and anoxygenic bacteria, and higher plants (Fig. 9.8). The sequence of *B. japonicum* was assigned as the outgroup. Although the bootstrap probabilities for individual branches, such as those for the fungi (0.58) and the cyanobacteria (0.87), are not significant, all internal branching probabilities are significant. The chloroplastic Rieske proteins cluster as one might have expected with a significant ($p < 0.05$) dichotomy between the cyanobacterial and higher plant sequences. From this analysis, it would appear that the Rieske

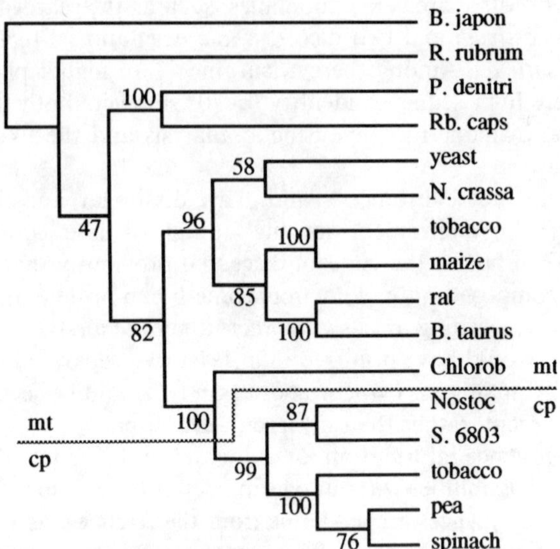

Figure 9.8   Majority-rule bootstrap tree of Rieske iron–sulfur protein sequences from the 16 species (see Table 9.6). Tree representations and construction methodology are described in Figure 9.5. mt, mitochondria; cp, chloroplasts.

iron–sulfur protein component of the cytochrome bc_1 and $b_6 f$ complexes is monophyletic. What is not clear from this analysis is whether the protein is evolving at the same rate as the b-cytochrome of the respective complex.

9.8 Conclusions

The ubiquity of cytochrome b of the bc_1 complex has been used as the basis of a wide variety of phylogenetic analyses [1–5]. More is becoming known about the evolution of the complexes themselves and their individual constituents. It now seems clear that the larger cytochrome b underwent a number of changes, including gene fission, to yield cytochrome b_6 and subunit IV. A better picture of the evolution of the Rieske iron–sulfur protein has resulted from a phylogenetic analysis of the distribution of quinone species within the Bacteria [51]. The two c-type cytochromes of the complexes, c_1 and f, are so dissimilar at the primary structural level that they may represent convergent evolution in terms of function. Implicit in such a view is that they would have been individually "recruited" to a simpler ancestral complex, consisting only of cytochrome bc_1 and a Rieske iron–sulfur protein, similar perhaps to that found in *S. acidocaldarius*. The general absence of a c-type cytochrome in the aerobic

Archaea further supports this view. Alternatively, resolution of a cytochrome c_1 structure may yet show that cytochromes c_1 and f are much more similar at the structural level than is currently thought. In this case, the great degree of sequence dissimilarity between cytochromes c_1 and f may be explained by selective pressures related to the cellular component in which they are encoded. In general, the evolution of the cytochrome bc_1/b_6f complexes and their subunits will remain poorly understood until more members of the Archaea have been surveyed for the presence of the complex, and many more cytochromes c_1 and f have been sequenced.

Accession Numbers

Cytochrome b_6: *Chlamydomonas reinhardtii*, D01036; *Nostoc*, J03855; *Marchantia polymorpha*, S01534; *Nicotiana tabacum* (tobacco), P06449; *Oenothera hookeri* (Hooker's evening primrose), P04658; *Oryza sativa* (rice), P07888; *Pisum sativum* (pea), X56315; *Spinacia oleracea* (spinach), P16013; *Synechococcus* sp. 7002 (*Agmenellum quadruplicatum*), P26293; *Synechocystis* sp. 6803, X58532; *Triticum aestivum* (wheat), P05151; *Vicia faba* (broad bean), P06669.

Subunit IV: *Agmenellum quadruplicatum*, X63049; *Chlamydomonas eugametos*, X14503; *Chlamydomonas reinhardtii*, X72919; *Chlorella protothecoides*, X15244; *Nostoc*, J03967; *Hordeum vulgare* (barley), X14107; *Zea mays*, X05422; *Marchantia polymorpha* (liverwort), X04465; *Nicotiana tabacum*, Z00044; *Oryza sativa* (rice), M35995; *Prochlorothrix hollandica*, X60313; *Spinacia oleracea*, X07106; *Synechocystis* sp. 6803, X58522; *Triticum aestivum*, X54751.

Cytochrome f: *Brassica campestris* (turnip), [29]; *Chlamydomonas reinhardtii*, D01036; *Nostoc*, J03855; *Marchantia polymorpha*, S01534; *Nicotiana tabacum* (tobacco), P06449; *Oenothera hookeri* (Hooker's evening primrose), P04658; *Oryza sativa* (rice), P07888; *Pisum sativum* (pea), X56315; *Spinacia oleracea* (spinach), P16013; *Synechococcus* sp. 7002 (*Agmenellum quadruplicatum*), P26293; *Synechocystis* sp. 6803, X58532; *Triticum aestivum* (wheat), P05151; *Vicia faba* (broad bean), P06669.

Cytochrome b: *Acipenser transmontanus* (white sturgeon), P11669; *Antilocapra americana* (pronghorn), P24992; *Artemia franciscana*, X69067; *Ascaris suum* (pig roundworm), P24878; *Aspergillus nidulans*, J01389; *Balaenoptera physalus* (finback whale), P24950; *Bos taurus*, J01394; *Bradyrhizobium japonicum*, J03176; *Caenorhabditis elegans* (nematode), P24890; *Camelus dromedarius*, X56281; *Capra hircus* (goat), X56289; *Chlamydomonas reinhardtii*, X59471; *Chlamydomonas smithii*, P23663; *Chlorobium limicola*, X73628; *Cyprinus carpio* (carp), P24951; *Dama dama* (deer), X56290; *Drosophila melanogaster* (fruit fly),

P18935; *Drosophila yakuba* (fruit fly), P07704; *Gallus gallus* (chicken), P18946; *Homo sapiens* (human), V00662; *Hystrix africaeaaustralis* (guinea pig), X70674; *Leishmania tarentolae*, P14548; *Marchantia polymorpha*, M68929; *Monodelphis domestica* (opossum), Q04911; *Mus musculus* (mouse), P00158; *Neurospora crassa*, M37324; *Oenothera bertiana* (Bertero's evening primrose), X07126; *Oryza sativa* (rice), P14833; *Paracentrotus lividus* (common sea urchin), P12778; *Paracoccus denitrificans*, X05799; *Paramecium aurelia*, X15917; *Paramecium tetraurelia*, P15585; *Phoca vitulina* (harbor seal), Q00530; *Plasmodium falciparum*, M99416; *Plasmodium yoelli* (*P. falciparum*, strain Malay Camp), M99416; *Podospora anserina* (fungus), Q02655; *Rattus rattus* (rat), J01436; *Rhodobacter capsulatus*, M18576; *Rhodobacter sphaeroides*, X56157; *Rhodospirillum rubrum*, X55387; *Saccharomyces cerevisiae*, V00696; *Schizosaccharomyces pombe*, X02819; *Solanum tuberosum* (potato), P29757; *Stenella longirostris* (pantropical spinner dolphin), P24962; *Sulfolobus acidocaldarius*, X62643; *Sus scrofa* (pig), P24964; *Tayassu tajacu* (collared peccary), P24966; *Theileria annulata*, M63015; *Tragulus napu* (balabac chevrotain), P24965; *Trypanosoma brucei*, X02618; *Vicia faba* (broad bean), P05718; *Xenopus laevis* (african clawed frog), P00160; *Zea mays* (maize), P04165.

Cytochrome c_1: *Bos taurus*, A00118; *Bradyrhizobium japonicum*, J03176; *Euglena gracilis*, P20114; *Homo sapiens* (human), X06994; *Neurospora crassa*, X05235; *Paracoccus denitrificans*, X05799; *Rhodobacter capsulatus*, X05630; *Rhodobacter sphaeroides, X03476; Rhodospirillum rubrum, X55387; Solanum tuberosum* (potato), S20014; *Saccharomyces cerevisiae* (yeast), X00791.

Rieske: *B. japonicum*, J03176; *C. limicola* X73628; *Bos taurus* (cattle), S58789; *Nostoc* J03855; *Neurospora crassa*, X02472; *Nicotiana tabacum* (cp), X64353; *Nicotiana tabacum* (mt), L16810; *Oryza sativa* (rice), D21116; *Paracoccus denitrificans* X05799; *Pisum sativum* (pea), X63605; Rat, M24542; *Rhodobacter capsulata*, X05630; *Rhodobacter sphaeroides*, X03476; *Rhodospirillum rubrum*, X55387; *Saccharomyces cerevisiae* (yeast), M24500; *Spinacia oleracea* (cp), S00454; *Synechocystis* sp. 6803, X58532; *Zea mays*, M77224.

Acknowledgments

The writing and work on this chapter, and the authors' studies incorporated therein, were supported by NIH Grant GM-38323. We thank F. Daldal, D. Huang, and B. L. Trumpower for Figures 9.1A to C, W. Nitschke and N. Sone for communication of information prior to publication, J. B. Heymann for digitizing gel data for Figures 9.1A and B, S. Martinez for Figure 9.3, K. Zeller for helpful discussions regarding the tree methods, N. Nelson and B. L. Trumpower for access to manuscripts prior to publication, and M. Saraste for a helpful discussion.

References

1. Meyer, A., Kocher, T. D., Basasibwaki, P., Wilson, A. C. *Nature* 1990; *347*, 550–553.

2. Irwin, D. M., Kocher, T. D., Wilson, A. C. *J. Mol. Evol.* 1991; *32*, 128–144.

3. Hedges, S. B., Bogart, J. P., Maxson, L. R. *Nature* 1992; *356*, 708–710.

4. Martin, A. P., Naylor, G. J. P., Palumbi, S. R. *Nature* 1992; *357*, 153–155.

5. Sturmbauer, C., Meyer, A. *Nature* 1992; *358*, 578–581.

6. Degli Esposti, M., DeVries, S., Crimi, M., Ghelli, A., Patarnello, T., Meyer, A. *Biochim. Biophys. Acta* 1993; *1143*, 243–271.

7. Crofts, A. R. In *The Enzymes of Biological Membranes*, Vol. 4, Martonosi, A. N. (ed.), 1985; pp. 347–382.

8. Rich, P. R. *Biochim. Biophys. Acta* 1988; *932*, 33–42.

9. Trumpower, B. L. *Microbiol. Rev.* 1990; *54*, 101–129.

10. Cramer, W. A., Furbacher, P. N., Szczepaniak, A., Tae, G.-S. In *Current Topics in Bioenergetics*, C. P. Lee (ed.), Academic Press, New York, 1991; pp. 179–222.

11. Gennis, R. B., Barquera, B., Hacker, B., Van Doren, S. R., Araud, S., Crofts, A. R., Davidson, E., Gray, K. A., Daldal, F. *J. Bioenerg. Biomembr.* 1993; *25*, 195–209.

12. Link, T. A., Haase, U., Brandt, U., von Jagow, G. *J. Bioenerg. Biomembr.* 1993; *25*, 221–232.

13. Trumpower, B. L., Gennis, R. B. *Annu. Rev. Biochem.* 1994; *63*, 675–716.

14. Wikström, M. K. F., Berden, J. A. *Biochim. Biophys. Acta* 1972; *283*, 403–420.

15. Kramer, D. M., Crofts, A. R. *Biochim. Biophys. Acta* 1994; *1184*, 193–201.

16. Takahashi, E., Wraight, C. A. *Biochemistry* 1992; *31*, 855–866.

17. Okamura, M. Y., Feher, G. *Annu. Rev. Biochem.* 1992; *61*, 861–896.

18. di Rago, J. P., Colson, A.-M. *J. Biol. Chem.* 1988; *263*, 12564–12570.

19. di Rago, J. P., Coppee, J. Y., Colson, A.-M. *J. Biol. Chem.* 1989; *264*, 14543–14548.

20. Howell, N., Gilbert, K. *J. Mol. Biol.* 1988; *203*, 607–618.

21. Daldal, F., Tokito, M. K., Davidson, E., Faham, M. *EMBO J.* 1989; *8*, 3951–3961.

22. Saraste, M. *FEBS Lett.* 1984; *166*, 367–372.

23. Widger, W. R., Cramer, W. A., Herrmann, R. G., Trebst, A. *Proc. Natl. Acad. Sci. USA* 1984; *81*, 674–678.

24. Gonzalez-Halphen, D., Lindorfer, M. A., Capaldi, R. A. *J. Biol. Chem.* 1988; *27*, 7021–7031.

25. Breyton, C., de Vitry, C., Popot, J.-L. *J. Biol. Chem.* 1994; *269*, 7597–7602.

26. Szczepaniak, A., Cramer, W. A. *J. Biol. Chem.* 1990; *265*, 17720–17726.

27. Haley, J., Bogorad, L. *Proc. Natl. Acad. Sci. USA* 1989; *86*, 1534–1538.

28. Popot, J. L., Pierre, Y., Breyton, C., Lemoine, Y., Takahashi, Y., Rochaix, J. D. In *Photosynthesis: from Light to Biosphere*, Mathis, P. (ed.), Kluwer, Dordrecht, 1995; Vol. II, pp. 509–512.

29. Gray, J. C. *Photosynth. Res.* 1992; *34*, 359–374.

30. Martinez, S. E., Huang, D., Szczepaniak, A., Cramer, W. A., Smith, J. L. *Structure* 1994; *2*, 95–105.

31. Cramer, W. A., Martinez, S. E., Furbacher, P. N., Huang, D., Smith, J. L. *Curr. Op. Struct. Biol.* 1994; *4*, 536–544.

32. Wheelis, M. L., Kandler, O., Woese, C. R. *Proc. Natl. Acad. Sci. USA* 1992; *89*, 2930–2934.

33. Lübben, M., Kolmerer, B., Saraste, M. *EMBO J.* 1992; *11*, 805–812.

34. Woese, C. R., Kandler, O., Wheelis, M. L. *Proc. Natl. Acad. Sci. USA* 1990; *87*, 4576–4579.

35. Gray, M. W., Cedergren, R., Abel, Y., Sankoff, D. *Proc. Natl. Acad. Sci. USA* 1989; *86*, 2267–2271.

36. Gray, M. W. *Trends Genet.* 1989; *5*, 294–299.

37. Gest, H. *FEMS Lett.* 1980; *7*, 73–77.

38. Schmidt, C. L., Anemüller, S., Teixeira, M., Schäffer, G. *FEBS Lett.* 1995; *359*, 239–243.

39. Anemüller, A., Schäffer, G. *Eur. J. Biochem.* 1990; *191*, 297–306.

40. Lübben, M. *Biochim. Biophys. Acta* 1995; *1229*, 1–22.

41. Sone, N., Go, S., Sone, T., Noguchi, S. *J. Biol. Chem.* 1995; *270*, 10612–10617.

42. Schütz, M., Zirngibl, S., le Coutre, J., Büttner, M., Xie, D.-L., Nelson, N., Deutzmann, R., Hauska, G. *Photosynth. Res.* 1994; *39*, 163–174.

43. Knaff, D. B., Malkin, R. *Biochim. Biophys. Acta* 1976; *430*, 244–252.

44. Lewis, R. J., Prince, R. C., Dutton, P. L., Knaff, D. B., Krulwich, T. A. *J. Biol. Chem.* 1981; *256*, 10543–10549.

45. Zannoni, D., Ingeldew, W. J. *FEBS Lett.* 1985; *193*, 93–98.

46. Kuila, D., Fee, J. A. *J. Biol. Chem.* 1986; *261*, 2768–2771.

47. Liebl, U., Rutherford, A. W., Nitschke, W. *FEBS Lett.* 1990; *261*, 427–430.

48. Liebl, U., Pezennec, S., Riedel, A., Kellner, E., Nitschke, W. *J. Biol. Chem.* 1992; *267*, 14068–14072.

49. Riedel, A., Kellner, E., Grodzitzki, D., Liebl, U., Hauska, G., Müller, A., Rutherford, A. W., Nitschke, W. *Biochim. Biophys. Acta* 1993; *1183*, 263–268.

50. Blankenship, R. E. *Photosynth. Res.* 1992; *33*, 91–111.

51. Nitschke, W., Kramer, D. M., Riedel, A., Liebl, U. In *Photosynthesis: from Light to Biosphere*, Mathis, P. (ed.), Kluwer, Dordrecht, 1995; Vol. I, pp. 945–950.

52. Cramer, W. A., Martinez, S., Huang, D., Tae, G.-S., Everly, R. M., Heymann, J. B., Cheng, R. H., Baker, T. S., Smith, J. L. *J. Bioenerg. Biomembr.* 1994; *26*, 31–47.

53. Graham, L. A., Brandt, U., Sargent, J. S., Trumpower, B. L. *J. Bioenerg. Biomembr.* 1993; *25*, 245–257.

54. Brown, W. M., Prager, E. M., Wang, A., Wilson, A. C. *J. Mol. Evol.* 1982; *18*, 225–239.

55. Palmer, J. D., Herbron, L. A. *J. Mol. Evol.* 1988; *28*, 87–97.

56. Edwards, S. V., Arctander, P., Wilson, A. C. *Proc. R. Soc. Lond. B* 1991; *243*, 99–107.

57. Swofford, D. L. PAUP: Phylogenetic Analysis Using Parsimony, version 3.0d. Illinois Natural History Survey, 1991.

58. Felsenstein, J. *J. Evolut.* 1985; *39*, 783–379.

59. Moore, G. R., Pettigrew, G. W. In *Cytochromes c.* Springer-Verlag, New York, 1990; pp. 255–307.

60. Simpkin, D., Palmer, G., Devlin, J., McKenna, M. C., Jensen, G. M., Stephens, P. J. *Biochemistry* 1989; *28*, 8033–8039.

61. Gray, K. A., Davidson, E., Daldal, F. *Biochemistry.* 1992; *35*, 11864–11873.

62. Ljungdahl, P. O., Pennoyer, J. D., Robertson, D. E., Trumpower, B. L. *Biochim. Biophys. Acta* 1987; *891*, 227–241.

63. Schägger, H., von Jagow, G. *Anal. Biochem.* 1987; *166*, 368–379.

Evolution of Cytochrome Oxidase

*Matti Saraste, Jose Castresana,
Desmond Higgins, Mathias Lübben,
and Matthias Wilmanns*

10.1 Introduction

Nucleotide sequences of genes and operons that encode cytochrome oxidases in bacteria have accumulated since 1986. These data have revealed a large superfamily of homologous heme–copper cytochrome oxidases that contains several structurally and functionally distinct members. The highly conserved cytochrome aa_3 of eukaryotes forms one branch of the family. Several aerobic bacteria also have this type of cytochrome c oxidase [1–3]. Another branch is formed by complexes, such as the *Escherichia coli* cytochrome *bo*, that use quinols as electron donors. Recently, a novel cytochrome c oxidase has been characterized in *Bradyrhizobium japonicum*, *Rhodobacter sphaeroides*, and *Paracoccus denitrificans* [3–6]. This enzyme, which we shall call the FixN complex (it is also called cytochrome cbb_3), appears to be the most distant member of the family. However, it still clearly shares the heme–copper active site with the other homologous cytochrome oxidases (Fig. 10.1).

The first crystal structures of cytochrome c oxidase were published in 1995. Hartmut Michel and his colleagues determined the structure of the *P. denitrificans* enzyme [7], and Shinya Yoshikawa and his co-workers solved the structure of the bovine mitochondrial oxidase [8]. The new structural data help us to understand common structural features as well as major differences between cytochrome oxidases in different branches of the superfamily. We first discuss the molecular anatomy of the enzyme's functional core, and then proceed to build an evolutionary tree of the superfamily using sequence alignments. It will become evident that quinol oxidases have evolved from

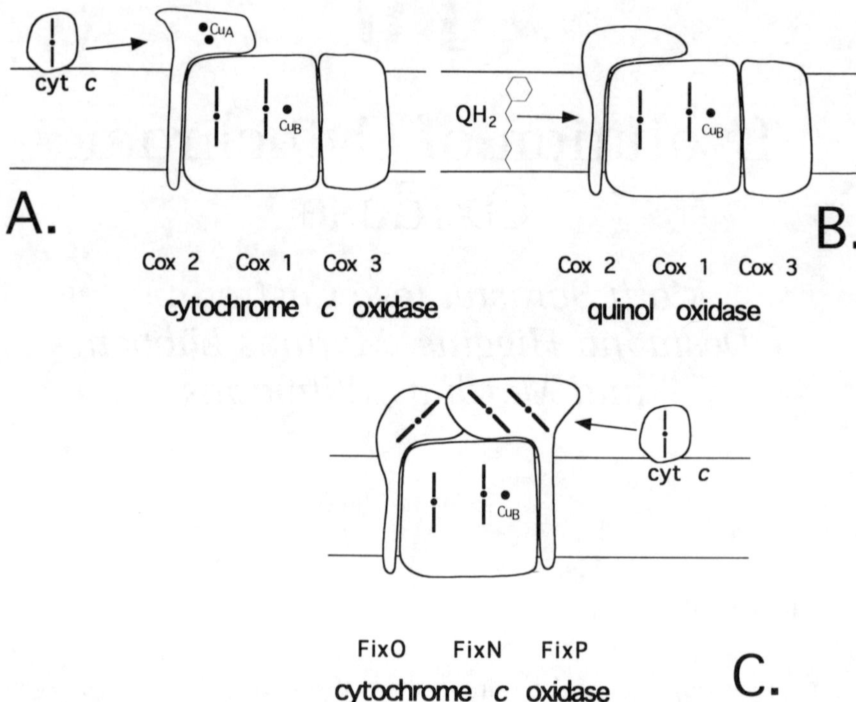

Figure 10.1   Three main members of the cytochrome oxidase superfamily. (A) Cu_A-containing cytochrome c oxidase such as the *Paracoccus denitrificans* cytochrome aa_3. (B) A quinol oxidase without a Cu_A center. (C) A FixN-type cytochrome c oxidase (cytochrome cbb_3). In the latter the cytochrome c binding and Cu_A-containing domains are replaced by two cytochromes c.

cytochrome c oxidases, and that the evolution of eubacterial oxidases may include lateral gene transfer. This tree will also indicate that the enzyme is older than atmospheric oxygen, and that respiration may have a monophyletic origin. There are also several links between denitrification and oxygen-based respiration. The first oxidase may have arisen from a denitrification enzyme.

10.2 Molecular Anatomy of Cytochrome Oxidase

Cytochrome oxidase reduces dioxygen to water using electrons coming from either ferrocytochrome c or quinol, depending on the type of enzyme:

$$O_2 + 4H^+ + 4e^- \Leftrightarrow 2H_2O$$

The liberated redox energy is in part employed to pump protons across the membrane. Proton pumping and the topological arrangement of the redox reaction generate membrane potential and ΔpH [9], and this electrochemical gradient is used to drive ATP synthesis.

Details of the oxygen reduction mechanism are emerging. This mechanism appears to be similar in quinol and cytochrome c reducing enzymes [10]. The active site where oxygen is reduced to water is formed by a pentacoordinated heme iron and a copper called Cu_B. This center seems to be ideally designed for trapping toxic intermediates, such as peroxide, superoxide, and hydroxyl radical, that are produced during the reduction of dioxygen [9, 10]. The current consensus view is that the secret of proton pumping is centered at the active site (see Refs. [10] and [11]) and that the mechanism may involve changes in the histidine coordination around the high-spin heme iron and Cu_B [7, 10a]. All cytochrome oxidases contain a hexacoordinated low-spin heme that is involved in electron transfer to the active site. Both hemes and Cu_B are bound to the largest subunit of cytochrome oxidase, subunit I, which we shall call Cox1 in this chapter.

With the exception of the FixN complex (Fig. 10.1) and one archaebacterial enzyme, all oxidases contain subunits II and III (Cox2 and Cox3). In cytochrome c oxidase, Cox2 has the binding site for the fourth metal center of the enzyme, a copper site called Cu_A. It is most likely the site that first receives electrons from cytochrome c [12]. In most eukaryotes, Cox1, Cox2, and Cox3 are encoded by the mitochondrial DNA, the leftover genome of an ancient endosymbiont that gave rise to this organelle. These three proteins form the functional core of the enzyme. A large number of "supernumerary" subunits (up to 10) is assembled with this core in the mitochondrial oxidases. In yeast, most of them [13], but not all [14], are needed for the assembly of the active enzyme. The bacterial oxidases may also contain one or two additional subunits but these are not related to the eukaryotic supernumerary proteins.

The discussion of the evolution of cytochrome oxidase will be limited to the common catalytic core. We shall not discuss the nuclear-encoded subunits of the eukaryotic enzyme [13, 15–17] or additional proteins found in the bacterial complexes, as their functional roles are not known in detail. However, the eukaryotic cytochrome oxidase may be regulated in a more complicated way than the prokaryotic enzymes. For instance, adenine and guanine nucleotides appear to have an effect on the oxidase activity in mitochondria [14, 16, 18].

The bacterial and yeast enzymes can be the target of mutagenesis. Recent data concerning the metal centers have been generated by a combination of site-directed and random mutagenesis, genetic complementation, activity assays, and various spectroscopic measurements [3, 11, 19–21]. These data have led to the identification of the ligands that bind the metal centers and to a model of the active site that was later confirmed by X-ray crystallography [7, 8].

Figure 10.2 shows the structure of the *P. denitrificans* cytochrome c oxidase complex [7]. In addition to Cox1, Cox2, and Cox3, this complex contains a hydrophobic peptide (subunit 4) with unknown function. Most of the protein is folded into long transmembrane helices connected by relatively short loops.

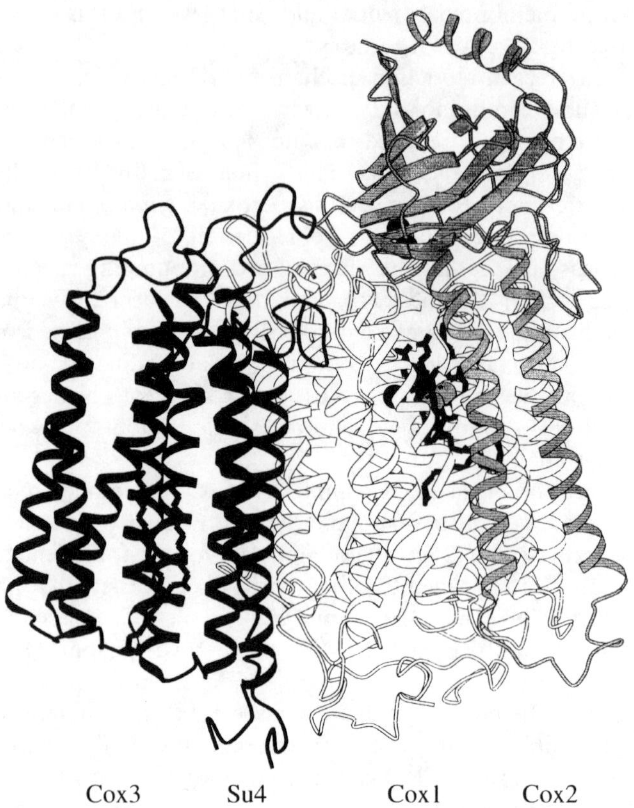

Figure 10.2   Structure of the cytochrome *c* oxidase from *P. denitrificans* [7]. Cox1, Cox2, and Cox3 subunits are labeled; the complex contains an additional small subunit (su 4). The metal centers in Cox1 and Cox2 as well as a phospholipid molecule bound to Cox3 are shown. (Courtesy of Drs. Iwata and Michel.)

Only the C-terminal domain of Cox2 is outside the bilayer. The core structure of the bovine mitochondrial oxidase [8] is identical to that of the bacterial enzyme.

10.2.1 Subunit I

As shown in Figure 10.3, the length of Cox1 in different oxidases is variable. However, in all cases the Cox1 proteins have a common core, consisting of 12 transmembrane helices. These helices were originally predicted by analysis of Cox1 sequences [2, 3, 22] and by membrane topology obtained with gene fusions [23]. The 12-helix structure was verified by crystallographic studies

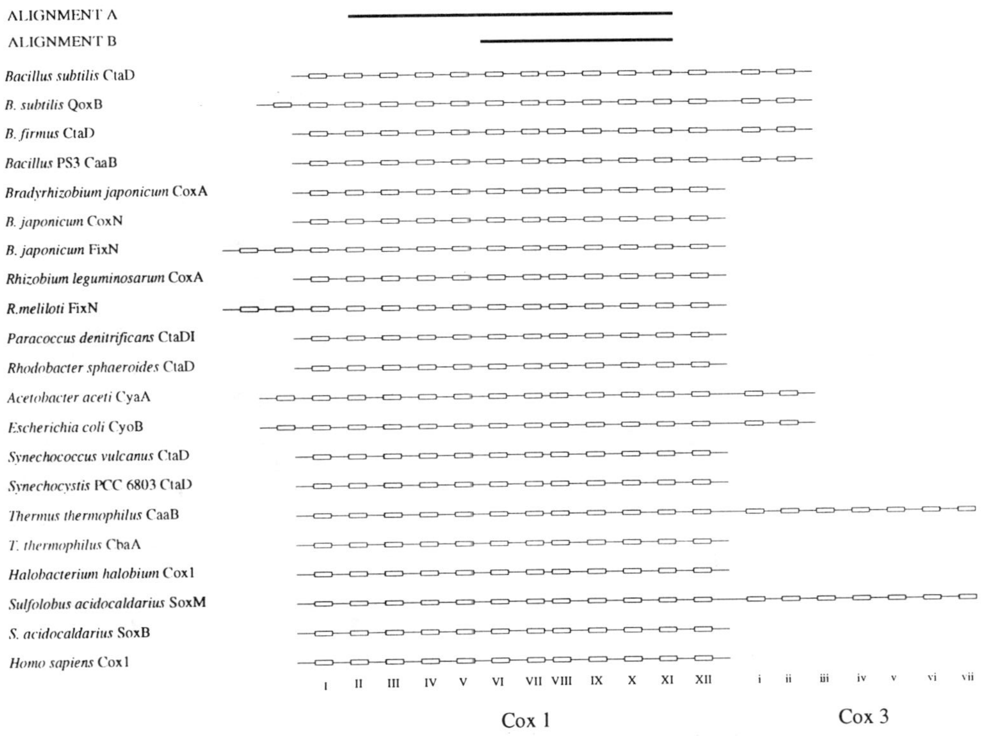

Figure 10.3   Schematic presentation of Cox1 sequences. The boxes represent transmembrane helices which are numbered with Roman numerals on the bottom. Membrane helices that normally belong to the separate Cox3 subunit are indicated by lowercase numerals. The common core of Cox1 contains helices I to XII, but some sequences have extension in the N- or C-terminus, or in both termini. The C-terminal extensions make a contribution to the total number of transmembrane helices that is normally made by Cox3. The bars at the top indicate two regions used for the alignments that are the basis of the evolutionary trees discussed in Section 10.4. (Adapted from Ref. [24].)

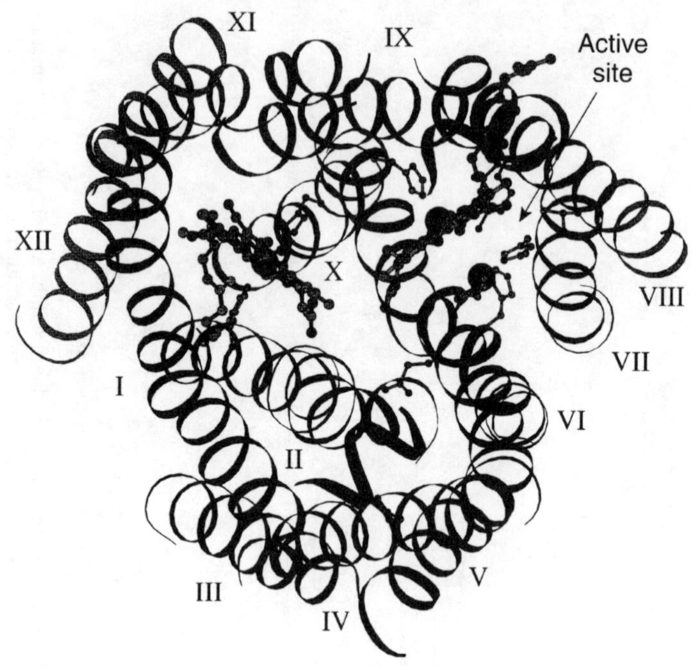

XI
IX
Active site
XII
X
VIII
VII
VI
I
II
III
IV
V

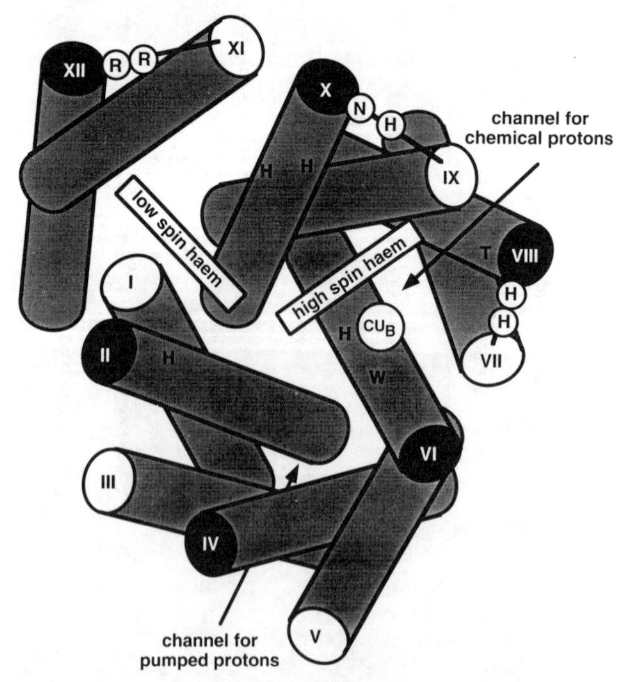

XII
R
R
XI
X
N
H
channel for
chemical protons
H
H
IX
low spin haem
high spin haem
T
VIII
I
H
H
II
H
Cu_B
H
VII
W
III
VI
IV
channel for
pumped protons
V

[7, 8] (Fig. 10.4). The variations in size of the Cox1 proteins and in the number of predicted transmembrane segments are noted only in the termini [24]. For instance, the Cox1 proteins of two oxidase complexes from *Bacillus subtilis* and the *E. coli* cytochrome *bo* complex have an extended C-terminus with two extra hydrophobic segments, whereas the FixN proteins contain two additional transmembrane segments in the N-termini (Fig. 10.3).

Cox1 has six invariant histidines that bind three metal centers [25–29]; the sequences surrounding these histidines are shown in Figure 10.5. Four histidines in helices VI, VII, and X are involved in binding of the iron–copper active site, and two histidines in helices II and X coordinate to the low-spin heme. In Figure 10.4A, they are shown in the skeletonized structure of the *P. denitrificans* Cox1 [7]. The basic arrangement of metal centers in Cox1 is probably identical in the quinol and cytochrome *c* oxidases. The low-spin heme is sandwiched between transmembrane helices II and X. The high-spin heme of the active site is located on the opposite side of helix X and coordinated to the histidine, which is separated by a single residue from the low-spin heme ligand. The residue between these two histidines is a conserved phenylalanine (Fig. 10.5) that may be involved in electron transfer between the hemes. The heme planes are perpendicular to the membrane and tilted 104 to 108° relative to each other [7, 8]. Cu_B is bound to three histidines. Two of these are adjacent in sequence in helix VII, and the third one is the residue in the conserved HPEVY sequence in helix VI (Fig. 10.5). The tyrosine residue of this sequence motif may form a hydrogen bond with the Cu_B-liganding histidine residue in helix VI [7].

The arrangement of metal centers leads to a rather simple topology of transmembrane helices (Fig. 10.4B). They wrap around the active site and the low-spin heme in a counterclockwise manner when viewed from the outer membrane surface as threefold semicircular arcs, each containing four helices [7]. Those helices that are adjacent in sequence are also adjacent in the three-dimensional structure. The same topological principle is found in bacterio-rhodopsin [30] and the bacterial photosynthetic reaction center [31], and it may be the simplest arrangement for folding and membrane insertion of a helical membrane protein.

The amino acid sequences of helices II, VI, VII, VIII, and X as well as those of the loops II > III and IX > X are shown in Figure 10.5 for the five most diverged Cox1 proteins — SoxB and SoxM of the archaeon *Sulfolobus acidocaldarius* [32, 33], CaaB and CbaA of the thermophilic eubacterium *Thermus*

Figure 10.4 Structure of Cox1. (A) The 12 transmembrane helices are viewed from the outer membrane surface. The metal binding sites and all six histidine ligands are included in the figure. (Courtesy of Drs. Iwata and Michel [7].) (B) A schematic illustration of the arrangement of 12 helices and the metal centers in Cox1. Approximate locations of some functionally important residues discussed in the text are indicated. (Adapted from Ref. [7].)

Helix II

```
            H-bond/low-spin haem          pH+        pH+
                     \                      \          \
E.c.  CyoB    HYDQIFTAHGVIMIFFVAMPF-VIGLMN
P.d.  CtaDII  LWNVMITYHGVLMMFFVVIPALFGGFGN        113
S.a.  SoxM    DYYDAVTLHGIFMIFFVVMPL-STGFAN
T.t.  CaaB    QYNQILTLHGATMLFFFIIQAGLTGFGN
S.a.  SoxB    LYYSALTIHGWAAMIAFVPMAAAAVIGF
T.t.  CbaA    SYYQGLTLHGVLNAIVFTQLFAQAAIMV
B.j.  FixN    SFGRLRPLHTSAVIFAFGGNVLIATSFY
P.s.  NorB    PFNVARMVHTNLLIVWLLFGFMGAAYYL
                                \/
                          Low-spin haem
```

HELIX VI

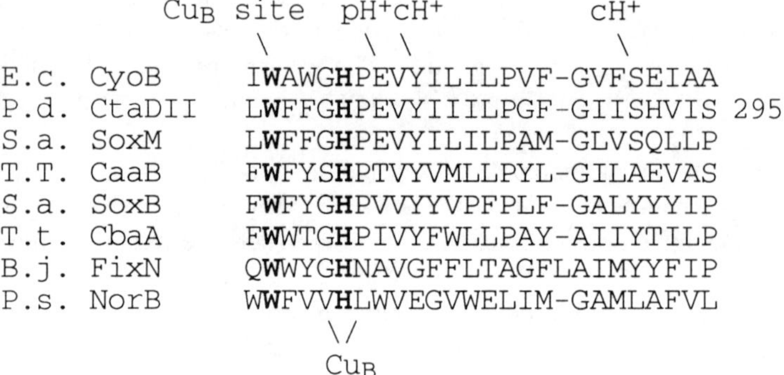

```
            CuB site  pH+cH+              cH+
                 \      \ \                \
E.c.  CyoB    IWAWGHPEVYILILPVF-GVFSEIAA
P.d.  CtaDII  LWFFGHPEVYIIILPGF-GIISHVIS  295
S.a.  SoxM    LWFFGHPEVYILILPAM-GLVSQLLP
T.T.  CaaB    FWFYSHPTVYVMLLPYL-GILAEVAS
S.a.  SoxB    FWFYGHPVVYYVPFPLF-GALYYYIP
T.t.  CbaA    FWWTGHPIVYFWLLPAY-AIIYTILP
B.j.  FixN    QWWYGHNAVGFFLTAGFLAIMYYFIP
P.s.  NorB    WWFVVHLWVEGVWELIM-GAMLAFVL
                         \/
                        CuB
```

Figure 10.5 Functionally important sequence segments of Cox1 proteins from *Paracoccus denitrificans* (CtaDII, [35]), *Sulfolobus acidocaldarius* (SoxB and SoxM, [32, 33], *Thermus thermophilus* (CaaB and CbaA, [34]), *Bradyrhizobium japonicum* (FixN [4]), *Escherichia coli* (CyoB, [36]), and NorB subunit of nitric oxide reductase [37]. Sequences in the regions corresponding to helices II, VI, VII, VIII, and X, as well as to the loops connecting helices II and III, IX and X, XI, and XII are shown. The residues discussed in the text are either bold or labeled with functional assigments. Heme–iron and copper ligands are shown in boldface and assigned. The positions of the histidine and aspartate residues that coordinate to a magnesium ion are labeled in loop IX > X. The label "Cu$_B$ site" above the sequences marks residues that stabilize the copper site. cH$^+$ and pH$^+$ indicate residues that align the channels for chemical and pumped protons, respectively [7].

Helix VII

```
E.c. CyoB     SLVWATVCITVLSFIVWLHHFFTMG
P.d. CtaDII   PMVLAMAAIGILGFVVWAHHMYTAG    331
S.a. SoxM     IALSSIAIAFLSALGVWMHHMFTAI
T.T. CaaB     QMVWAQMGIVVLGTMVWAHHMFTVG
S.a. SoxB     WARWNIYLLAIGTMGVWVHHLQTWP
T.t. CbaA     MARLAFLLFLLLSTPVGFHHQFADP
B.j. FixN     LSIIHFWALIFLYIWAGPHHLHYTA
P.s. NorB     WLYVIIAMALITGIIGTGHHFFWIG
                                      \/
                                      Cu_B
```

Helix VIII

```
                  Cu_B site cH+ cH+
                    \       \   \
E.c. CyoB     FFGITTMIIAIPTGVKIFNWLFTM
P.d. CtaDII   YFMLATMTIAVPTGIKVFSWIATM    362
S.a. SoxM     VSSATTMAIAIPSGVKVLNWTATL
T.T. CaaB     AFAFFTALIAVPTGVKLFNIIGTL
S.a. SoxB     WVNLSTLILATGSGLTVLNLGLTI
T.t. CbaA     IIISVLTLFVAVPSLMTAFTVAASL
B.j. FixN     LGMTFSIMLWMPSWGGMINGLMTL
P.s. NorB     VGSIFSALEPLPFFAMVLFALNMV
```

Helix X

```
                High-spin haem
                  /\
E.c. CyoB     LIAHFHNVIIGGVVFGCFAGMTY
P.d. CtaDII   VVAHFHYVMSLGAVFGIFAGVYY    430
S.a. SoxM     VVGHFHYMVYAILYALLGALFYY
T.t. CaaB     VVAHFHNVLMAGSGFGAFAGLYY
S.a. SoxB     VVGHFHLMIWTLIIMGYTTVFLD
T.t. CbaA     VPGHFHLQVASLVTLTAMGSLYW
B.j. FixN     TIGHVHSGALGWVGFVSFGALYC
P.s. NorB     TAAHGHLAFYGAYAMIVMTMISY
                  \/
                Low-spin haem
```

Figure 10.5 (*Continued*).

Loop II>III

```
                          pH+         pH+
                           \           \
E.c.  CyoB      PLQI GARDVAFPF LNNLS
P.d.  CtaDII    PLHI GAPDMAFPR LNNLS    134
S.a.  SoxM      PRMI GAHDLYWPK INALS
T.t.  CaaB      PLML GARDVALPR VNAFS
S.a.  SoxB      SLYK SKLSIIHTK QMAIF
T.t.  CbaA      MVYL PARELNMRP NMGLM
B.j.  FixN      SFYV VQKSCRVRL AGDLA
P.s.  NorB      IPEE SDCELHSPK LAIIL
```

Loop IX>X

```
                       pH+
                        \
E.c.  CyoB      AV PGADFVLH NSLF
P.d.  CtaDII    SQ APLDRVYH DTYY    407
S.a.  SoxM      PV LPIDYALN GTYF
T.t.  CaaB      SM TPLDYQFH DSYF
S.a.  SoxB      PE NSINPLFH NSYY
T.t.  CbaA      AS FTLDYVVH NTAW
B.j.  FixN      SI KVVNSLSH YTDW
P.s.  NorB      TL APVNYYTH GSQL
                             \/\/
                             Mg2+
```

Loop XI>XII

```
                  H-bond/high-spin haem
                       \
E.c.  CyoB      GMTRRLSQQIDPQF
P.d.  CtaDII    QMPRRYIDYPVEFA    481
S.a.  SoxM      GMPRRYAVIPSPIY
T.t.  CaaB      GMPRRYYTYNADIA
S.a.  SoxB      GFLRRMIAYPVIFQ
T.t.  CbaA      NVPRRYIAQVPDAY
B.j.  FixN      GLMWRAYTSLGFLE
P.s.  NorB      IWLQRIPADGAAMS
                       \
                  H-bond/low-spin haem
```

Figure 10.5   (*Continued*).

thermophilus [34], and FixN of *B. japonicum* [4]. For comparison, we have included sequences of CtaDII from *P. denitrificans* [35], the "free-living mitochondrion," and of the *E. coli* CyoB [36] as well as those from the NorB subunit of nitric oxide reductase (NOR) of *Pseudomonas stuzeri* [37], which also belongs to the superfamily (Section 10.5.4).

Only about 10 amino acid residues within ca. 500 positions of the common alignment have remained totally invariant in the entire superfamily. Apart from the six histidine residues, the highly conserved residues include a tryptophan in helix VI. This residue, together with a threonine (or serine) in helix VIII, is involved in stabilization of the Cu_B site, the latter also possibly in the mechanism of proton pumping [7]. Another invariant residue is an arginine in loop XII > XII. This residue makes a hydrogen bond to one propionate of the low-spin heme. The adjacent arginine in the PRRY motif (Fig. 10.5) makes a similar bond to the high-spin heme [7]. It is remarkable that the strongest conservation directly relates to the structure of the metal binding sites. In contrast, none of the residues that have been proposed to participate in proton translocation (see Ref. [7]) is strictly invariant.

10.2.1.1 *Proton Channels*

The metal centers in Cox1 are rather close to the outer surface of the membrane [7, 8] (Fig. 10.2). This means that one of the substrates, a proton, which is taken from the cytoplasmic (inner) side of the membrane [9], has to traverse the bilayer. The enzyme contains a hydrophilic channel or pore that connects the active site to the cytoplasmic side of the bacterial plasma membrane [7]. This pore is ligned by helices VI, VII, and VIII (Fig. 10.4B). The main building element of this channel is the transmembrane helix VIII, which is strongly amphipathic and has retained this character throughout evolution. It contains conserved hydrophilic residues (asparagines, lysines, serines, and threonines in the positions shown in boldface in Fig. 10.5) in a pattern that places them on the same side of an helix. Substitutions of these hydrophilic residues in helix VIII have been found to inactivate the oxidase activity [3].

The crystal structure of the *P. denitrificans* oxidase [7] indicates that the conserved lysine and its preceding threonine residue in helix VIII (Fig. 10.5) could be involved in the transport of the chemical protons to the active site. In addition, a serine residue in helix VI could be part of the channel that would terminate with the tyrosine residue of the PEVY motif in helix VI and to the hydroxyl group of heme A. The hydroxyl group of tyrosine might be the donor of proton to oxygen during the reduction cycle [7]. This tyrosine is conserved neither in the FixN proteins nor in NorB (Fig. 10.5).

The key lysine residue of helix VIII within the pore for chemical protons can be substituted with threonine. It is replaced by glycine in the FixN oxidases, suggesting that the channel may have a different structure. In NorB, the chemical channel along helix VIII appears to be blocked, which is in accordance with the evidence suggesting that no membrane potential is generated during NO reduction when ascorbate is the electron donor (see Ref. [38]). The chemical proton used in this reaction may originate from the outer side of the membrane.

A major problem concerning the function of cytochrome oxidase remains the coupling between electron transfer and proton translocation. Before structural data were obtained, mutagenesis studies identified only a few residues that could be involved in translocation of the pumped proton. These residues were located in the connecting loop II > III [3, 39]. In particular, an aspartic residue within this loop (in the GARD motif; see Fig. 10.5) was found to be important for proper coupling. The crystal structure indicates that loop II > III can be the inlet of the pore for pumped protons which begins at the inner surface and ends at the low-spin heme [7]. This pore would continue along helices II, III, and IV toward the glutamic acids of the PEVY motif in helix VI. The residues predicted to assemble this pore are highly conserved although not invariant. In particular, the sequences corresponding to the loop II > III are impossible to align (Fig. 10.5).

The *Thermus* CbaA and *Sulfolobus* SoxB (which form a distant branch of the family; see Sections 10.5.1 and 10.5.2) together with the FixN proteins are particularly deviant with respect to the residues in the channel for pumped protons. This may reflect a deviation of the functional properties of these enzymes from those of the majority of oxidases. Some cytochrome oxidases may not be proton pumps.

10.2.2 Subunit II

Cox2 is normally anchored to the membrane with two membrane-spanning segments in its N-terminus (Fig. 10.2), of which only the second one has conserved amino acids [2]. The first and variable membrane-spanning segment is absent from the *S. acidocaldarius* SoxA and *T. thermophilus* CbaB [32, 34]. The C-terminal domain of Cox2 is outside the membrane (Fig. 10.6). This domain hosts the Cu_A center and contains a part of the cytochrome c binding site [22, 40, 41].

The redox centers in Cox1 are conserved in all types of cytochrome oxidases, and the six histidines and their surrounding sequences are diagnostic for the entire superfamily. In contrast, the Cu_A center in Cox2 is specific for cytochrome c oxidases [42]. All known quinol oxidases belonging to the family have lost the ligands that bind this center. As the evolution of quinol oxidases has apparently required removal of the Cu_A site, a part of the binding site for the new substrate may reside within the lost copper center (Section 10.5.3.).

The sequence connecting helices IX and X in Cox1 on the outer side of the membrane contains a conserved HDTY motif (Fig. 10.5). The crystal structures and mutagenesis [7, 8, 43] show that the histidine and aspartate residues coordinate to a magnesium ion that is located at the interface between Cox1 and Cox2. The third amino acid ligand of Mg^{2+} is glutamic acid, which is present in the Cu_A loop in Cox2 (Fig. 10.7). As this glutamate is not normally

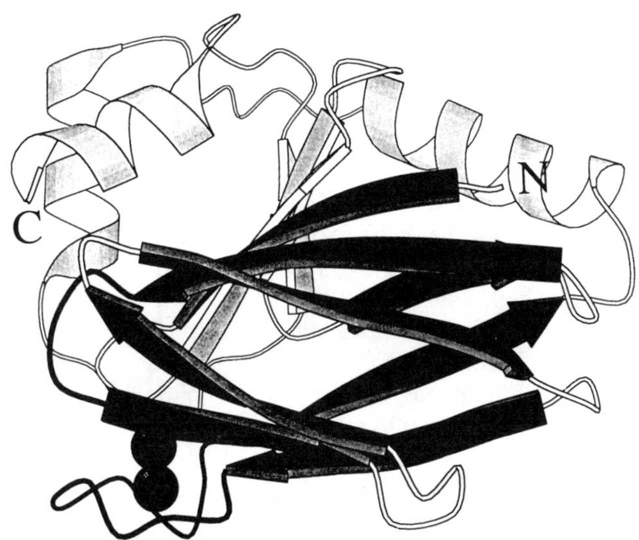

Figure 10.6 Structure of the membrane-exposed CyoA domain with an engineered copper center. The β-sandwich has an extended cupredoxin fold. The location of the dinuclear Cu_A site is shown in the orientation that it has in the complex (Fig. 10.2), that is, the Cu–Cu axis pointing toward the membrane. (Adapted from Ref. [48].)

```
                                     Cyt c   CuA
                                       \   /
P.s.  N2OR    EFTVKQGDEVTV----TITNIDQIEDVSHGFVVV--NHGV
P.d.  CtaC    EYLLATDNPVVVPVGKKVLVQVTATDVIHAWTIP--AFAV
T.t.  CbaB    FAFGYQPNPIEVPQGAEIVFKITSPDVIHGFHVE--GTNI
S.a.  SoxA    TQYKWTPDLIVVNKSEPVVLIINSPQVDTGFYLRTPDGVI
E.c.  CyoA    EQGIATVNEIAFPANTPVYFKVTSNSVMNSFFIP--RLGS
              bbbb bbbbb   bbbbbbbbb    bbbbbb       b

              Cyt c                            CuA
                \                         / / /   \  \
P.s.  N2OR    SMEISPQQTSSITFVADKGPLHWYYCSWFCHALHMEMVGRMMVEPA*
P.d.  CtaC    KQDAVPGRIAQLWFSVDQEGVYFGQCSELCGINHAYMPIVVKAVS..
T.t.  CbaB    NVEVLPGEVSTVRYTFKRPGEYRIICNQYCGLGHQNMFGTIVVKE*
S.a.  SoxA    NLNNVAGITSYAYFVINQPGNYTWRDAEYAGYNSSYMTGTVEVVG*
E.c.  CyoA    QIYAMAGMQTRLHLIANEPGTYDGISASYSGPGFSGMKFKAIATP..
              bbbbbb bbbbbbbbb    bbbbbbb  \          bbbbbbb
                                         Mg2+
```

Figure 10.7 Partial sequences of selected Cox2 proteins as well as a nitrous oxide reductase in a region within the cupredoxin fold. β-Strands in CyoA [48] are shown by underlining with "b." The only two invariant residues (a valine and a methionine) that are present in this segment are shown in boldface. The positions of Cu_A ligands are also highlighted. Two positions that contain conserved carboxylic acid residues which are probably involved in binding of cytochrome c in cytochrome c oxidases [41] are marked. The * denotes the C-termini. Sequences from *Pseudomonas stutzeri* N_2OR [51], *Paracoccus denitrificans* CtaC [50], *Thermus thermophilus* CbaB [34], *Sulfolobus acidocaldarius* SoxA [32], and *Escherichia coli* CyoA [36] are shown.

present in quinol oxidases, the magnesium binding site appears to be characteristic of the Cu_A-containing cytochrome c oxidases. The Cu_A site resides close to the Cox1–Cox2 interface.

10.2.2.1 Cupredoxin Fold

Sequence alignment studies of Cox2 proteins [2] have suggested that the membrane-exposed part of the protein contains a domain that has a characteristic β-barrel structure typical of the blue copper proteins, cupredoxins [44]. The similarity to blue copper proteins was noticed very early [45] and has been the basis of model building and protein engineering experiments [46, 47]. Comparison of the crystal structures with the protein database confirms that the extramembrane domain has an extended cupredoxin fold that is also conserved in quinol oxidases [7, 48] (Figs. 10.2 and 10.6).

The copper binding site in Cox2 has the same location as the metal site in blue copper proteins. However, the Cu_A binding loop is longer, as two coppers rather than one have to fit into it. Recent experiments have shown that the Cox2 domain can be engineered into a blue copper protein [47], and conversely, blue copper proteins can be engineered into CuA proteins [49].

Spectroscopic results have indicated that the only copper center that is similar to Cu_A is the purple copper center A in nitrous oxide reductase (N_2OR, a multicopper protein involved in denitrification; Section 10.5.4) [52–55]. The sequences of Cox2 and the C-terminus of N_2OR are distantly related in the regions corresponding to the cupredoxin domain [56]. This may imply that N_2OR contains a C-terminal domain that also has the cupredoxin fold. The copper binding ligands in these two proteins are conserved (Fig. 10.7).

10.2.2.2 The Structure of the Cu_A Site

A seven-line hyperfine splitting, which is seen in the electron paramagnetic resonance (EPR) spectrum of this copper site, has been interpreted to rise from a dinuclear mixed-valence $Cu^{1.5}/Cu^{1.5}$ center [52, 55]. The dinuclear model has received experimental support from a series of studies on the isolated membrane-exposed portion of Cox2. This has been expressed in a soluble form, and the copper center has been studied using mutagenesis in combination with biochemical and spectroscopic techniques as well as by X-ray crystallography [47, 48, 57–59]. The structure of the Cu_A center is shown in Figure 10.8. The chemical environment of two mixed-valence coppers has to be very similar to avoid valence trapping. The symmetry of coordination is the key feature of the site: two cysteines bridge the copper atoms which are at ca. 2.5 Å distance from each other, and two histidines are the terminal ligands. Furthermore, the additional weaker ligands — a methionine and the carbonyl oxygen — are also

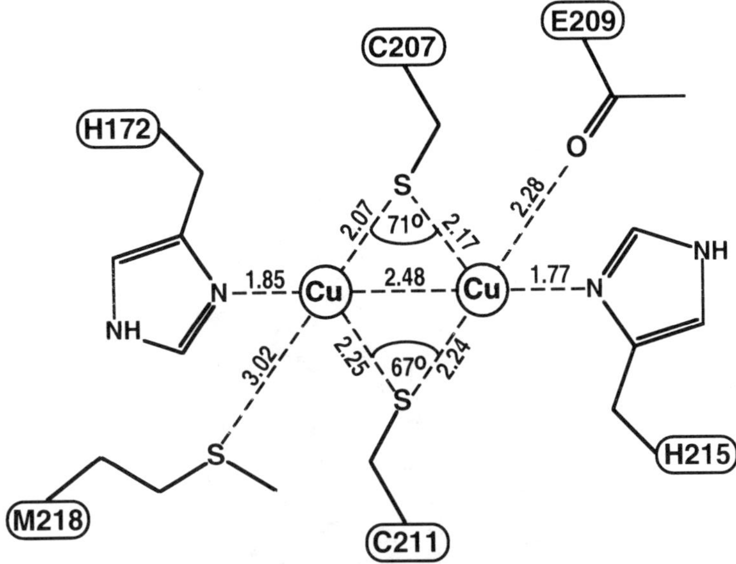

Figure 10.8 The dinuclear Cu_A center. Six ligands — two cysteines and two histidines, a main chain carbonyl, and a methionine — are symmetrically arranged around two coppers. The figure represents the engineered Cu_A center and the residue numbers refer to the CyoA sequence [48]. The bond lengths are indicated.

placed in symmetric positions with respect to the Cu–Cu axis [7, 8, 48] (Fig. 10.8).

10.2.3 Subunit III

Cox3 has no redox-active metal centers that would be involved in the electron transfer function of cytochrome oxidase. Deletion of the gene encoding Cox3 in *P. denitrificans* has shown that the protein has to be present during the assembly of the cytochrome oxidase complex [60]. In its absence, it is difficult for Cox1 and Cox2 to make a functional complex. Several mutations of Cox3 in yeast led to inactive enzyme [19, 20]. For a long time, it was anticipated that Cox3 would have a role in the proton translocation (see Refs. [2, 9, 22, 61]). However, site-directed mutagenesis of Cox3 and the distruption of its gene in *P. denitrificans* indicate that this may not be the case [62].

Figure 10.9 shows the structure of Cox3 of the *P. denitrificans* enzyme with seven transmembrane helices. The topology of these helices is more complex than that found in Cox1 as the C-terminal helix VII folds back between helices III and IV [7]. The protein has a cavity that contains a bound phospholipid molecule. The same cavity and bound phospholipids are also present in the

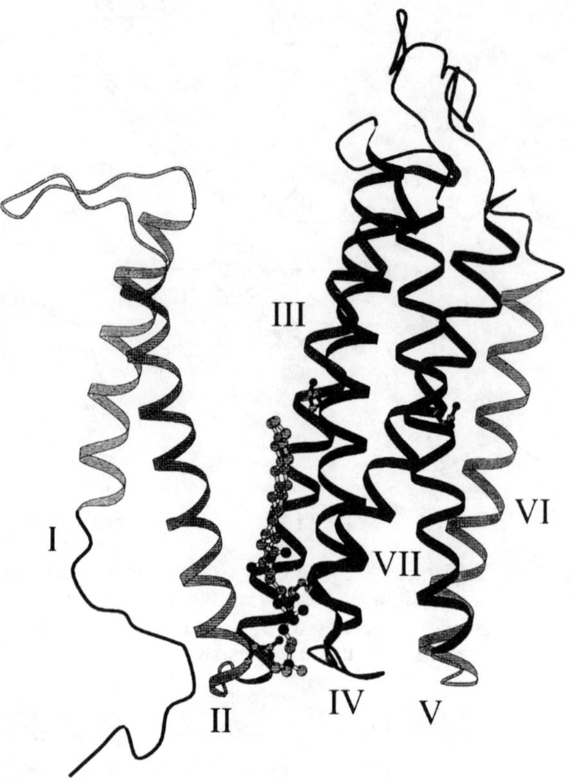

Figure 10.9   Structure of Cox3. Seven helices are shown from the side and labeled. The bound phosphatidylcholine [7] is included in the picture, and two conserved carboxylic acid residue in helices III and VII are also shown. (Courtesy of Drs. Iwata and Michel.)

mitochondrial enzyme [8]. The conserved carboxylic residue in helix III that reacts with dicyclohexyl carbodiimide [61] appears to have a structural role: it forms a hydrogen bond with a conserved histidine [7].

10.2.3.1 Cox1 and Cox3, and Their Fusions

The topological features of Cox1 and Cox3 are helpful for the interpretation of the evolution of cytochrome oxidases. There are three versions of Cox3 in cytochrome c and quinol oxidases. The version that is typical of the eukaryotic enzyme has seven transmembrane segments. In a number of bacterial enzymes, Cox3 has lost two N-terminal transmembrane helices. These have been transferred to the C-terminus of Cox1 (Fig. 10.3). The two "mobile" transmembrane helices are separated from the other five by the phopholipid-binding cleft

(Fig. 10.9). The movement of the N-terminal helices of Cox3 to Cox1 may alter the lipid binding site.

The third type of Cox3 is a fusion protein where Cox1 and Cox3 are encoded by a single gene. The fusion between Cox1 and Cox3 has been found in the SoxM protein of *S. acidocaldarius* as well as in the CaaB subunit of the cytochrome *caa₃* complex of *T. thermophilus*. Since SoxM is a part of cytochrome oxidase complex in an archaeon [33, 63] and CaaB belongs to a cytochrome oxidase in a thermophilic eubacterium [34], it is possible that these gene fusions have occurred independently. This is feasible because the *cox*3 gene follows *cox*1 in most cytochrome oxidase operons [2, 42, 64–68].

An experimental simulation of such an evolutionary gene fusion event has been carried out by Gennis and co-workers, who have fused the reading frames encoding subunits of the *E. coli* cytochrome *bo* complex. Three fusions were made. In one of them, Cox2, Cox1, and Cox3 were all joined, and the two others were pairwise fusions of Cox2 and Cox1, and of Cox1 and Cox3 [69]. The rationale for this experiment is the topology of the three proteins in the membrane. CyoB (Cox1) of the *E. coli* complex has an extra N-terminal transmembrane segment (Fig. 10.3) which brings the terminus to the periplasmic side of the membrane. The C-terminus of CyoA (Cox2) is located on this side as well, so it can be joined to the N-terminus of CyoB. Similarly, the C-terminus of CyoB and the N-terminus of CyoC (Cox3) are on the inside of the membrane, and their fusion does not change the membrane topology of the two proteins. All fused enzymes were active and able to complement a *cyo⁻* deletion strain [69].

10.3. Variant Heme Structures

One of the recent surprises in oxidase research has been the discovery of a novel heme structure that is a prosthetic group of the cytochrome *bo* [70]. This heme is an intermediate between hemes A and B, and it was called heme O on the basis of its source. Heme A, the prosthetic group of the mitochondrial cytochrome oxidase, has two substituents that distinguish it from protoheme [71]. In position 2 of the tetrapyrrole ring it has a hydroxyethyl farnesyl side chain and in position 8 it has a formyl group (Fig. 10.10). Heme O has the 8-methyl group as does the protoheme and 2-hydroxyethyl farnesyl as heme A [72].

At least two steps are involved in the biosynthesis of heme A [73]. The farnesyl-containing side chain is produced by a farnesyl transferase. In *E. coli*, this is the CyoE protein [74], the product of the last gene of the *cyo* operon [3]. In *B. subtilis*, the homologous protein is coded by *cta*B [65, 73], and in yeast, it is the product of *cox*10 [75]. The enzyme that oxidizes the 8-methyl to a formyl group is probably the product of the *cta*A gene in *B. subtilis*. Expression of this gene in *E. coli* leads to the appearance of heme A which this organism does not normally synthesize [73].

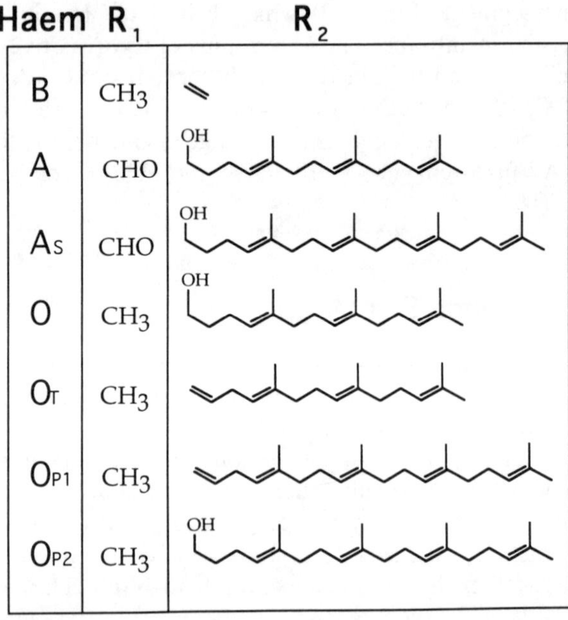

Figure 10.10 Novel heme structures in archaea and eubacteria. The substituents in positions 8 (R_1) and 2 (R_2) of the tetrapyrrole ring are shown in the box. In these positions, heme B (protoheme) has a methyl and a vinyl group and heme A has a formyl and a hydroxyethyl farnesyl side chain [71], whereas heme O has a methyl and a hydroxyethyl farnesyl side chain [72]. Novel hemes A_S, O_T, O_{P1}, and O_{P2} have variably a methyl or formyl group in the R_1 position, and hydroxyethyl farnesyl or geranylgeranyl, or dehydrated forms of these chains in the position R_2. (From Ref. [77].)

Table 10.1 Heme Composition[a] of Eubacterial and Archaebacterial Membranes

Organism	A	A_S	O	O_T	O_{P1}	O_{P2}	Characteristic
Archea							
Sulfolobus acidocaldarius	−	+	−	−	−	−	Thermoacidophile
Desulfurolobus ambivalens	−	+	−	−	−	−	Thermoacidophile
Halobacterium salinarium	−	+	−	−	−	−	Halophile
Thermoplasma acidophilum	−	−	−	+	−	−	Thermoacidophile
Methanosarcina barkeri	−	−	−	−	−	−	Methanogen
Pyrobaculum aerophilum	−	+	−	−	+	+	Thermophile
Eubacteria							
Thermus thermophilus	−	+	−	−	−	−	Thermophile
Aquifex pyrophilus	+	−	−	−	−	−	Thermophile
Bacillus subtilis	+	−	+	−	−	−	Mesophile
Escherichia coli	−	−	+	−	−	−	Mesophile
Paracoccus denitrificans	+	−	−	−	−	−	Mesophile

[a]In all samples heme B (= protoheme IX) was present; hemes C and D could not be determined.

10.3.1 Novel Hemes in Archaea

We have found that the terminal oxidases in the archaeon *S. acidocaldarius* contain a novel heme A [76]. It is spectroscopically similar to the mitochondrial heme A but contains a different isoprenoid side chain. Instead of a farnesyl-containing chain, this heme has a tail with one more isoprenoid unit, that is, an hydroxyethyl geranylgeranyl chain. The discovery of a new heme prompted a further study of archaean hemes. The results of this search are shown in Table 10.1 [77].

So far, four novel variations of hemes A and O have been found in archea (Fig. 10.10). Heme A_S, the geranylgeranyl-containing A-type heme, is present in *Sulfolobus, Desulfurolobus, Halobacterium,* and *Pyrobaculum,* and also in the thermophilic eubacterium *Thermus.* Heme O_T, in which the hydroxyethyl farnesyl has been dehydrated to ethenyl farnesyl, has been found in *Thermoplasma. Pyrobaculum* contains two more variations of heme O, in which the prenyl side chain is either ethenyl or hydroxyethyl geranylgeranyl (Fig. 10.10; [77]).

10.3.2 Heterogeneity of Heme Content

Another new aspect of cytochrome oxidases is the heterogeneity of heme content. The heme content of some enzymes may vary, depending on growth conditions or the level of expression [78–80]. For instance, heme O can replace heme A in the active site of cytochrome c oxidase of the thermophilic bacillus PS3 [78], and the low-spin heme in the *E. coli* cytochrome bo can be either a protoheme or heme O [80].

This heterogeneity, and the gallery of different hemes that can be accommodated by homologous oxidases, show that the heme-binding sites have plasticity. However, specific binding to the active site of some complexes seems to be restricted to hemes that have a long isoprenoid chain. The bimetallic active site is not formed in an *E. coli* strain that carries the *cta*E deletion and does not form heme O [74]. A protoheme cannot substitute for heme O in this site. The requirement for a isoprenoid side chain in the high-spin heme position of the active site seems to be a general characteristic of oxidases. An exception is the FixN oxidase which contains protohemes both in the low-spin position and the active site [5, 6].

The chemical structure of hemes is not an indication of enzyme activity — it is not a distinction between cytochrome c and quinol oxidases. For instance, *B. subtilis* has both a quinol oxidase and a cytochrome c oxidase which contain two A-type hemes (cytochromes aa_3 and caa_3, respectively [65, 81, 82]).

10.4 Cytochrome Oxidases in Archaea

Terminal oxidases belonging to the superfamily have been found in archaeal species [32, 33, 63, 76, 83–85]. *S. acidocaldarius* contains two different terminal oxidases, both of which are probably quinol oxidases (there is no cytochrome c in this organism) and use a sulfur-containing caldariellaquinol as electron donor [83]. Cloning of the oxidase genes and purification of these enzymes [32, 33, 63, 76] revealed that both contain components of the cytochrome bc_1 complex (see Chapter 9).

The SoxABCD terminal oxidase complex has a typical Cox1 subunit (SoxB) and a Cox2 (SoxA) subunit with a single transmembrane span. SoxC is related to cytochrome b of the bc_1 complex. However, it has a hydrophobic C-terminal extension [32]. SoxD is a small hydrophobic subunit, possibly the "lost helix" of SoxA (Fig. 10.11). The complex contains four hemes which are all A_s type (Fig. 10.7); SoxC, a "cytochrome b," is actually an a-type cytochrome. The other terminal oxidase, the SoxM complex, contains the Cox1/Cox3 fusion protein SoxM, and Cox2 with a probable Cu_A binding site (SoxH), another homologue of cytochrome b (SoxG), which probably again binds two hemes A_s, and a Rieske Fe–S protein (SoxF) [33, 63]. Furthermore, it is possible that a blue copper protein (SoxE) also associates with the SoxM complex. This could be a replacement of cytochrome c_1 that is normally found in the cytochrome bc_1 and involved in electron transfer to the oxidase.

Two of the core components of complex III (cytochrome b and Rieske Fe–S protein) have been scrambled with the oxidase subunits during the evolution of *Sulfolobus*. In one case, a complex of cytochrome b and Fe–S protein has joined the oxidase subunits SoxM and SoxH. In the other case, Cox1 and Cox2

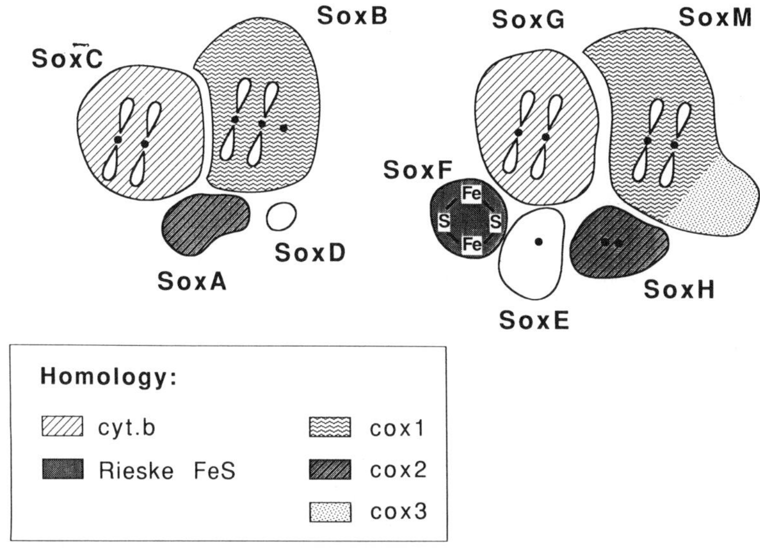

Figure 10.11 A cartoon showing two terminal oxidases of *Sulfolobus acidocaldarius*, SoxABCD (left) and SoxM (or SoxEFGHM) (right). CQ is the electron donating caldariella quinol. SoxE is a blue copper protein which may be involved in electron transfer from SoxG and SoxF to the oxidase subunits SoxH and SoxM. The shading indicates homologies to the subunits of cytochromes bc_1 and aa_3.

(SoxB and SoxA) have joined a cytochrome *b* (SoxC). The SoxC and SoxG proteins have two novelties. First, they preferentially bind heme A_S although the "normal" prosthetic group, protoheme, is also present. Second, they are apparently able to make complexes with the oxidase subunits which they cannot normally do.

We have pointed out a mechanistic corollary of these peculiar archaeal cytochrome complexes [32]. If the Q cycle, which is an inherent property of the arrangement of hemes and quinol binding sites in cytochrome *b*, operates in the SoxABCD and SoxM complexes, and if their oxidase parts function as proton pumps, they are able to conserve energy as effectively as the mitochondrial cytochrome *c* reductase and oxidase working together, and to do this with a minimal number of protein subunits.

10.5 Evolution of Cytochrome Oxidases

In this section, we analyze the molecular evolution of cytochrome oxidases using alignments of Cox1 and Cox2 sequences. The details of these alignments can be found in Ref. [24] and in the legends of Figures 10.12 to 10.14. A

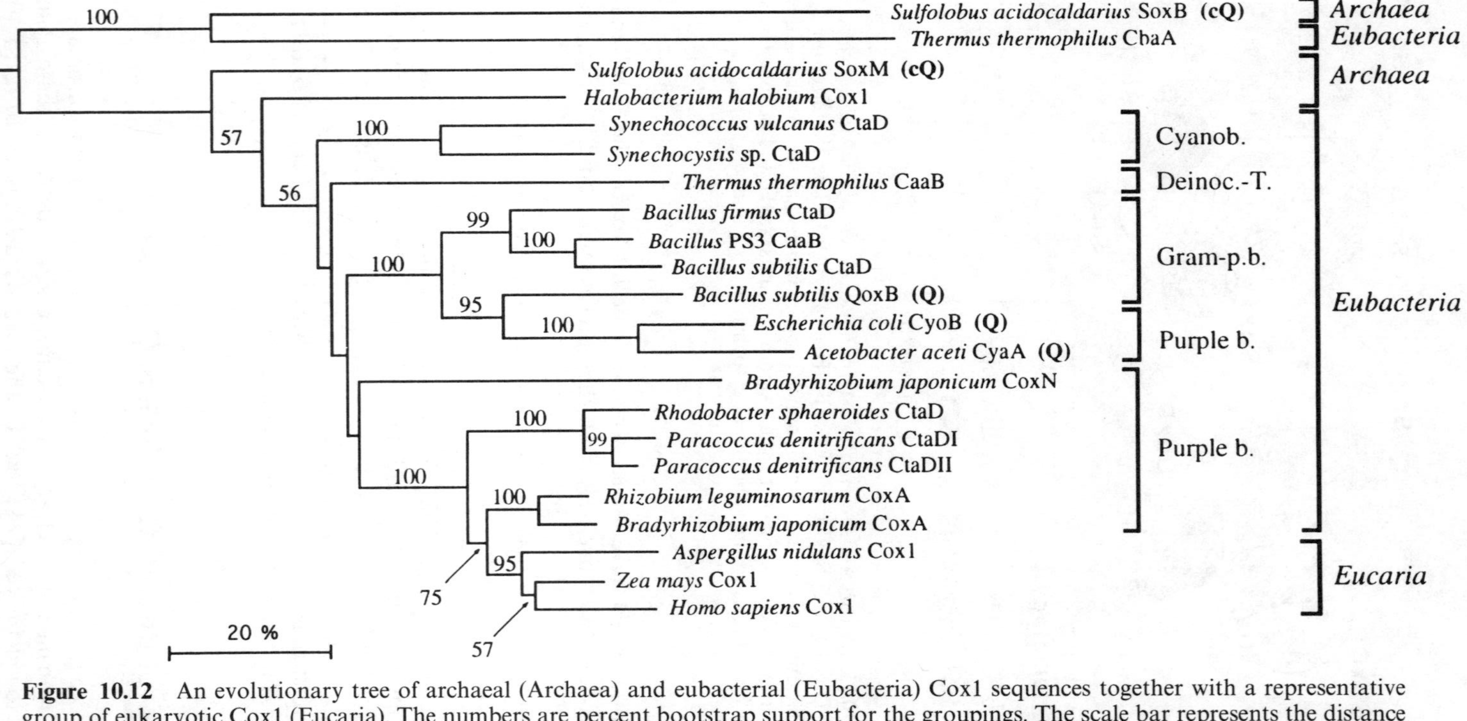

Figure 10.12   An evolutionary tree of archaeal (Archaea) and eubacterial (Eubacteria) Cox1 sequences together with a representative group of eukaryotic Cox1 (Eucaria). The numbers are percent bootstrap support for the groupings. The scale bar represents the distance of 20%. The tree is based on an alignment of Cox1 sequences covering helices II to XI. Species belonging to Cyanobacteria, Deinococceae, and Thermus; Gram-positive bacteria; and purple bacteria are labeled with corresponding abbreviations. (From Ref. [24].) For the details of calculations, as well as for the sources of sequences, see this reference.

number of general conclusions can be made from the evolutionary analysis. In addition, the question of "sequence signatures" (mechanistically important residues that are, for instance, involved in proton pumping) was addressed in Section 10.2.1.1.

10.5.1 Evolutionary Trees

Figure 10.12 shows a phylogenetic tree of cytochrome oxidases, based on the analysis of aligned Cox1 sequences that span from helix II to helix XI (marked by the longer bar at the top of Fig. 10.2). It excludes the FixN sequences, because these are the most divergent and most difficult to align with the others. Figure 10.13 shows a similar analysis of a smaller set of the most deviant Cox1 proteins, now including FixN. This tree is based on an alignment that covers only the sequence spanning from helix VI to helix XI (shorter bar in Fig. 10.3). Figure 10.14 shows a tree calculated using the Cox2 sequences. The Cox1 trees are statistically more stable than the Cox2 tree [24]. However, the Cox2 tree supports the main features of the Cox1 trees.

The Cox1 tree of Figure 10.12 includes only a few sequences of mitochondrial Cox1s, as these are mutually very similar and add little information. The eukaryotic Cox1 sequences cluster with those of the purple bacteria. This is consistent with the generally accepted idea that the endosymbiont that was engulfed by an early eukaryote and became the mitochondrion is related to contemporary purple bacteria [86]. The tree was rooted in the middle with the assumption that the evolutionary rate of bacterial genomes is roughly constant [87, 88].

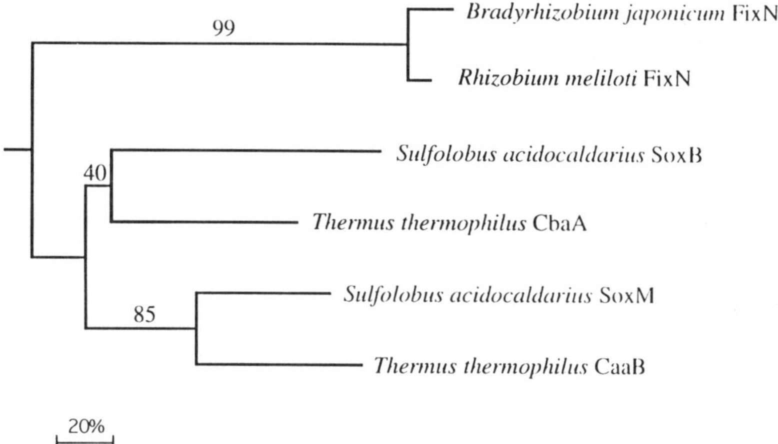

Figure 10.13 An evolutionary tree of a smaller sample of Cox1 sequences, which are the most diverged within the family. This tree was calculated on the basis of a shorter alignment corresponding to the helix VI to IX span of Cox1 (From Ref. [24].)

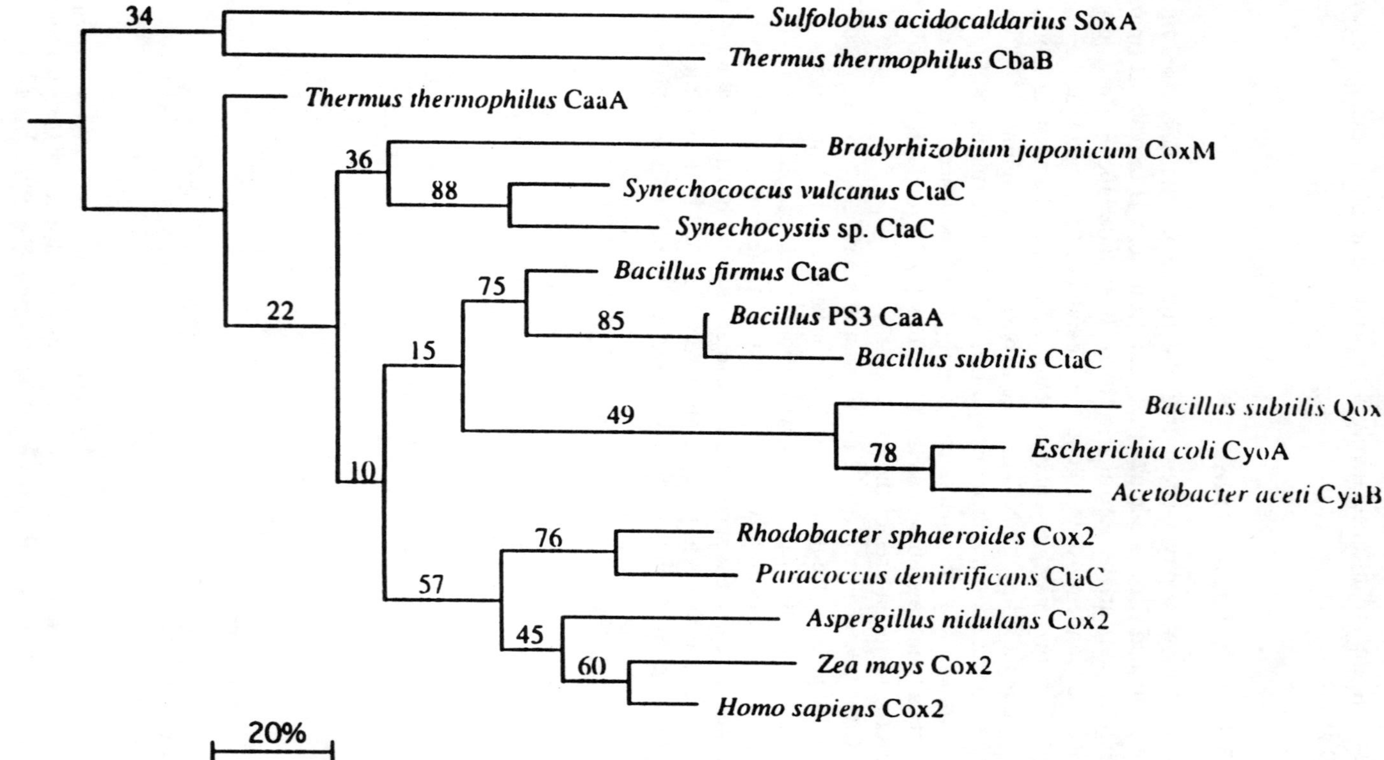

Figure 10.14 An evolutionary tree calculated using the Cox2 sequences. Only the sequence span from the beginning of the second transmembrane helix to the end of the cupredoxin-like domain (see Ref. [2] and Fig. 10.5) was used in the alignment. (From Ref. [24].)

The main feature of the Cox1 tree is that it does not follow the phylogenic trees based on rRNA [89, 90]. This discrepancy must be the result of several gene duplications or lateral gene transfer during evolution. It is obvious that gene duplications have taken place during the evolution of cytochrome oxidases because most of the studied bacterial species contain multiple homologous oxidases.

10.5.2 FixN Descends from the Uroxidase

The earliest gene duplication separated FixN-type complexes from the others (Fig. 10.13). This duplication predates the split between archaea and eubacteria. FixN-type complexes have been found in a number of purple bacteria [4–6, 91], and we can predict that they might be present in archaeal species, unless their genes have been lost in this lineage.

The gene content of the *fixNOQP* operon in *B. japonicum* [4] suggests the complex contains two membrane-attached *c*-type cytochromes, one (FixO) with a single heme and the other (FixP) with two hemes, and the major catalytic FixN (Cox1) subunit (Fig. 10.1). This view is consistent with the biochemical data. The complex purified from *R. sphaeroides* contains three subunits, and heme C, protoheme, and copper in a ratio of $3:2:1$. It has a heme–copper binuclear center but no Cu_A, and it oxidizes cytochrome *c* with rapid kinetics [5]. There is no indication of the presence of Cox2 or Cox3 subunits in the complex.

Although the binuclear heme–copper active site is present in FixN, the sequence comparison suggests that its functional properties may be distinct from those of the other oxidases. For instance, the FixN subunit lacks the otherwise invariant tryrosine of helix VI, and the sequence around the two Cu_B ligands in helix VII has undergone profound changes (the conserved WXHH motif is not present; see Fig. 10.5). Such changes may reflect a different mechanism in the details of oxygen reduction, or in proton translocation. Contrary to the experimental evidence ([91a], but see discussion in Ref. [91b]), the absence of key residues in the proposed channel for pumped protons suggests that the FixN/cytochrome cbb_3 oxidase is either not a proton pump, or it has a different channel.

B, japonicum contains several cytochrome oxidases [92]. The FixN complex is expressed in the bacteroids that inhabit the root nodules and participate in nitrogen fixation. The bacteroids synthesize ATP via oxidative phosphorylation and need an active respiratory chain. However, the oxygen tension in root nodules is very low and buffered by leghemmoglobin, partly to protect nitrogenase which is sensitive to oxidation. Under these conditions, the terminal oxidase must have a high affinity for oxygen, and it is believed that this high-affinity oxidase is FixN (see Refs. [4] and [93]).

FixN seems to be an oxidase that is adapted to function in an environment containing a very low concentration of oxygen. There are no experimental data concerning the oxygen affinity of the other FixN-type (*co*-type) oxidases, and it is not at all clear that they would have a high affinity for oxygen or be expressed under microaerobic conditions. However, the fact that FixN is the bacteroid oxidase in rhizobia, that its sequence is the most deviant among the Cox1 proteins, and that the subunit composition of the FixN complex is unique, support the primordial character of FixN and its probable similarity to the ancestral oxidase (uroxidase). Further evidence for this comes from the relationship of FixN to a denitrification enzyme (Section 10.5.4).

10.5.3 Quinol Oxidases Descend from Cytochrome *c* Oxidases

The second duplication during the evolution of cytochrome oxidases separated a group composed of SoxA and SoxB, together with CbaA and CbaB, from the rest of the oxidases (Figs. 10.12 and 10.13). This duplication also happened before the divergence of eubacteria and archaea, as the SoxA and SoxB are subunits of an oxidase complex in the archaeon *Sulfolobus*, and CbaA and CbaB are the subunits of the cytochrome ba_3 of the eubacterium *Thermus*. The SoxABCD complex is a quinol oxidase [76, 83] whereas cytochrome ba_3 is probably a cytochrome *c* oxidase [94]. Therefore, it is likely that the functional specificity of one of these enzymes has changed during evolution.

A number of arguments support the idea that quinol oxidases have evolved from cytochrome *c* oxidases and that this change occurred independently in the different evolutionary lines. The simplest argument is that the ancestor of FixN was probably a cytochrome *c* oxidase, and that all quinol oxidases may have descended from it. The second argument deals with the evolution of oxidases in bacilli (Fig. 10.12). Cytochrome *c* oxidases in three genera of *Bacillus* are more closely related to quinol oxidases in *B. subtilis*, *E. coli*, and *Acetobacter aceti* than to the cytochrome *c* oxidases of purple bacteria. This means that a gene duplication that occurred during the evolution of bacilli gave rise to these quinol oxidases. The original operon should have encoded a cytochrome *c* oxidase, as all eubacterial oxidases that have diverged prior to this duplication are cytochrome *c* oxidases.

The third argument derives from the structural similarity of the membrane-exposed Cox2 domains in quinol and cytochrome *c* oxidases [7, 8, 48]. The cupredoxin fold is present even if the copper ligands are absent; the metal site has been lost. Cu_A is the hallmark for the majority of cytochrome *c* oxidases and lost in all quinol oxidases studied until now. Therefore, quinol oxidases have evolved from cytochrome *c* oxidases.

Why has the copper site been lost? Perhaps a prerequisite for the evolution of a quinol binding site was the removal the copper center. The empty copper binding loop can perhaps be used in the creation of the new substrate binding site. This metal center is along the route of electron entry, and it would also be natural to use this location for the release of electrons from a new substrate. The molecular adaptation from cytochrome *c* oxidation to quinol oxidation, which occurred independently during evolution, possibly followed, at least in part, the same underlying mechanism.

Since the discovery of the superfamily, a general trend has been to emphasize the principal similarities of cytochrome c and quinol oxidases. A countercurrent has emerged more recently. Musser et al. [95] have proposed that there may be fundamental differences in the mechanisms of electron transfer and proton translocation by cytochrome c and quinol oxidases. These would be due to the different redox spans available for the reaction and to the chemical nature of quinols as electroneutral proton carriers in the membrane. This suggestion led to the discussion on the evolution of cytochrome oxidases [96, 97]. In our opinion the mechanistic differences are probably restricted to the electron transfer to the active site. It is likely that reduction of oxygen in the active site as well as the basic mechanism of proton pumping are shared by quinol- and cytochrome c-oxidizing branches of the superfamily [10]. However, the SoxB/CbaA and FixN branches can be possible exceptions because they may not function as proton pumps.

10.5.3.1 *Lateral Gene Transfer*

A gene duplication during the evolution of bacilli resulted in the evolution of quinol oxidase in these species (Fig. 10.12). The quinol oxidases of $E.$ $coli$ and $A.$ $aceti$, which are both purple bacteria, are more closely related to the quinol and cytochrome c oxidases of bacilli than to the cytochrome c oxidases of other purple bacteria. The close similarity of the quinol oxidases to the cytochrome c oxidases of bacilli, together with the long stem that separates this group from cytochrome c oxidases of purple bacteria, exclude the possibility that genes encoding a quinol oxidase were present before the purple bacteria and bacilli diverged. Therefore, we are left with the possibility that quinol oxidases of purple bacteria have evolved after lateral gene transfer from bacilli [24].

This suggestion is supported further by the molecular anatomy of the Cox1 subunits of quinol oxidases. The Cox1 proteins of $B.$ $subtilis$, $E.$ $coli$, and $A.$ $aceti$ all have an extended C-termini with two extra transmembrane segments and a single extra N-terminal hydrophobic segment (Fig. 10.3). The latter feature is unique. These C-terminally extended Cox1 proteins are accompanied by Cox3 subunits that have lost two N-terminal transmembrane segments (they contain five instead of seven helices). Swapping two helices from Cox3 to Cox1 first requires fusion of the reading frames and, second, a new split of the fused frame and creation of a new translational initiation site. Such a succession of events has probably happened only once during evolution It should also be pointed out that the organization of structural genes in the operons encoding eubacterial quinol oxidases and cytochrome c oxidases of bacilli is the same. This supports the idea that an entire operon was transferred from bacilli to purple bacteria [24].

10.5.4 Links Between Denitrification and Respiration

Denitrification is an anaerobic respiratory process in which nitrous salts are reduced to dinitrogen by a series of metalloprotein catalysts. It can produce energy and support cell growth. Cytochrome bc_1, cytochromes c, and quinols are involved in this electron transfer pathway [98, 99].

There are molecular links between the evolution of denitrification enzymes and cytochrome oxidases. Nitrous oxide reductase is a soluble multicopper protein that contains a binuclear active site, center Z, and the purple center A related to Cu_A (Section 10.2.2.2). The local sequence homology [46] (Fig. 10.7) and the likely structural similarity of the copper centers [52, 55] are an argument for the idea that N_2OR and Cox2 are related via divergent evolution. N_2OR catalyzes the final step in denitrification:

$$N_2O + 2H^+ + 2e^- \Leftrightarrow N_2 + H_2O$$

In analogy to cytochrome c oxidase, N_2OR seems to use cytochrome c as electron donor [98, 99].

The second link between denitrification and oxygen-based respiration deals with nitric oxide reductase. NOR is a membrane protein complex containing b- and c-type cytochromes. It has two subunits: NorC, which is a membrane-anchored cytochrome c, and NorB, which contains protoheme [37, 100–102]. NOR catalyzes the reaction in which the N–N bond is formed during denitrification:

$$2NO + 2H^+ + 2e^- \Leftrightarrow N_2O + H_2O$$

using electrons that are probably donated by cytochrome c [98, 99].

It has been shown that the NO-reducing cytochrome bc complex is biochemically distinct from the cytochrome co terminal oxidase (reminiscent of a FixN complex) in *P. stutzeri* [100]. The first sequence data [37] showed that NorB is homologous to FixN, and consequently to Cox1 of other cytochrome oxidases [103]. The NOR complex belongs to the superfamily. The major subunit of NOR (NorB) contains 12 transmembrane segments, and all six metal binding histidines (Fig. 10.5) are present at the correct locations in its sequence. However, the bimetallic active site of NOR may contain two irons (a heme iron and a nonheme iron) rather than a heme and a copper [100]. Another difference is the lack of the residues involved in the channel of chemical protons that connects the active site to the inner surface of the membrane (Fig. 10.5; see Section 10.2.1.1).

The third connection is the close structural relatedness of the Cox2 domain with one of the three cupredoxin domains in the multicopper nitrite reductase. Although the Cox2 domain matches other cupredoxins as well, this structure gives the top score in structural comparisons [48].

Denitrifying and oxygen-reducing respiratory chains have several common features, and they differ only in the terminal part of the chain. The links between denitrification and oxygen-based respiration suggest that the latter may have evolved from the former. The first oxidase (FixN) was an enzymic adaptation that begun to reduce dioxygen instead of joining two nitric oxides into nitrous oxide. Later on this oxidase acquired the Cu_A center from N_2OR. This, together with the loss of cytochrome c domains, has led to the modern mitochondrial cytochrome c oxidase (Fig. 10.1).

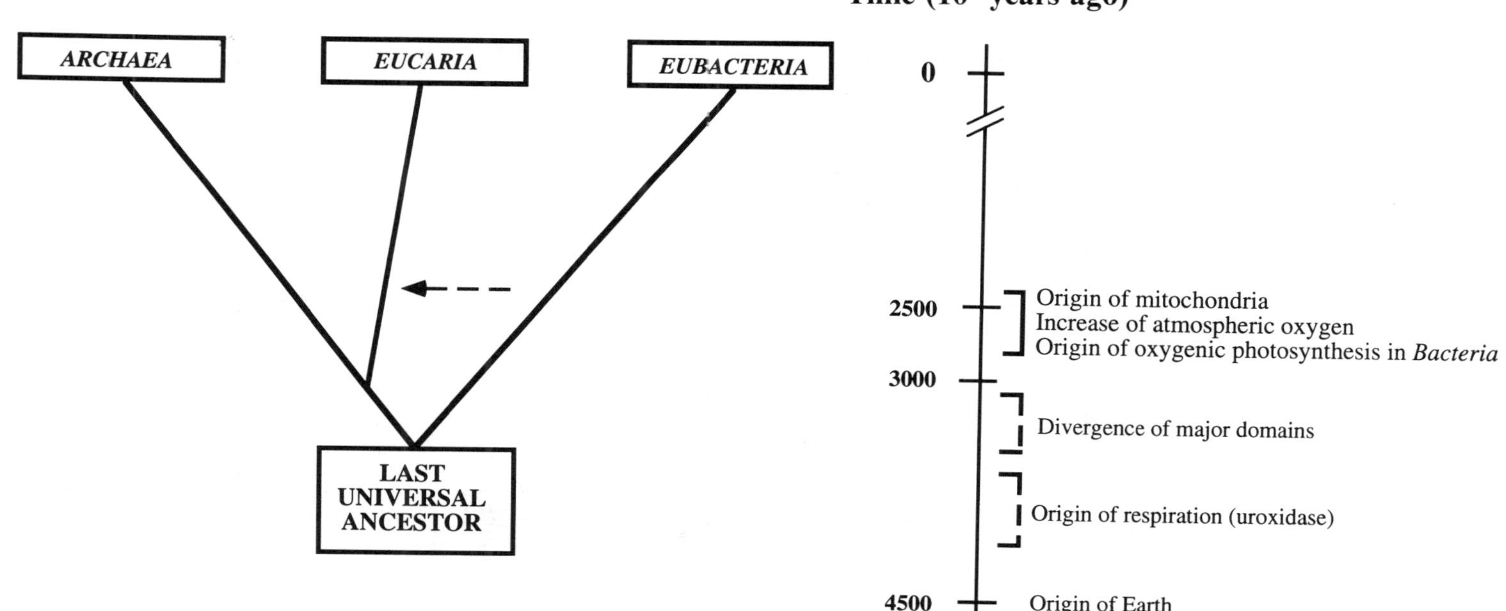

Figure 10.15 A scenario for the evolution of respiration. The first cytochrome oxidase is predicted to have existed in the common ancestor of archaea and eubacteria, before the emergence of atmospheric oxygen. The horizontal arrow marks the endosymbiontic origin of mitochondria. The dates for the divergence of the major domains and the appearance of the uroxidase are not known. Only the order of events is predicted. (From Ref. [24].)

10.6 Cytochrome Oxidase Is Older Than Atmospheric Oxygen

It has frequently been proposed that the evolution of oxygen-supported respiration was possible only after the emergence of oxygenic photosynthesis, and subsequent release of oxygen to the atmosphere [104–106]. This hypothesis leads to a view that respiration has evolved several times independently in the already diverged evolutionary lines [86, 105, 107, 108]. Our analysis of cytochrome oxidase sequences [24] does not support this polyphyletic view.

It is believed that the level of atmospheric oxygen began to rise about 2400 to 2600 million years ago [109], when the red beds containing oxidized iron started to appear, and that the present concentration of free oxygen was reached only about 500 million years ago (see Ref. [98]). The first organisms capable of performing oxygenic photosynthesis were probably ancestors of the present-day cyanobacteria [110]. Therefore, photosynthesis developed in the eubacterial lineage after the split between archaea and eubacteria (Fig. 10.15). No photosynthetic archaeon is known. In contrast, cytochrome oxidase (as well as core subunits of the cytochrome bc_1 complex) are present both in archaea and eubacteria. This indicates that the uroxidase predates the first photosynthetic systems [98, 111].

Small amounts of oxygen were generated from photolysis of water by ultraviolet radiation before the atmospheric oxygen and ozone shielded this out [112, 113]. Oxygen has a particular fitness as a terminal electron acceptor for respiration [9, 10], and oxygen-based respiration could have emerged very early, possibly evolving from anaerobic, denitrifying respiratory chain. This idea is feasible in light of the remarkable conservation of current cytochrome oxidases in archaea, eubacteria, and mitochondria. The basic structure of the active site in all oxidases may have evolved to its present-day configuration before the split between archaea and eubacteria, indicating that the function of the uroxidase was similar to the function of contemporary oxidases [114]. This proposal has to be tested by further experimental studies on the functional mechanisms of most diverged oxidases. If it is found true, it supports a monophyletic origin of respiration.

Acknowledgments

We thank Drs. So Iwata and Hartmut Michel for Figures 10.2, 10.4, and 10.9.

References

1. Ludwig, B. *FEMS Microbiol. Rev.* 1987; *46*, 41–56.

2. Saraste, M. *Q. Rev. Biophys.* 1990; *23*, 331–366.

3. Gargia-Horsman, J. A., Barquera, B., Rumbley, J., Ma, J., Gennis, R. B. *J. Bacteriol.* 1994, *176*, 5587–5600; Calhoun, M. W., Thomas, J. W., Gennis, R. B. *Trends Biochem. Sci.* 1994; *19*, 325–330.

4. Preisig, O., Anthamatten, D., Hennecke, H. *Proc. Natl. Acad. Sci. USA* 1993; *90*, 3309–3313.

5. Garcia-Horsman, J. A., Berry, E., Shapleigh, J. P., Alben, J. O., Gennis, R. B. *Biochemistry* 1994; *33*, 3113–3119; Gray, K. A., Grooms, M., Myllykallio, H., Moomaw, C., Slaughter, C., Daldal, F. *Biochemistry* 1994; *33*, 3120–3127.

6. De Gier, J-W. L., Lübben, M., Reijnders, W. N. M., Tipker, C. A., Van Spanning, R. J. M., Stouthamer, A. H., Van der Oost. *J. Mol. Microbiol.* 1994; *13*, 183–196.

7. Iwata, S., Ostermeier, C., Ludwig, B., Michel, H. *Nature* 1995; *376*, 660–669.

8. Tsukihara, T., Aoyama, H., Yamashita, E., Tomizaki, T., Yamaguchi, H., Shinzawa-Itoh, K., Nakashima, R., Yaono, R., Yoshikawa, H. *Science* 1995; *269*, 1069–1074.

9. Wikström, M., Krab, K., Saraste, M. *Cytochrome Oxidase—A Synthesis;* Academic Press: London, 1981; Haltia, T., Wikström, M. In *Molecular Mechanisms in Bioenergetics*, Ernster, L. (ed.), Elsevier, Amsterdam, 1992; pp. 217–239.

10. Babcock, G. T., Wikström, M. *Nature* 1992; *356*, 301–309.

10a. Wikström, M., Bogachev, A., Finel, M., Morgan, J. E., Puustinen, A., Raitio, M., Verhovskaya, M., Verhovsky, M. I. *Biochim. Biophys. Acta* 1994; *1187*, 106–111.

11. Brown, S., Moody, A. J., Mitchell, R., Rich, P. R. *FEBS Lett.* 1993; *316*, 216–223; Rich, P. R. *Aust. J. Plant Physiol.* 1995; *22*, 479–486.

12. Hill, B. C. *J. Bioenerg. Biomembr.* 1993; *25*, 115–120.

13. Tzagoloff, A., Dieckmann, C. L. *Microbiol. Rev.* 1990; *54*, 211–225.

14. Taanman, J.-W., Capaldi, R. A. *J. Biol. Chem.* 1993; *268*, 18754–18761.

15. Cooper, C. E., Nicholls, P., Freedman, J. A. *Biochem. Cell Biol.* 1991; *69*, 585–607.

16. Kadenbach, B., Stroh, A., Hüther, F.-J., Reimann, A., Steverding, D. *J. Bioenerg. Biomembr.* 1991; *23*, 321–334.

17. Taanman, J.-W., Capaldi, R. A. *J. Biol. Chem.* 1992; *267*, 22481–22485.

18. Anthony, G., Reimann, A., Kadenbach, B. *Proc. Natl. Acad. Sci. USA* 1993; *90*, 1652–1656.

19. Meunier, B., Lemarre, P., Colson, A.-M. *Eur. J. Biochem.* 1993; *213*, 129–135.

20. Brown, S., Colson, A.-M., Meunier, B., Rich, P. R. *Eur. J. Biochem.* 1993; *213*, 129–135.

21. Meunier, B., Coster, F., Lemarre, P., Colson, A.-M. *FEBS Lett.* 1993; *321*, 159–162.

22. Capaldi, R. A. *Annu. Rev. Biochem.* 1990; *59*, 569–596.

23. Chepuri, V., Gennis, R. B. *J. Biol. Chem.* 1990; *265*, 12978–12986.

24. Castresana, J., Lübben, M., Saraste, M., Higgins, D. G. *EMBO J.* 1994; *13*, 2516–2525.

25. Lemieux, L. J., Calhoun, M. W., Thomas, J. W., Ingledew, W. J., Gennis, R. B. *J. Biol. Chem.* 1992; *267*, 2105–2113.

26. Minagawa, J., Mogi, T., Gennis, R. B., Anraku, Y. *J. Biol. Chem.* 1992; *267*, 2096–2104.

27. Shapleigh, J. P., Hosler, J. P., Tecklenburg, M. M. J., Kim, Y., Babcock, G. T., Gennis, R. B., Ferguson-Miller, S. *Proc. Natl. Acad. Sci. USA* 1992; *89*, 4786–4790.

28. Calhoun, M. W., Hill, J. J., Lemieux, L. J., Ingledew, W. J., Alben, J. O., Gennis, R. B. *Biochemistry* 1993, *32*, 11524–11529.

29. Uno, T., Mogi, T., Tsubaki, M., Nishimura, Y., Anraku, Y. *J. Biol. Chem.* 1994; *269*, 11912–11920.

30. Henderson, R., Baldwin, J. M., Ceska, T. A., Zemlin, F., Beckmann, E., Downing, K. H. *J. Mol. Biol.* 1990; *213*, 899–929.

31. Deisenhofer, J., Michel, H. *EMBO J.* 1989; *8*, 2149–2169.

32. Lübben, M., Kolmerer, B., Saraste, M. *EMBO J.* 1992; *11*, 805–812.

33. Lübben, M., Arnaud, S., Castresana, J., Warne, A., Albracht, S., Saraste, M. *Eur. J. Biochem.* 1994; *224*, 151–159.

34. Mather, M. W., Springer, P., Hensel, S., Buse, G., Fee, J. A. *J. Biol. Chem.* 1993; *268*, 5359–5408. The accession number of the unpublished CbaA and CbaB sequences is L09121 in the EMBL data library.

35. Raitio, M., Pispa, J. T., Metso, T., Saraste, M. *FEBS Lett.* 1990; *261*, 431–435.

36. Chepuri, V., Lemieux, L., Au, D. C.-T., Gennis, R. B. *J. Biol. Chem.* 1990; *265*, 11185–11192.

37. Zumft, W. G., Braun, C., Cuypers, H. *Eur. J. Biochem.* 1994; *219*, 481–490.

38. Bell, L. C., Richardson, D. J., Ferguson, S. J. *J. Gen. Microbiol.* 1992; *138*, 437–443.

39. Thomas, J. W., Puustinen, A., Alben, J. O., Gennis, R. B., Wikström, M. *Biochemistry* 1993; *32*, 10923–10928; Garcia-Horsman, J. A., Puustinen, A., Gennis, R. B., Wikström, M. *Biochemistry* 1995; *34*, 4428–4433.

40. Taha, S. M. T., Ferguson-Miller, S. *Biochemistry* 1992; *31*, 9090–9097.

41. Lappalainen, P., Watmough, N. J., Greenwood, C., Saraste, M. *Biochemistry* 1995; *34*, 5824–5830.

42. Saraste, M., Holm, L., Lemieux, L., Lübben, M., Van der Oost, J. *Biochem. Soc. Trans.* 1991; *19*, 608–612.

43. Hosler, J. P., Espe, M. P., Zhen, Y., Babcock, G. T., Ferguson-Miller, S. *Biochemistry* 1995; *34*, 7586–7592.

44. Adman, E. T. *Adv. Prot. Chem.* 1991; *42*, 145–197.

45. Steffens, G. J., Buse, G. *Hoppe Seylers Z. Physiol. Chem.* 1979; *360*, 613–619.

46. Holm, L., Saraste, M., Wikström, M. *EMBO J.* 1987; *6*, 2819–2823.

47. Van der Oost, J., Lappalainen, P., Musacchio, A., Warne, A., Lemieux, L., Rumbley, J., Gennis, R. B., Aasa, R., Pascher, T., Malmström, B. G., Saraste, M. *EMBO J.* 1992; *11*, 3209–3217.

48. Wilmanns, M., Lappalainen, P., Kelly, M., Sauer-Eriksson, E., Saraste, M. *Proc Natl. Acad. Sci. USA* 1995; *92*, 11955–11959.

49. Dennison, C., Vijgenboom, E., de Vries, S., van der Oost, J., Canters, G. W. *FEBS Lett.* 1995; *365*, 92–94.

50. Raitio, M., Jalli, T., Saraste, M. *EMBO J.* 1987; *6*, 2825–2833.

51. Viebrock, A., Zumft, W. G. *J. Bacteriol.* 1988; *170*, 4658–4668.

52. Malmström, B. G., Aasa, R. *FEBS Lett.* 1993; *325*, 49–52; Lappalainen, P., Saraste, M. *Biochim. Biophys. Acta* 1994; *1187*, 222–225.

53. Kroneck, P. M. H., Antholine, W. A., Riester, J., Zumft, W. G. *FEBS Lett.* 1988; *242*, 70–74.

54. Scott, R. A., Zumft, W. G., Coyle, C. L., Dooley, D. M. *Proc. Natl. Acad. Sci. USA* 1989; *86*, 4082–4086.

55. Antholine, W. E., Kastrau, D. H. W., Steffens, G. C. M., Buse, G., Zumft, W. G., Kroneck, P. M. H. *Eur. J. Biochem.* 1992, 209, 875–881; Farrar, J. A., Lappalainen, P., Zumft, W. G., Saraste, M., Thomson, A. *Eur. J. Biochem.* 1995; *232*, 294–303.

56. Zumft, W. G., Dreutsch, A., Löchelt, S., Cuypers, H., Friedrich, B., Schneider, B. *Eur. J. Biochem.* 1992; *208*, 31–40.

57. Von Wachenfeldt, C., de Vries, S., van der Oost, J. *FEBS Lett.* 1994; *340*, 109–113.

58. Kelly, M., Lappalainen, P., Talbo, G., Haltia, T., Van der Oost, J., Saraste, M. *J. Biol. Chem.* 1993; *268*, 16781–16787.

59. Lappalainen, P., Aasa, R., Malmström, B. G., Saraste, M. *J. Biol. Chem.* 1993; *268*, 26416–26421.

60. Haltia, T., Finel, M., Harms, N., Nakari, T., Raitio, M., Wikström, M., Saraste, M. *EMBO J.* 1989; *8*, 3571–3579.

61. Brunori, M., Antonini, G., Malatesta, F., Sarti, P., Wilson, M. T. *Eur. J. Biochem.* 1987; *169*, 1–8.

62. Haltia, T., Saraste, M., Wikström, M. *EMBO J.* 1991; *10*, 2015–2021.

63. Castresana, J., Lübben, M., Saraste, M. *J. Mol. Biol.* 1995; *250*, 202–210.

64. Ishizuka, M., Machida, K., Shimada, S., Mogi, A., Tsuchiya, T., Ohmori, T., Souma, Y., Gonda, M., Sone, N. *J. Biochem.* 1990; *108*, 866–873.

65. Saraste, M., Metso, T., Nakari, T., Jalli, T., Lauraeus, M., van der Oost, J. *Eur. J. Biochem.* 1991; *195*, 517–525; Santana, M., Kunst, F., Hullo, M. F., Rapoport, G., Danchin, A., Glaser, P. *J. Biol. Chem.* 1992; *267*, 10225–10231.

66. Quirk, P. G., Hicks, D. B., Krulwich, T. A. *J. Biol. Chem.* 1993; *268*, 678–685.

67. Alge, D., Peschek, G. A. *Biochim. Biophys. Res. Commun.* 1993; *191*, 9–17.

68. Sone, N., Tano, H., Ishizuka, M. *Biochim. Biophys. Acta* 1993; *1183*, 130–138.

69. Ma, J., Lemiuex, L., Gennis, R. B. *Biochemistry* 1993; *32*, 7692–7697.

70. Puustinen, A., Finel, M., Haltia, T., Gennis, R. B., Wikström, M. *Biochemistry* 1991; *30*, 3936–3942.

71. Caughey, W. S., Smythe, G. A., O'Keefe, D. H., Maskasky, J. E., Smith, M. L. *J. Biol. Chem.* 1975; *250*, 7602–7622.

72. Wu, W., Chang, C. K., Varotsis, C., Babcock, G. T., Puustinen, A., Wikström, M. *J. Am. Chem. Soc.* 1992; *114*, 1182–1187.

73. Svensson, B., Lübben, M., Hederstedt, L. *Mol. Microbiol.* 1993; *10*, 193–201; Mogi, T., Saiki, K., Anraku, Y. *Mol. Microbiol.* 1994; *14*, 391–398.

74. Saiki, K., Mogi, T., Anraku, Y. *Biochem. Biophys. Res. Commun.* 1992; *189*, 1491–1497.

75. Nobrega, M. P., Nobrega, F. G., Tzagoloff, A. *J. Biol. Chem.* 1990; *265*, 14220–14226.

76. Lübben, M., Warne, A., Albracht, S., Saraste, M. *Mol. Microbiol.* 1994; *13*, 327–335.

77. Lübben, M., Morand, K. *J. Biol. Chem.* 1994; *269*, 21473–21479.

78. Sone, N., Fujiwara, Y. *FEBS Lett.* 1991; *288*, 154–158.

79. Matsushita, K., Ebisuya, H., Adachi, O. *J. Biol. Chem.* 1992; *267*, 24748–24753.

80. Puustinen, A., Morgan, J. E., Verkhovsky, M., Thomas, J. W., Gennis, R. B., Wikström, M. *Biochemistry* 1992; *31*, 10363–10369.

81. Lauraeus, M., Haltia, T., Saraste, M., Wikström, M. *Eur. J. Biochem.* 1991; *197*, 699–705.

82. Van der Oost, J., von Wachenfedt, C., Hederstedt, L., Saraste, M. *Mol. Microbiol.* 1991; *5*, 2063–2072.

83. Anemüller, S., Schäfer, G. *Eur. J. Biochem.* 1990; *191*, 297–305.

84. Denda, K., Fujiwara, T., Seki, M., Yoshida, M., Fukumori, Y., Yamanaka, T. *Biochem. Biophys. Res. Comm.* 1991; *181*, 316–322.

85. Anemüller, S., Bill, E., Schäfer, G., Trautwein, A. X., Teixeira, M. *Eur. J. Biochem.* 1992; *210*, 133–138.

86. Margulis, L. *Symbiosis in Cell Evolution. Life and Its Environment on The Early Earth*, Freeman, San Francisco, 1981.

87. Ochman, H., Wilson, A. C. *J. Mol. Evol.* 1987; *26*, 74–86.

88. Moran, N. A., Munson, M. A., Baumann, P., Ishikawa, H. *Proc. R. Soc. Lond. B* 1993; *253*, 167–171.

89. Woese, C. R. *Microbiol. Rev.* 1987; *51*, 221–271.

90. Cedergren, R., Gray, M. W., Abel, Y., Sankoff, D. *J. Mol. Evol.* 1988; *28*, 98–112.

91. Batut, J., Daveran-Mingot, M.-L., David, M., Jacobs, J., Garnerone, A. M., Kahn, D. *EMBO J.* 1989; *8*, 1279–1286.

91a. Raitio, M., Wikström, M. *Biochim. Biophys. Acta* 1994; *1186*, 100–106.

91b. De Gier, J.-W. PhD Thesis, Free University Amsterdam, 1995.

92. Bott, M., Preisig, O., Hennecke, H. *Arch. Microbiol.* 1992; *158*, 335–343.

93. Bergersen, F. R. S., Turner, G. L. *Proc. R. Soc. Lond. B* 1990; *238*, 295–320.

94. Zimmerman, B. H., Nitsche, C., Fee, J. A., Rusnak, F., Muenck, E. *Proc. Natl. Acad. Sci. USA* 1988; *85*, 5779–5783.

95. Musser, S. M., Stowell, M. H. B., Chan, S. I. *FEBS Lett.* 1993; *327*, 131–136.

96. Haltia, T. *FEBS Lett.* 1993; *335*, 294–295.

97. Musser, S. M., Stowell, M. H. B., Chan, S. I. *FEBS Lett.* 1993; *335*, 296–298.

98. Stouthamer, A. H. *Antonie van Leeuwenhoek*, 1992; *61*, 1–33.

99. Nicholls, D. G., Ferguson, S. J. *Bioenergetics 2*, Academic Press, London, 1992.

100. Heiss, B., Frunzke, K., Zumft, W. G. *J. Bacteriol.* 1989; *171*, 3288–3297.

101. Carr, G. J., Ferguson, S. J. *Biochem. J.* 1990; *269*, 423–429.

102. Dermantia, M., Turk, T., Hollocher, T. C. *J. Biol. Chem.* 1991; *266*, 10899–10905.

103. Saraste, M., Castresana, J. *FEBS Lett.* 1994; *341*, 1–4; Van der Oost, J., de Boer, A. P. N., de Gier, J.-W., Zumft, W. G., Stouthamer, A. H., van Spanning, R. J. M. *FEMS Microbiol. Lett.* 1994; *121*, 1–10.

104. Dickerson, R. E., Timkovich, R., Almassy, R. J. *J. Mol. Biol.* 1976; *100*, 473–491.

105. Broda, E., Peschek, G. A. *J. Theor. Biol.* 1979; *81*, 201–212.

106. Moore, G. R., Pettigrew, G. W. *Cytochromes c. Evolutionary, Structural and Physiochemical Aspects*, Springer-Verlag, Berlin, 1990.

107. Fox, G. E., Stackebrandt, E., Hespell, R. B., Gibson, J., Maniloff, J., Dyer, T. A., Wolfe, R. S., Balch, W. E., Tanner, R. S., Magrum, L. J., Zablen, L. B., Blackmore, R., Gupta, R., Bonen,

L., Lewis, B. J., Stahl, D. A., Luehrsen, K. R., Chen, K. N., Woese, C. R. *Science* 1980; *209*, 457–463.

108. Buse, G., Steffens, G. C. M. *J. Bioenerg. Biomembr.* 1991; *23*, 269–289.

109. Knoll, A. H. *Science* 1992; *256*, 622–627.

110. Margulis, L., Walker, J. C. G., Rambler, M. *Nature* 1976; *264*, 620–624.

111. Buse, G., Steffens, G. C. M., Fee, J. D. *Eur. J. Biochem.* 1989; *181*, 261–268; Blankenship, R. E. *Photosynth. Res.* 1992; *33*, 91–111.

112. Sagan, C., Thompson, W. R., Carlson, R., Gurnett, D., Hord, C. *Nature* 1993; *365*, 715–721.

113. Kasting, J. F. *Science* 1993; *259*, 920–926.

114. Castresana, J., Saraste, M. *Trends Biochem. Sci.* 1995; *20*, 443–448.

Evolution of V- and F-ATPases

*Lincoln Taiz and Nathan Nelson**

11.1 Introduction

Proton ATPases are among the most fundamental proteins that function in biological energy transduction. There are two related families of H^+-ATPases, V-ATPases and F-ATPases, that share similar structure and function. They are composed of distinct catalytic and membrane sectors, each containing multiple subunits [1–4]. They are present in every living cell and have the potential of functioning in ATP synthesis at the expense of energy provided by respiration and photosynthesis. F-ATPases function in the plasma membrane of eubacteria and in the mitochondria and chloroplasts of eukaryotes. V-ATPases function in the plasma membrane of archaebacteria and in the vacuolar system of eukaryotes. Because V- and F-ATPases are present in every known living cell they can be used as evolutionary markers along with ribosomal RNA and RNA polymerases. Conversely, understanding the evolution of the three kingdoms of eubacteria, archaebacteria, and eukaryota may shed light on the function of these proton pumps in the wide variety of organelles and membranes that they energize. The antiquity of these enzymes helps us glimpse into early events in their evolution and in the evolution of the different organelles in eukaryotic cells.

11.2 Structure–Function Relations Among the Various V- and F-ATPases

Figure 11.1 depicts the minimal structure of F- and V-ATPases. The *E. coli* enzyme is taken as a model for F-ATPases and the model for V-ATPases contains subunits that are likely to be present in all the enzymes of this family including the plasma membrane of archaebacteria and eukaryotes as well as the vacuolar system of eukaryota. The catalytic sectors of both families are composed of five different subunits. The catalytic site is located on the β-subunit and the α-subunit contains an additional ATP binding site, the function of which is unclear. The γ, δ- and ε-subunits function in the proper attachment of the catalytic sector onto the membrane and are vital for coupling between the ATPase and proton conduction activities of the enzyme [5–8]. The γ-subunit is likely to play a key role in energy coupling but its mechanism of action is still an enigma. In V-ATPases the A-subunit contains the catalytic site of the enzyme and the B-subunit contains the additional ATP binding site [9–14]. The roles of subunits C, D, and E have not been defined as yet and even the essentiality of subunit D for the function of V-ATPases has not been proven [15–17]. While subunits α and β of F-ATPases are functionally and structurally related to the respective B- and A-subunits, respectively, of the V-ATPases, the relationships between the other subunits are not apparent

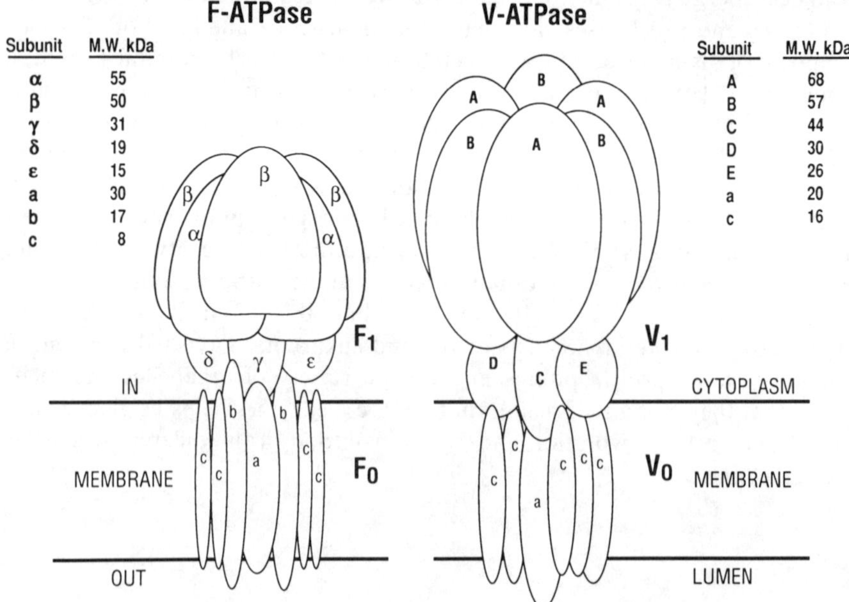

Figure 11.1 Schematic depiction of the minimal subunit structure of F- and V-ATPases.

[1–3]. The stoichiometry of the various subunits of F- and V-ATPases is similar and the respective catalytic sectors are composed of $\alpha_3, \beta_3, \gamma, \delta, \varepsilon$ or A_3, B_3, C, D, E [10, 18].

The structure, function, and subunit composition of the membrane sectors are much more complicated. The membrane sector of F-ATPase from *E. coli* contains three different subunits denoted as a, b, and c (proteolipid). Subunits a and c are directly involved in proton conduction across the membrane and subunit b is more involved in connecting the two sectors and energy coupling [19]. The stoichiometry of these subunits was determined to be ab_2c_{9-12} [20, 21]. The membrane sector of the mammalian enzyme may contain about six additional subunits [22, 23]. The structure and function of individual subunits in the membrane sector of V-ATPases is less clear. The only subunit proven to be functional is the proteolipid (subunit c) [24–26]. Additional subunits with molecular weights of 39 and 115 kDa have been identified in most preparations of V-ATPases from eukaryotic cells [27–29]. A hydrophobic subunit of about 20 kDa was identified in some mammalian enzymes [10, 15–17], and two smaller polypeptides, of 12 kDa and 13 kDa, have been detected in plant V-ATPases [30]. The proteolipid is almost certain to function in proton conduction. The proteolipids of F- and V-ATPases are structurally and functionally related [31, 32].

The main function of F-ATPases is to synthesize ATP at the expense of the protonmotive force (pmf) generated by the photosynthetic and respiratory electron transport chains. However, under anaerobic conditions and certain metabolic circumstances F-ATPases may function as an H^+-pumping ATPase, generating pmf at the expense of ATP. A similar if not identical function is fulfilled by V-ATPases in archaebacteria. Since F-ATPases function in chloroplasts and mitochondria we assume that they are more efficient in an oxygenic environment than the V-ATPases. The reason for the presence of a V-ATPase in archaebacteria for energy conversion is not known. This may have to do with the fact that they are not as successful in an evolutionary sense as eubacteria [33]. On the other hand, V-ATPase may be advantageous under extreme conditions of low pH, high salinity, and elevated temperatures [34, 35]. The vacuolar system of eukaryotic cells may have exploited the relative inefficiency of V-ATPases to gain better control over the extent of acidification of the various internal organelles [2, 36]. We proposed that this control is governed by a proton slip resulting in the inability of these V-ATPases to synthesize ATP [37].

11.3 Early Events in the Evolution of F- and V-ATPases

It is logical to assume that the ancestral proton-ATPase that evolved into F- and V-ATPases was composed of a single gene product [4]. Because pseudosixfold symmetry is maintained in all current enzymes of these families, we suggested that a hexamer of that ancestral protein comprised the functional enzyme [2–4]. This protein should have contained three structural elements: a globular part with six ATP binding sites, each subunit containing one catalytic site; a membrane sector capable of conducting protons across the membrane; and a connecting element functioning in energy transduction. In constructing the membrane component, the most important consideration would be whether the proteolipid by itself was capable of fulfilling the function of the entire channel. The current F-ATPases have a minimum of three subunits in the membrane sector, all of which are required for proton conduction across the membrane [19]. However, the known operons encoding V-ATPases in archaebacteria contain six genes, only one of which is hydrophobic enough to function in the membrane sector [38]. If there are no other genes encoding V-ATPase subunits, the proteolipid may be the only subunit in the membrane sector of the archaebacterial enzyme. However, we believe that at least one more subunit analogous to subunit a of the *E. coli* F-ATPase may be functional in V-ATPases.

Figure 11.2 shows a schematic proposal for a single gene product containing all of the above features: a globular component with six ATP binding sites, a connecting segment and a membrane sector containing the features of the current proteolipids, and charged elements of subunit a. In this scheme the original gene subdivided early in evolution into six separate catalytic subunits, several "stalk" subunits, and a complex of six hydrophobic subunits comprising the channel.

Because the catalytic (A, β) and noncatalytic (B, α) subunits of the current V- and F- ATPases are related to each other, that is, derived from a gene

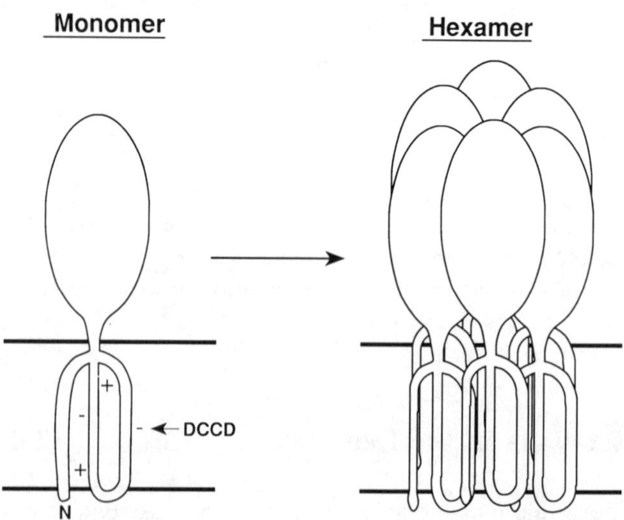

Figure 11.2   A proposed single gene product functions as an ancestral ATP-dependent proton pump.

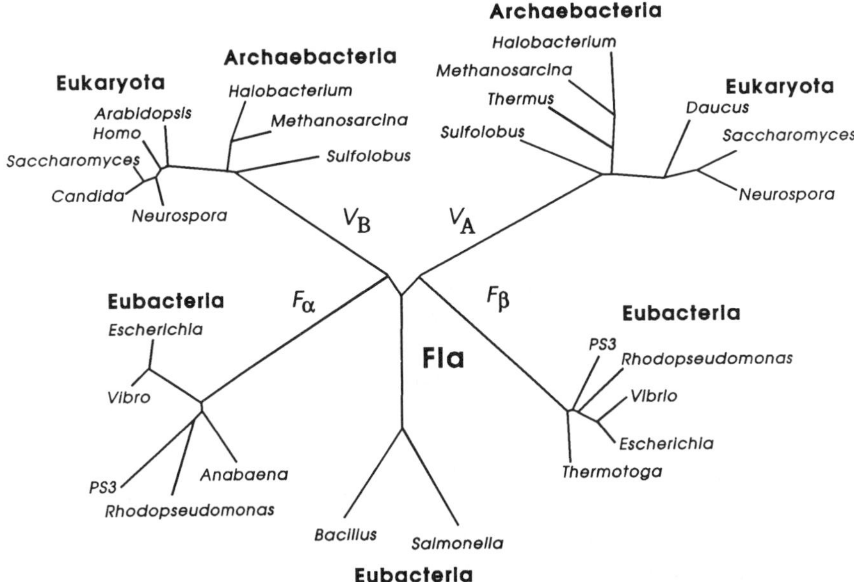

Figure 11.3 Phylogenetic relationships between the catalytic and noncatalytic subunits of the F-, V-, and archaebacterial-type H^+-ATPase-like flagellar protein of eubacteria. The unrooted tree was derived from pairwise comparisons of the amino acid sequences and calculated from a distance matrix. V_A and V_B are the catalytic and noncatalytic subunits, respectively, of the V- and archaebacterial type ATPases; F_α and F_β are the noncatalytic and catalytic subunits, respectively, of the F-ATPases; Fla is the ATPase-like flagellar protein of eubacteria. [From Gogarten et al. (1992).]

duplication, we can assume that the ancestral ATPase contained six identical catalytic subunits encoded by a single gene. Following the gene duplication, one of the subunits lost its catalytic function and became a regulatory subunit. The resulting stoichiometry of three catalytic: three noncatalytic subunits preserved the total mass of the complex. The universality of this arrangment of the complex suggests that the gene duplication ocurred prior to the divergence from the common ancestor. This feature of the ATPases has been exploited in evolutionary studies to root the tree of life (Fig. 11.3). The tree can be rooted at the point where the two "halftrees" for the catalytic and noncatalytic subunits join [39]. Since the archaebacterial subunits clearly group with their eukaryotic counterparts, the tree is rooted between the eubacteria on one side and the common ancestor of the eukaryotes and archaebacteria on the other. The sequence analysis is further supported by the presence in the catalytic subunits of archaebacteria and eukaryotes of a "nonhomologous region" that is completely absent from the eubacteria [11]. This region thus appears to represent an evolutionary insert that extended the length of the catalytic subunit by about 20 kDa, making it larger than the regulatory subunit. In the F-ATPases, the catalytic subunit is shorter than the regulatory subunit.

Gene duplication has also played an important role in the evolution of the *c* subunit. The *c* subunit of eubacteria is an 8-kDa proteolipid consisting of two membrane-spanning helices [40, 41]. An aspartate residue on the second helix binds the inhibitor DCCD and is thought to undergo protonation and deprotonation during H^+ transport. The c subunit (proteolipid) of the eukaryotic V-ATPase is a 16-kDa proteolipid composed of four membrane-spanning helices [31, 32]. A total of six copies are present per enzyme, preserving the overall mass of the complex [24]. Both the N-terminal and C-terminal halves are homologous to F-ATPase c subunits and are related to each other, but only the fourth helix contains the essential DCCD-reactive residue E^{139}. In contrast, the c-subunit of *Sulfolobus* consists of only two membrane-spanning helices [42, 43]. Thus it appears that the gene duplication event that gave rise to the 16-kDa c-subunit is associated with the divergence of eukaryotes from the archaebacteria [31, 32, 44, 45]. Figure 11.4 shows the evolutionary tree derived from the amino acid sequences of proteolipids from various sources, using the N-terminal and C-terminal halves of the 16-kDa proteolipid as separate entries for tree construction. It is shown that the gene duplication and fusion event occurred somewhat later than the divergence of eubacteria and archaebacteria. The N-terminal half of the V-ATPase c-subunit apparently evolved at a faster rate than the C-terminal half, as indicated by the longer branch length. The C-terminal half, with its essential DCCD binding

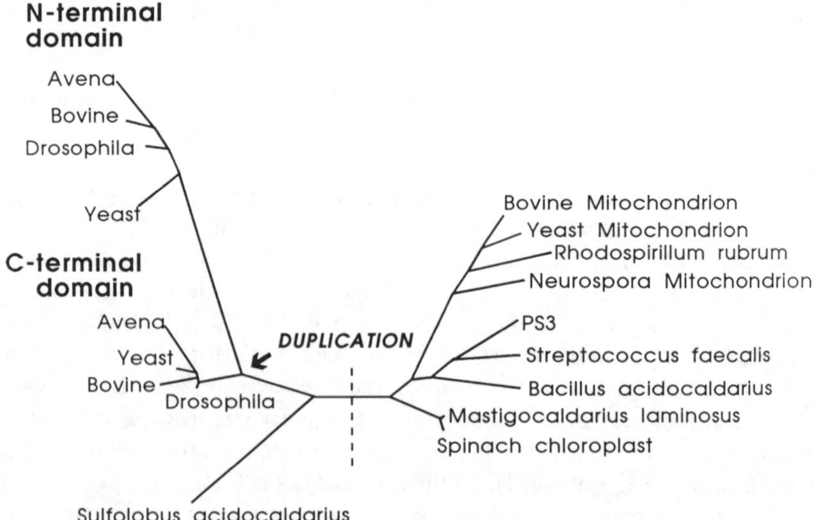

Figure 11.4 Phylogenetic tree of the proteolipids of the F-, V-, and archaebacterial ATPases. The gene duplication event that gave rise to N- and C-terminal halves of the V-type 16-kDa proteolipid is indicated by the arrow. [From Gogarten et al. (1992).]

glutamate residue on the fourth helix, is probably more intimately involved in proton transport and therefore under greater selection pressure.

The evidence that gene duplication has played a central role in the evolution of the ATPases is overwhelming. The evolutionary driving force or selection pressure for these changes is more difficult to explain. In this regard, it has been pointed out that the enzyme had undergone two significant reversals in function that roughly coincide with the two gene duplications [46]. It first evolved under anaerobic conditions and functioned as an ATP-driven proton pump. Following the evolution of photosynthesis, it reversed directions and became an ATP synthase. Finally, it reverted to being a proton pump when it became internalized on the vacuoles of eukaryotes.

For an ATPase to function as a pump, the ΔG of ATP hydrolysis must be greater (more negative) than the quantity $-n(\Delta\mu H^+)$, where n is the H^+/ATP stoichiometry of the enzyme. Because of this relationship, a small n favors a pump, whereas a large n favors a synthase. We assume that n was low in the original ATPase, increased when the ATPase became a synthase, and was reduced again when it reverted to being a pump. This assumption is consistent with the known H^+/ATP stoichiometries of the contemporary enzymes: three or four for the F-ATPases and two for the eukaryotic V-ATPases [1–8]. It was further suggested that the H^+/ATP stoichiometry can be approximated by the ratio of the number of proteolipid subunits to the number of catalytic subunits [46]. The original ATPase, with its six proteolipid subunits and six catalytic subunits, would have an $n = 1$; the F-ATPases and archaebacterial ATPases, with 10 to 12 c-subunits and three catalytic subunits, would have $n = 4$; whereas the vacuolar ATPase, with its six c-subunits and three catalytic subunits, would again have an $n = 2$. In this way, gene duplications may have served to alter the ratios of the two types of subunits, shifting the direction of the reaction, as required, toward either proton pumping or ATP synthesis. Regardless of the precision of these numbers the general correlation between the evolution of the membrane sector and the function of the enzyme in ATP synthesis or proton pumping is overwhelming.

11.4 Lateral Gene Transfer

Although the vast majority of eubacteria thus far examined contain F-ATPases, at least one exception has been reported. *Thermus thermophilus*, an authentic eubacterium based on rRNA analyses, has been shown to contain a V-type ATPase [47]. An alternative model for ATPase evolution was proposed on the basis of this finding. It was argued that the common ancestor contained both the F-type and the V-type ATPases, and that one or the other was selectively lost during evolution [47]. However, phylogenetic analysis of the A-subunit of this enzyme demonstrated that it clearly groups with the archaebacteria [44]. If the V-ATPase of *Thermus* were very ancient, that is, present in the last common ancestor, it should be equally related to archaebacteria and eukaryotes. Because it groups with the archaebacteria, the V-ATPase of *Thermus* seems to be a clear case of lateral gene transfer from an archaebacterium to a eubacterium.

Another apparent exception to the rule, *Enterococcus* [48], appears to contain both an F-type ATPase and a V-type enzyme. The latter is believed to function as a sodium pump. Other exceptions are bound to be discovered. It will be important to subject these to rigorous phylogenetic analysis to distingush between the two models of ATPase evolution.

11.5 Relations Among Other Subunits of F- and V-ATPases in Prokaryotes and Eukaryotes

While the A, B, and proteolipid subunits of archaebacterial V-ATPases show an apparent common ancestry with the corresponding subunits in the vacuolar system of eukaryotes, the remaining three subunits show no sequence similarity to any subunit of the eukaryotic V-ATPases [38, 49–52]. This may have resulted from two different evolutionary pathways: either the subunits of prokaryotes evolved independently from those of the eukaryotes, or they evolved from common ancestors that underwent rapid evolutionary changes, eliminating any sequence similarity among the functionally related subunits. Our view on this matter depends on the origin of the vacuolar system. If the vacuolar system evolved as early as the divergence between archaebacteria and eukaryota, it is more likely that those V-ATPases subunits evolved independently of each other. However, if a symbiotic archaebacterium contributed to the development of the vacuolar system, it is likely that some of these subunits of archaebacteria and eukaryota had common ancestors. It is worth pointing out that the analogous subunits in F-ATPases are also less well conserved, presumably because they have structural rather than catalytic function. However there is enough sequence homology for recognizing the corresponding subunit in eubacteria, chloroplasts, and mitochondria [5–8, 23].

It is widely accepted that chloroplasts and mitochondria evolved from symbiotic eubacteria [53, 54]. Therefore, their F-ATPases are clearly related and sequence similarities exist among functionally related subunits of eubacteria, chloroplasts, and mitochondria [4–8]. Studies on the evolutionary relationships among F-ATPase subunits have provided insights into the evolution of the organelles. A prime example for such a pathway is the evolution of the genes encoding subunits I and II of the membrane sector (CF_0) of the chloroplast F-ATPase. It was demonstrated that whereas subunit I is a chloroplast gene product, subunit II is synthesized on cytoplasmic ribosomes as a larger precursor and transported into the chloroplasts before assembling into the enzyme [55]. Sequencing information revealed that both subunits I and II of chloroplasts are related to subunit b of *E. coli* and most probably evolved from a common ancestor [56–58]. Cloning and sequencing the operons encoding F-ATPases in purple and cyanobacteria provided the information for the intermediate stage of this evolutionary pathway [57]. In these bacteria there are two genes encoding subunit b-like proteins. They probably evolved by internal gene duplication within the operon. The chloroplast subunit I evolved from a gene related to the cyanobacterial b and subunit II from a gene related to subunit b' [57, 58]. Subsequently the gene encoding subunit II was transferred from the chloroplast to the nucleus of plants and algae. This suggests that the transfer of the b gene from the chloroplasts to the nucleus took part prior to the segregation of plants and algae. Mechanistically this advancement in evolution conferred the *E. coli*-like enzyme with a potential twofold symmetry into an inherently asymmetric enzyme in cyanobacteria and chloroplasts. This deviation from symmetry may have evolved for gaining better control over partial activities of the enzyme.

11.6 Structure and Evolution of Operons

Only one operon containing genes encoding V-ATPase subunits has been cloned and sequenced [38]. The operon of *Sulfolobus acidocaldarius* contains six open reading frames encoding potential subunits of the enzyme (Fig. 11.5).

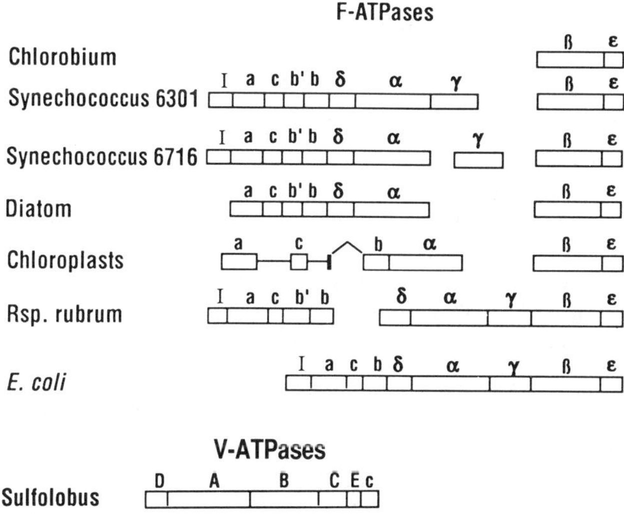

Figure 11.5 Structure of operons encoding F- and V-ATPases in different organisms.

Three of these genes encode proteins homologous to subunits A, B and the proteolipid of the enzyme of eukaryotic cells [38]. The order of the genes in this operon has no relation to the order of the homologous genes in operons encoding F-ATPase subunits, and therefore it is likely that the archaebacterial operons evolved independently from the eubacterial ones. On the other hand, the order of F-ATPase genes in the operons of eubacteria and chloroplasts has been strictly maintained [8]. Thus it is likely that all of these operons evolved from a common ancestor. This assumption implies that the F-ATPase operon(s) are very old and became organized during the early evolution of the eubacterial kingdom. In *E. coli* all genes encoding the enzyme subunits are organized into a single operon [6, 7, 60]. The gene order reflects the structure of the complex such that genes encoding subunits of the membrane sector (F_0) are clustered in the beginning of the operon, followed by genes encoding the catalytic sector (F_1). The same arrangement was found in operons encoding F-ATPases of other eubacteria such as PS3 [61], *Bacillus megaterium* [62], *Bacillus firmus* [63], and *Vibrio alginolyticus* [64]. In purple or cyanobacteria the genes are present in two separate operons [57, 58]. Although in the purple nonsulfur bacterium *Rhodospirillum rubrum* the two operons are divided according to the function of the gene products in the membrane or the catalytic sector, in cyanobacteria *Synechococcus* 6301, *Anabaena* 7120, and *Synechocystis* 6803 one of the operons contains the genes for the β- and ε-subunits and the other operon harbors the remainder of the genes [57, 58, 65–67]. Separate operons for the β- and ε-subunits can also be found in chloroplasts and in the green bacterium *Chlorobium limicola* [68]. This arrangement of a separate operon encoding β- and ε-subunits is present in all organisms having a photosystem I — like reaction center [69]. There is no apparent explanation for this correlation. In chloroplasts of higher plants the genes encoding subunit II of the membrane sector and subunits γ and δ of the catalytic sector are situated in the nucleus [8]. Intermediate steps in the gene transfer were identified in cyanobacteria and the plastids of red alga and diatoms. Recently it was shown that the thermophilic cyanobacterium *Synechococcus* 6716 contains three operons encoding F-ATPase subunits [70]. In this cyanobacterium the gene encoding the γ-subunit is present on a separate operon. This evolutionary event may represent the first step in the transfer of the γ-subunit gene. In the plastid of the red alga *Antithamnion* an operon encoding the a, cb′-, bδ- and α-subunits was identified [71]. A similar operon was found in the plastid of the diatom *Odontella sinensis* along with a second operon encoding the β- and ε-subunits [72]. This is an apparent evolutionary intermediate in which the γ-subunit is probably encoded by a nuclear gene but the δ-subunit is still encoded on plastid DNA. Further sequencing of operons encoding V- and F-ATPase subunits are likely to reveal more intermediate steps in the evolution of their corresponding operons. The question of the origin of these operons may be addressed by cloning and sequencing the genes encoding F-ATPase subunits in ancient thermophilic eubacteria recently discovered [73].

11.7 Evolution of Mitochondrial F-ATPase

While F-ATPase of *E. coli* is composed of eight subunits, enzyme preparations from mammalian mitochondria contain up to 14 polypeptides [6, 7, 22, 23]. Unlike the chloroplast F-ATPase the mitochondrial enzyme did not leave apparent evolutionary marks and there are no operons encoding its subunits in mitochondria. It is not known whether the operon structure was lost during the evolution of mitochondria or the first mitochondria preceded the operon formation in eubacteria. We will compare the various polypeptide compositions of the F-ATPase from *E. coli*, chloroplasts, and bovine mitochondria. Subunits α, β, and γ of the two enzymes show sufficient similarity to assume a common ancestor for each one [23]. However, the δ-subunit of *E. coli* has no homology to the bovine δ-subunit, and it shows similarity to OSCP that also has a similar function to the δ-subunits of the *E. coli* and chloroplast F-ATPases [23]. The mitochondrial δ-subunit may be related to the ε-subunit of the *E. coli* enzyme but not to the chloroplast one [8, 23]. The ε-subunit of the chloroplast F-ATPase functions as an ATPase inhibitor and converts the enzyme into a latent ATPase preventing excessive ATP hydrolysis in the dark [8]. A specific polypeptide (ATPase inhibitor) functions in mitochondria, but has no sequence homology to the chloroplast ε-subunit, and they probably evolved independently [74]. Thus the ε-subunit of F-ATPases may have been a later addition to the various enzymes conferring specificity to their function in the different organisms and organelles.

The location of the genes encoding the various F-ATPase subunits is also indicative of evolutionary events. It is believed that mitochondria and chloroplasts originated from prokaryotic ancestors. Some or all genes encoding the catalytic sector of F-ATPases were transferred to the nucleus in the course of evolution. The frequency of gene transfer from organelles to nucleus is relatively high, and therefore there was ample time for all to be translocated to the nucleus [75]. This being the case, why are all mammalian genes encoding subunits of the catalytic sector situated in the nucleus, whereas the α-subunit of plant mitochondria is organellar and in most chloroplasts only the genes encoding γ- and δ-subunits were transferred to the nucleus? There is no apparent answer to this question.

The most unique part of F-ATPases is their membrane sector. Whereas prokaryotic and chloroplast enzymes contain three or four subunits, the mitochondrial membrane sector may contain up to eight different polypeptides [22, 23]. The proteolipid is the only subunit with a clear homology and a common ancestry in all F-ATPases. Subunits a (ATPase-6) and b exhibit very poor identity with the bacterial and chloroplast corresponding subunits, leaving some doubts about their common ancestors [23]. The origin of OSCP may have been the same as that of the δ-subunit, but all other subunits of the mitochondrial membrane sectors A6L, d, and e may have evolved from independent sources. The conservation of the amino acid sequences of proteolipids in F-ATPases from various sources may be in correlation with the number of subunits in the membrane sector. Whereas the amino acid sequences of proteolipids from cyanobacteria and higher plant chloroplasts are highly conserved (85% to 90% identity), those of proteolipids from eubacteria and mitochondria are poorly conserved (20% to 25% identity). This may be the result of the added subunits to the mitochondrial membrane sector that relaxed the strict conservation among the proteolipids. The requirement of greater versatility for the mitochondrial F-ATPase may have been the evolutionary pressure for adding more subunits to the membrane sector. In addition there are gene products necessary for the correct assembly of the mitochondrial enzyme. These proteins are present in substoichiometric amounts and therefore are not considered subunits [76].

So far more than 15 gene products were identified as subunits of mitochondrial F-ATPase and/or specifically involved in its assembly. Thus on the molecular level there is little doubt that the advancement in evolution makes multisubunit membrane proteins more complex. Within the protein complex the subunits directly involved in the catalytic activity are strictly conserved. Other subunits are much less conserved but allow tracing of common ancestors. Some of the subunits evolved independently or were modified beyond recognition. Therefore, tracing evolutionary trends among these subunits in F-ATPases from various sources is very difficult. It is apparent that lateral gene transfer did not cease in latter stages of evolution. It may have dominated the events of early evolution, but continued in higher organisms through viral infections and other vehicles. That way many of the numerous pseudogenes were introduced to the mammals genome. A pseudogene of the F-ATPase proteolipid in sheep was estimated to be only 8 million years old [77]. The lack of operons encoding mitochondrial F-ATPase subunits leave distance matrix analyses as the method of choice for evolutionary tracing.

11.8 Conclusion

In conclusion, the energy-transducing H^+-ATPases appear to have evolved from relatively simple proteins comprised of one to several subunits, to highly complex protein complexes involving as many as 15 different subunits. Gene splitting may have given rise to the assembly of smaller subunits, whereas gene duplication events resulted in the paralogous catalytic and noncatalytic subunits, as well as the larger proteolipid of the V-ATPases. Specialization for function has undoubtedly been the driving force for evolution. Initially functioning as a proton pump on the plasma membrane of the prokaryotic common ancestor living under reducing conditions, the enzyme reversed direction and became an ATP synthase. Finally, the ATP synthase reverted to being a proton pump when it was internalized on the vacuolar membranes of eukaryotes. It is likely that the structural changes associated with these reversals in function affected the H^+/ATP stoichiometry of the ATPase. Other evolutionary changes must have been driven by the need for regulatory functions in their particular organelles, which probably accounts for the higher degree of diversity among the minor subunits than among the major ones. Since relatively few organisms have been characterized as yet, particularly among the prokaryotes, further complexities in the evolution of the F/V-ATPase superfamily can be expected.

References

1. Nelson, N., Taiz, L. *Trends Biochem. Sci.* 1989; *14*, 113–116.

2. Nelson, N. *J. Bioenerg. Biomembr.* 1989; *21*, 553–571.

3. Nelson, N. *Curr. Opin. Cell Biol.* 1992; *4*, 654–660.

4. Nelson, N. *Biochim. Biophys. Acta* 1992; *1100*, 109–124.

5. Penefsky, H. S., Cross, R. L. In *Advances in Enzymology and Related Areas of Molecular Biology*, Vol. 64, Meister, A. (ed.), John Wiley & Sons, New York, 1991; pp. 173–214.

6. Futai, M., Noumi, T., Maeda, M. *Annu. Rev. Biochem.* 1989; *58*, 111–136.

7. Senior, A. E. *Annu. Rev. Biophys. Biophys. Chem.* 1990; *19*, 7–41.

8. Nalin, C. M., Nelson, N. *Curr. Topics Bioenerget.* 1987; *15*, 273–294.

9. Moriyama, Y., Nelson, N. *J. Biol. Chem.* 1987; *262*, 14723–14729.

10. Arai, H., Terres, G., Pink, S., Forgac, M. *J. Biol. Chem.* 1988; *263*, 8796–8802.

11. Zimniak, L., Dittrich, P., Gogarten, J. P., Kibak, H., Taiz, L. *J. Biol. Chem.* 1988; *263*, 9102–9112.

12. Manolson, M. F., Ouellette, B. F. F., Filion, M., Poole, R. J. *J. Biol. Chem.* 1988; *263*, 17987–17994.

13. Bowman, J. B., Allen, R., Wechser, M. A., Bowman, E. J. *J. Biol. Chem.* 1988; *263*, 14002–14007.

14. Bowman, E. J., Tenney, K., Bowman, B. J. *J. Biol. Chem.* 1988; *263*, 13994–14001.

15. Moriyama, Y., Nelson, N. *J. Biol. Chem.* 1989; *264*, 3577–3582.

16. Moriyama, Y., Nelson, N. *Biochim. Biophys. Acta* 1989; *980*, 241–247.

17. Moriyama, Y., Nelson, N. *J. Biol. Chem.* 1989; *264*, 18445–18450.

18. Lai, S., Randall, S. K., Sze, H. *J. Biol. Chem.* 1988; *263*, 16731–16737.

19. Schneider, E., Altendorf, K. *Microbiol. Rev.* 1987; *51*, 477–497.

20. Foster, D. L., Fillingame, R. H. *J. Biol. Chem.* 1982; *264*, 6797–6803.

21. Fillingame, R. H. *J. Bioenerg. Biomembr.* 1992; *24*, 485–491.

22. Papa, S., Guerrieri, F., Zanotti, F., Capozza, G., Fiermonte, M., Cocco, T., Altendorf, K., Deckers-Hebersteit, G. In *Ion-Motive ATPases: Structure, Function, and Regulation*, Vol. 671, Scarpa, A., Carafoli, E., Papa, S. (eds)., The New York Academy of Sciences, New York, 1992; pp. 345–358.

23. Walker, J. E., Lutter, R., Dupuis, A., Runswick, M. J. *Biochemistry* 1991; *30*, 5369–5378.

24. Arai, H., Berne, M., Forgac, M. *J. Biol. Chem.* 1987; *262*, 11006–11011.

25. Nelson, H., Nelson, N. *Proc. Natl. Acad. Sci. USA* 1990; *87*, 3503–3507.

26. Ward, J. M., Sze, H. *Plant Physiol.* 1992; *99*, 925–931.

27. Wang, S.-Y., Moriyama, Y., Mandel, M., Hulmes, J. D., Pan, Y.-C. E., Danho, W., Nelson, H., Nelson, N. *J. Biol. Chem.* 1989; *263*, 17638–17642.

28. Perin, M. S., Fried, V. A., Stone, D. K., Xie, X.-S., Südhof, T. C. *J. Biol. Chem.* 1991; *266*, 3877–3881.

29. Manolson, M. F., Proteau, D., Preston, R. A., Stenbit, A., Roberts, T., Hoyt, M. Andrew, Preuss, D., Mulholland, J., Botstein, D., Jones, E. W. *J. Biol. Chem.* 1992; *267*, 14294–14303.

30. Sze, H., Ward, J. M., Lai, S. *J. Bioenerg. Biomembr.* 1992; *24*, 371–381.

31. Mandel, M., Moriyama, Y., Hulmes, J. D., Pan, Y.-C. E., Nelson, H., Nelson, N. *Proc. Natl. Acad. Sci. USA* 1988; *85*, 5521–5524.

32. Nelson, H., Nelson, N. *FEBS Lett.* 1989; *247*, 147–153.

33. Nelson, N. *Plant Physiol.* 1988; *86*, 1–3.

34. Nanba, T., Mukohata, Y. *J. Biochem.* 1987; *102*, 591–598.

35. Lübben, M., Lünsdorf, H., Schäfer, G. *Eur. J. Biochem.* 1987; *167*, 211–219.

36. Nicolls, D., Attwell, D. *Trends Pharmacol. Sci.* 1990; *11*, 462–468.

37. Moriyama, Y., Nelson, N. in *The Ion Pumps, Structure, Function and Regulation*, Stein, W. D., (ed.), Alan R. Liss, New York, 1988; pp. 387–394.

38. Denda, K., Konishi, J., Hajiro, K., Oshima, T., Date, T., Yoshida, M. *J. Biol. Chem.* 1990; *265*, 21509–21513.

39. Gogarten, J. P., Kibak, H., Dittrich, P., Taiz, L., Bowman, E. J., Bowman, B. J., Manolson, M. F., Poole, R. J., Date, T., Oshima, T., Konishi, J., Denda, K., Yoshida, M. *Proc. Natl. Acad. Sci. USA* 1989; *86*, 6661–6665.

40. Fraga, D., Fillingame, R. H. *J. Bacteriol.* 1991; *173*, 2639–2643.

41. Miller, M. J., Oldenburg, M., Fillingame, R. H. *Proc. Natl. Acad. Sci. USA* 1990; *87*, 4900–4904.

42. Denda, K., Konishi, J., Oshima, T., Date, T., Yoshida, M. *J. Biol. Chem.* 1989; *264*, 7119–7121.

43. Schäfer, G., Meyering-Vos, M. In *Ion-Motive ATPases: Structure, Function, and Regulation*, Vol. 671, Scarpa, A., Carafoli, E., Papa, S. (eds.), The New York Academy of Sciences, New York, 1992; pp. 293–309.

44. Kibak, H., Taiz, L., Starke, T., Bernasconi, P., Gogarten, J. P. *J. Bioenerg. Biomembr.* 1992; *24*, 415–424.

45. Gogarten, J. P. and Starke, T., Kibak, H., Fishmann, J., Taiz, L. *J. Exp. Biol.* 1992; *172*, 137–147.

46. Cross, R. L., Taiz, L. *FEBS Lett.* 1990; *259*, 227–229.

47. Yokoyama, K., Oshima, T., Yoshida, M. *J. Biol. Chem.* 1990; *265*, 21946–21950.

48. Takase, K., Yamato, I., Kakinuma, Y. *J. Biol. Chem.* 1993; *268*, 11610–11616.

49. Hirsch, S., Strauss, A., Masood, K., Lee, S., Sukhatme, V., Gluck, S. *Proc. Natl. Acad. Sci. USA* 1988; *85*, 3004–3008.

50. Foury, F. *J. Biol. Chem.* 1990; *265*, 18554–18560.

51. Nelson, H., Mandiyan, S., Noumi, T., Moriyama, Y., Miedel, M. C., Nelson, N. *J. Biol. Chem.* 1990; *265*, 20390–20393.

52. Beltrán, C., Kopecky, J., Pan, Y.-C. E., Nelson, H., Nelson, N. *J. Biol. Chem.* 1992; *267*, 774–779.

53. Maloney, P. C., Wilson, T. H. *Bioscience* 1985; *35*, 43–48.

54. Margulis, L. In *Sybiosis in Cell Evolution*, W. H. Freeman, New York, 1987; pp. 37–62.

55. Nelson, N., Cidon, S. *J. Bioenerg. Biomembr.* 1984; *16*, 11–36.

56. Hennig, J., Herrmann, R. G. *Mol. Gen. Genet.* 1986; *203*, 117–128.

57. Cozens, A. L., Walker, J. E. *J. Mol. Biol.* 1987; *194*, 359–383.

58. Lill, H., Nelson, N. *Plant Molec. Biol.* 1991; *17*, 641–652.

59. Walker, J. E., Saraste, M., Gay, N. J. *Biochim. Biophys. Acta* 1984; *768*, 164–200.

60. Herrmann, R. G., Steppuhn, J., Herrmann, G. S., Nelson, N. *FEBS Lett.* 1993; *326*, 192–198.

61. Ohta, S., Yohda, M., Ishizuka, M., Hirata, H., Hamamoto, T., Otawara-Hamamoto, Y., Matsuda, K., Kagawa, Y. *Biochim. Biophys. Acta* 1985; *933*, 141–155.

62. Brusilow, W. S. A., Scarpetta, M. A., Hawthorne, C. A., Clark, W. P. *J. Biol. Chem.* 1989; *264*, 1528–1533.

63. Ivey, D. M., Krulwich, T. A. *Mol. Gen. Genet.* 1991; *229*, 292–300.

64. Krumholz, L. R., Esser, U., Simoni, R. D. *Nucleic Acids Res.* 1989; *17*, 7993.

65. McCarn, D. F., Whitaker, R. A., Alam, J., Vrba, J. M., Curtis, S. E. *J. Bacteriol.* 1988; *170*, 3448–34586.

66. Falk, G., Walker, J. E. *Biochem. J.* 1992; *229*, 663–668.

67. Falk, G., Walker, J. E. *Biochem. J.* 1988; *254*, 109–122.

68. Xie, D.-L., Lill, H., Hauska, G., Maeda, M., Futai, M., Nelson, N. *Biochim. Biophys. Acta* 1993; *1172*, 267–273.

69. Büttner, M., Xie, D.-L., Nelson, H., Pinther, W., Hauska, G., Nelson, N. *Proc. Natl. Acad. Sci. USA* 1992; *89*, 8135–8139.

70. Van Walraven, H. S., Lutter, R., Walker, J. E. *Biochem. J.* 1993; *294*, 239–251.

71. Kostrzewa, M., Zetsche, K. *J. Mol. Biol.* 1992; *227*, 961–970.

72. Pancic, P. G., Strotmann, H., Kowallik, K. V. *J. Mol. Biol.* 1992; *224*, 529–536.

73. Huber, R., Langworthy, T. A., Konig, H., Thomm, M., Woese, C. R., Sleytr, U. B., Stetter, K. O. *Arch. Microbiol.* 1986; *144*, 324–333.

74. Pedersen, P. L., Schwerzmann, K., Cintrón, N. *Curr. Topics Biognerget.* 1981; *11*, 149–199.

75. Thorsness, P. E., Fox, T. D. *Nature* 1990; *346*, 376–379.

76. Ackerman, S. H., Tzagoloff, A. *Proc. Natl. Acad. Sci. USA* 1990; *87*, 4986–4990.

77. Medd, S. M., Walker, J. E., Jolly, R. D. *Biochem. J.* 1993; *293*, 65–73.

Index